Aspects of Protein Biosynthesis

Part A

Contributors

VINCENT G. ALLFREY

CHARLES J. EPSTEIN

E. PETER GEIDUSCHEK

NICHOLAS M. KREDICH

MARSHALL NIRENBERG

G. M. TOMKINS

GUNTER VON EHRENSTEIN

Aspects of Protein Biosynthesis

Edited by

C. B. Anfinsen, Jr. *Boehmer*

LABORATORY OF CHEMICAL BIOLOGY
NIAMD
NATIONAL INSTITUTES OF HEALTH
BETHESDA, MARYLAND

Part A

ACADEMIC PRESS   New York and London   1970

ACADEMIC PRESS, INC.
111 Fifth Avenue, New York, New York 10003

United Kingdom Edition published by
ACADEMIC PRESS, INC. (LONDON) LTD.
Berkeley Square House, London W1X 6BA

LIBRARY OF CONGRESS CATALOG CARD NUMBER: 69-18355

PRINTED IN THE UNITED STATES OF AMERICA

List of Contributors

Numbers in parentheses indicate the pages on which the authors' contributions begin.

VINCENT G. ALLFREY (247), The Rockefeller University, New York, New York

CHARLES J. EPSTEIN (367), Department of Pediatrics, University of California Medical School, San Francisco, California

E. PETER GEIDUSCHEK (43), Department of Biophysics, University of Chicago, Chicago, Illinois

NICHOLAS M. KREDICH (1), Department of Medicine, Duke University Hospital, Durham, North Carolina

MARSHALL NIRENBERG (215), Laboratory of Biochemical Genetics, National Heart Institute, National Institutes of Health, Bethesda, Maryland

G. M. TOMKINS (1), Laboratory of Molecular Biology, National Institute of Arthritis and Metabolic Diseases, National Institutes of Health, Bethesda, Maryland

GUNTER VON EHRENSTEIN* (139), Department of Biophysics, Johns Hopkins University, School of Medicine, Baltimore, Maryland

* Present address: Department of Molecular Biology, Max-Planck-Institute for Experimental Medicine, Göttingen, Germany.

Preface

The term "protein biosynthesis" has come to include a very large part of what is otherwise known as "molecular biology." Indeed, in a recent, randomly selected issue of the *Proceedings of the National Academy of Sciences of the U. S.*, 17 of the 35 articles in the Biological Section dealt with some aspect of this problem. As the chapter titles of this book indicate, it is necessary to include the whole range of topics from bacterial genetics and regulatory mechanisms, through the chemistry and biology of RNA and DNA, to evolution, and the hereditary diseases of man.

The actual synthesis of peptide bonds during the elaboration of polypeptide chains remains the area of greatest uncertainty. The subjects of chain initiation, translocation and termination, as well as ribosome and polyribosome structure and formation, are undergoing intense investigation at the present time, and we may safely assume that well-informed and dependable discussions will be possible in the near future. These areas of research will form the subject matter of Part B, now in preparation, of this two-volume treatise of protein biosynthesis.

Although a definitive work on protein synthesis cannot yet be written, the group of individuals who became authors of these chapters agreed that a book that summarized those aspects now generally accepted as fact or near-fact would be of great value to students as well as to scientists working in unrelated disciplines. Speculation has been held to a minimum. A discussion of the historical development of the field has not been included since this has been treated fairly recently from a technical point of view in a volume edited by P. N. Campbell and J. R. Sargent ("Techniques in Protein Biosynthesis," Academic Press, 1967) and in a more narrative manner by R. Hendler ("Protein Biosynthesis and Membrane Biochemistry," John Wiley and Sons, 1968).

C. B. Anfinsen, Jr.

Bethesda, Maryland
November, 1969

Contents

I. GENETIC CONTROL OF PROTEIN STRUCTURE AND THE
    REGULATION OF PROTEIN SYNTHESIS

G. M. Tomkins and Nicholas M. Kredich

II. SYNTHESIS OF DNA AND ITS TRANSCRIPTION AS RNA

E. Peter Geiduschek

III.   TRANSFER RNA AND AMINO ACID ACTIVATION

        Gunter von Ehrenstein

IV.   THE FLOW OF INFORMATION FROM GENE TO PROTEIN

    Marshall Nirenberg

V.   BIOSYNTHETIC REACTIONS IN THE CELL NUCLEUS

    Vincent G. Allfrey

VI. THE THREE-DIMENSIONAL STRUCTURE AND EVOLUTION OF PROTEINS

Charles J. Epstein

Contents of Part B (Tentative)

I

Genetic Control of Protein Structure
and the Regulation of Protein Synthesis

G. M. Tomkins and Nicholas M. Kredich

I. NATURE OF THE GENETIC DETERMINANTS
OF PROTEIN STRUCTURE

A. IDENTITY OF NUCLEIC ACIDS AS THE GENETIC SUBSTANCE

1. *Deoxyribonucleic Acid*

The genetic substance must have two distinct properties: a capacity for self-replication and an ability to determine cellular activity. In view of the overwhelming evidence implicating DNA in both these regards, it is difficult to realize that, until fairly recently, the chemical nature of genetic determinants was not known, and that proteins, rather than nucleic acids, seemed the most likely candidates. However, in 1944, Avery and associates (7) demonstrated conclusively that DNA, in the complete absence of protein, could carry genetic information. In these beautiful experiments, a genetic substance was isolated from one strain of pneumococcus which, when added to a culture of a second, conferred on the recipient the antigenic properties of the donor strain capsule. This transformation of the recipient persisted for many generations. An analysis of the chemical composition of the transforming substance showed it to correspond exactly to that of DNA, and its ultraviolet absorption spectrum was found to be identical to that of DNA. DNase, but not RNase or proteolytic enzymes, inactivated the transforming substance. These studies solidly established that DNA can transfer genetic information from one cell to another. Since the recipient organism had acquired the ability to synthesize a new type of capsular polysaccharide, the authors suggested that the action of genes must be to control either enzyme formation or function.

Further proof that DNA, and not protein, is the hereditary material was presented by Hershey and Chase in a classic paper on the nature of the genetic determinants in bacteriophage T4 (49). This elegant work established the general mechanism by which the genes of a bacterial virus are introduced into, and replicate within, the host cell (Fig. 1). The investigators prepared phage labeled with either ^{32}P, which is incorporated into DNA, or ^{35}S, which occurs only in the viral proteins. Then phage particles, labeled in either the DNA or protein portion, were allowed to attach to bacteria for a given length of time after which the virus–host complex was separated by mechanical agitation. Under these conditions, where virus multiplication occurred normally, almost all the ^{35}S-labeled phage protein was removed from the bacterial host, but the

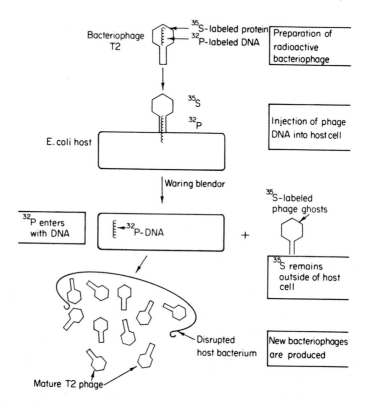

FIG. 1. Schematic representation of the Hershey–Chase experiment (49).

[32]P remained. This result established that phage DNA alone can direct the synthesis of new viruses, while the proteins of the phage particle act as a kind of syringe, which injects the information-containing nucleic acid into a bacterium.

The experiments of Avery *et al.* and Hershey and Chase show that DNA carries genetic information both in bacteria and their viruses. Many more recent experiments have substantiated these conclusions and have shown that this is true in almost all living things. For example, highly purified DNA can also transform *Bacillus subtilis* (*86*). This bacterium is more suitable for genetic studies than is the *Diplococcus pneumoniae* originally used by Avery *et al.* (*7*), since it can grow on a completely defined simple medium, to permit the easy demonstration of nutritional requirements. A genetic map of the *B. subtilis* chromosome, based on transformation experiments and other techniques, has been compiled by Dubnau *et al.* (*25*). Transformation can also be shown in the classic organism of bacterial genetics *Escherichia coli*. In this case purified DNA from the lysogenic bacteriophage (lambda) can transform the cells, provided a normal phage (helper phage) is used to facilitate the entry of the nucleic acid into the bacteria (*60*). Purified DNA from certain mammalian viruses can infect animal cells with the consequent production of virus particles or other manifestations of viral infection (*24, 37*). The single-strand circular DNA molecule from the virus ϕX174 (containing 5500 nucleotides) can also be taken up and expressed in spheroplasts of *E. coli* (*46*).

Using the latter system, Goulian, Kornberg, and Sinsheimer have provided perhaps the most elegant demonstration to date that DNA carries genetic information (*42*). These workers used purified single-strand DNA [called the (+) strand], extracted from ϕX174, as a template for the *in vitro* enzymic synthesis of its complementary, (−) strand (see Chapter II), using purified *E. coli* DNA polymerase. The synthesis requires the triphosphates of the four normal deoxyribonucleosides: deoxyadenosine (dATP), deoxycytidine (dCTP), deoxyguanosine (dGTP), and deoxythymidine (dTTP). In this experiment, however, the dTTP was substituted for by its denser, synthetic analog, bromo-deoxyuridine triphosphate (dBrUTP). Light circular (+) strand was used as a template, and the dense (−) complementary strand was enzymically synthesized, with its free ends joined by the recently discovered DNA ligase enzyme (*36*). The double helical complex of light template and its dense complement was dissociated, and the heavier (−) strand, synthesized *in vitro*, was isolated by centrifugation in a cesium chloride density gradient. The newly synthesized (−) strand, after separation from its (+) template, was next used as a template for the *in vitro* synthesis of a light (+) strand by substituting dTTP for dBrUTP in the reaction mixture. Its free ends were joined by the ligase enzyme. In this reaction a completely synthetic duplex circular molecule was formed.

Both the separated circular (+) and (−) molecules, products of *in vitro* enzymic synthesis, as well as the synthetic duplex circle, were then used to infect spheroplasts of *E. coli*, and, in all cases, normal virus particles were formed by the host. One evident consequence of this classic experiment is that DNA indeed carries genetic information.

2. Ribonucleic Acid

After the general acceptance of the idea that DNA served the genetic function, it was assumed for some time that this was the only molecule that did so. The discovery of self-replicating viruses containing only RNA suggested, however, that RNA in some instances must also transmit genetic information. Fraenkel-Conrat and Williams (*29*) have clearly demonstrated the genetic role of the ribonucleic acid in the RNA-containing virus, tobacco mosaic virus (TMV). The rod-shaped virus particles consist of a single RNA chain of approximately 6400 nucleotides, surrounded by a protein coat composed of about 2100 protein units, each of molecular weight 18,000. By separating and purifying the RNA and protein components of the virus, it has been possible to demonstrate that the infectivity, and, hence, the genetic determinants of the virus, fractionates with the RNA and not the protein. The infectivity of the purified RNA is only 0.1% that of the whole virus, but can be increased several hundredfold by reconstituting the virus particle from its purified RNA and protein. The increased infectivity of the RNA–protein particle over that of the pure RNA may be attributed to the ability of the protein coat to protect the RNA from nucleases found in the tobacco leaf.

If a virus particle is reconstituted from the RNA of one strain and the coat protein of another, the progeny obtained from infection are those of the strain from which the RNA was derived (*30*). Valuable information concerning the nature of mutagenesis has been obtained by chemically modifying purified TMV–RNA, reconstituting it with normal untreated coat protein and studying the mutant progeny produced by such hybrids (*31, 93*). These reconstitution experiments strengthen the contention that nucleic acid, rather than protein, is the genetic material.

Spiegelman *et al.* (*85*) have successfully replicated the *E. coli* RNA virus Qβ in a cellfree *in vitro* system utilizing the RNA-dependent RNA replicase induced by this bacteriophage. It has been shown that this newly synthesized RNA is capable of infecting *E. coli* spheroplasts to give rise to an apparently normal bacteriophage.

Although these experiments with RNA viruses conclusively establish the ability of RNA to carry genetic determinants, DNA appears to comprise most if not all the genetic material of bacteria and higher organisms.

B. PHYSICAL AND CHEMICAL STRUCTURE OF GENES AND CHROMOSOMES

1. *Structure of Deoxyribonucleic Acid*

The first account of the isolation of DNA was published in 1871 by Friedrich Miescher (*68*), who purified the substance from pus cells. By the 1940's its primary structure was generally recognized to be that of a polymer, composed of deoxyribonucleotides linked together in a linear fashion by 3',5'-phosphodiester bonds (Fig. 2). The purines, adenine and guanine, and the pyrimidines, cytosine and thymine, appear to constitute almost the entire base content of DNA from all sources examined, except for small amounts of 5-methylcytosine in plants and animals, and 5-hydroxymethylcytosine, which occurs instead of cytosine in certain bacteriophage. The occurrence of other rare bases cannot be excluded by present analytical techniques.

In 1953, Watson and Crick (*94*) proposed a model for the secondary structure of DNA, which, now generally accepted as correct, has been a powerful stimulus to the growth of molecular genetics and our understanding of the genetic control of protein synthesis. Based on the X-ray diffraction data of Wilkins and his co-workers (*101*), and Franklin and Gosling (*32*), Watson and Crick postulated that DNA consists of two helical polynucleotide chains, each coiled about the same axis. The deoxyribose phosphate chain comprises the backbone of each helix, while the bases lie in planes perpendicular to the longitudinal axis of the molecule. While both chains are right-hand helices, the deoxyribose phosphate backbones run in opposite directions to one another with the bases from one chain sharing the same planes with bases of the opposing chain (Fig. 3). This allows specific hydrogen bonding between bases (Fig. 4) to hold the entire structure together. In the proposed structure the repeat distance between planes of nucleotide pairs is 3.4 Å, while the helix is so pitched as to repeat itself every 34 Å (every 10 nucleotide pairs); both distances were derived from X-ray diffraction data. According to the model, the diameter of the molecule is 20 Å, which also agrees with the X-ray diffraction data and with measurements performed on electron microphotographs of DNA (*9*) (Fig. 5).

From the biological standpoint, the most critical feature of this

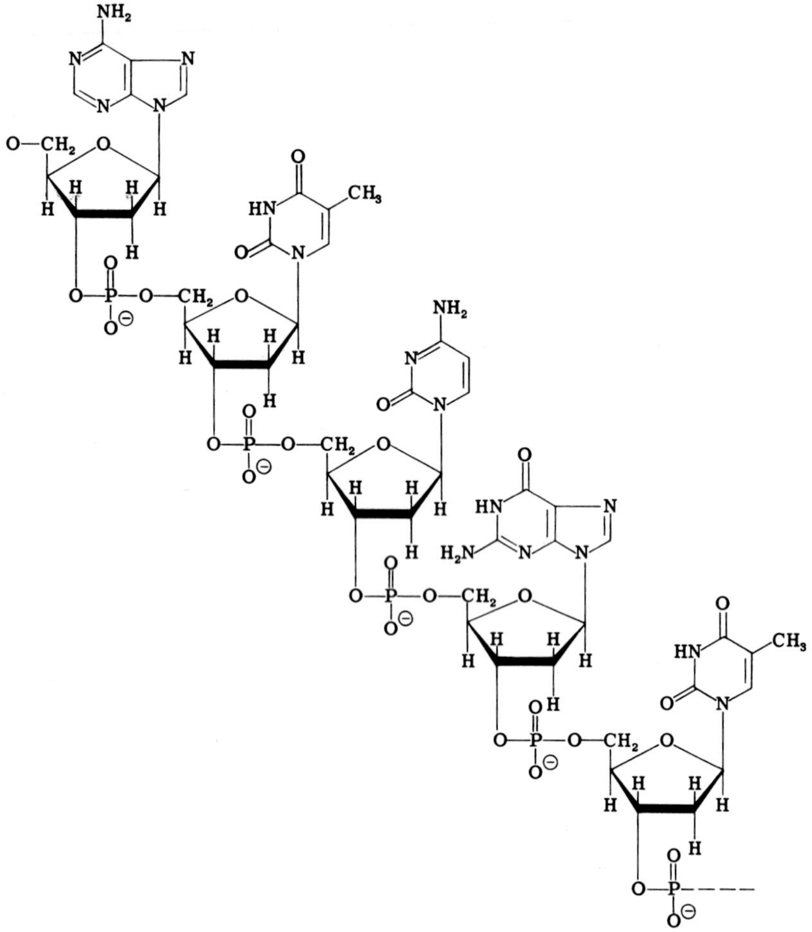

Fig. 2. The primary structure of DNA (from *60a*). Note the purine and pyrimidine bases attached to the polydeoxyribose phosphate backbone. The base sequence, adenine-thymine-cytosine-guanine-thymine, is hypothetical.

model is that adenine on one chain can base-pair only with thymine on the other chain, and guanine can pair only with cytosine (or 5-hydroxymethylcytosine or 5-methylcytosine). This is due to the fact that the distance between glycosidic bonds of opposite chains is a constant one allowing only these two possible combinations to fit within this distance. Thus, the two chains are complementary to one another in their base sequence, an A, G, G, T, C sequence in one requiring a T, C, C, A, G sequence in the other. Although the model imposes these

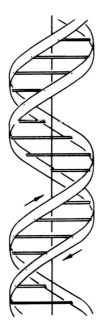

FIG. 3. A diagrammatic representation of the Watson–Crick model (*94*) for double-strand DNA. The two ribbons symbolize the two deoxyribose phosphate chains, and the horizontal rods the pairs of bases holding the chains together. The vertical line marks the fiber axis.

restrictions on the pairing of bases between chains, there are no restrictions on the order of base-pairs along the length of the molecule.

X-Ray diffraction data and electron microscopy support the helical nature and dimensions of the DNA model, while the base-pairing features are consistent with chemical data. Although the base compositions of DNA from different sources vary considerably, the adenine-to-thymine and guanine-to-cytosine ratios for a given species are always very near unity (*20*). In DNA containing appreciable amounts of 5-methylcytosine or 5-hydroxymethylcytosine, the sum of these bases and cytosine is equal to the guanine content, suggesting that all three bases pair with guanine. An exception to this finding is in single-strand DNA viruses (i.e., ϕX174) (*84*), where these ratios are not, and are not expected to be, near unity.

The Watson–Crick model further serves to explain certain other physical and chemical characteristics of DNA. Sedimentation, diffusion, light-scattering, and viscosity determinations indicate that DNA is a very long thin molecule with a surprising amount of rigidity, considering the fact that the free rotation expected about the phosphodiester

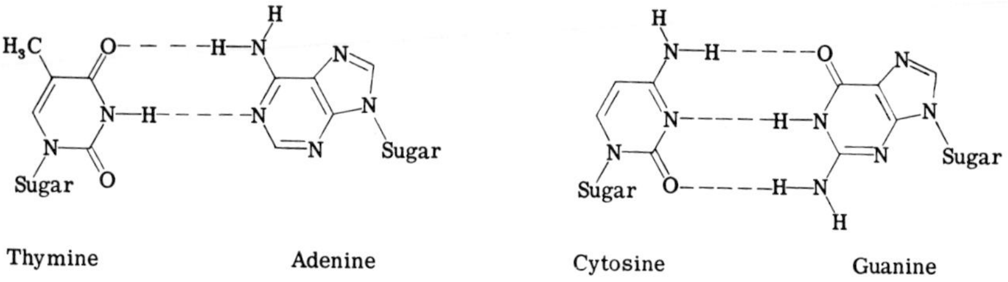

Thymine   Adenine   Cytosine   Guanine

FIG. 4. A representation of the hydrogen bonding between adenine and thymine, and guanine and cytosine, which occurs in double-strand DNA (from *60a*).

bonds and deoxyribose carbon–carbon bonds should give the polymer a great deal of flexibility. The constraint imposed upon these bonds by the hydrogen-bonded double helix offers an explanation for the observed rigidity. Physical measurements indicate that the DNA molecule loses its rigidity at pH extremes and elevated temperatures, presumably by breaking the hydrogen bonds between bases and allowing the two chains to separate (melt out). After such treatment, new titratable groups are found which appear to correspond to groups on the bases which were unavailable before the treatment because of their involvement in hydrogen bonding.

It is evident that base-pairing between complementary chains offers an attractive mechanism for DNA replication and hence the perpetuation of genetic information. Each strand of the double helix may serve as a template for the synthesis of its complementary strand, the restriction on pairing between bases ensuring a faithful reconstitution of the original base sequences. This hypothesis has been verified by enzymic studies on the *in vitro* replication of DNA (*62*). It is also important to note that the only irregular feature of the Watson–Crick DNA model is the sequence of base-pairs along the length of the molecule. Succeeding chapters will consider the manner in which this sequence of bases directs the linear order of amino acids in polypeptide chains.

2. Genes and Chromosomes

a. Genes and Polypeptide Chains. We consider next the mechanism of gene expression. Interestingly enough, the mere introduction of the word gene raises a number of questions (see Pontecorvo, *74*). Originally, the term was used to designate the hereditary element which controls a single character of the particular organism used for genetic study. However, the complexity of a gene, defined in this way, would obviously depend on the complexity of the character determined by it. Thus, at the time a given gene is identified, it may be technically impossible to

FIG. 5. A DNA molecule extracted from bacteriophage T4, stained with uranyl acetate, and photographed with an electron microscope (150,000×). (Courtesy of H. Erickson and M. Beer, Johns Hopkins University.)

resolve its phenotypic counterpart into smaller units, but, as **analysis** proceeds, this may become possible. For example, the genetic determinants for an entire metabolic pathway are often provisionally considered a single character, but when the various biochemical reactions leading to a metabolite are elucidated, a single gene is identified as the element controlling a single biochemical step. These considerations led to the "one-gene–one-enzyme" hypothesis of Beadle and Tatum (*8*). Complications arose, however, when it was found that some proteins are made up of combinations of nonidentical polypeptide chains and that genes can be reduced to simpler functional units.

b. Substructure of the rII Locus in Bacteriophage T4. Genetic substructure was originally explored by Seymour Benzer in a series of ingenious experiments using bacteriophage T4 (*10*). He concentrated entirely on the single gene *rII* of the T4 chromosome. Normally when these viruses burst from infected host bacteria grown as a confluent layer in a Petri dish, they produce a clear zone of lysis called a plaque, which is easily visible against the background lawn of intact cells. The *r* (rapid lysis) mutants were so called because the appearance of the plaques they produced differed from that of the wild type. Although several different genetic loci control the rapid lysis character, Benzer confined his studies to mutations in only one of these, the *rII*.

Taking advantage of the fact that *rII* mutants can multiply (with the formation of an abnormal plaque) in strain B of *E. coli* but not in strain K, he used the ability of the *rII* mutant to produce plaques on strain K as an indication that it had reverted to wild type. Benzer also discovered that when strain K bacteria were doubly infected with certain pairs of different *rII* mutants, these viruses could cooperate within the cytoplasm of the host cell to produce wild-type plaques by a process not involving genetic recombination (Fig. 6). This phenomenon is called *complementation.* Although two different mutations can complement one another when on separate viral chromosomes in the same bacterial cell (in the trans position), complementation does not occur when both mutations are on a single chromosome (in the cis position). An examination of a large number of *rII* point mutants (see below) revealed that they all fell into either of two complementation groups, or cistrons, A and B such that all mutants in the A cistron complemented all B cistron mutants, while two mutants within the same cistron would not complement one another. This technique for assigning mutants into several groups according to their ability to complement one another has been useful in many other genetic analyses and is called the cis–trans test. Complementation shows that it is not necessary that both nonmutant cistrons be on the same chromosome. A reasonable (although not necessarily

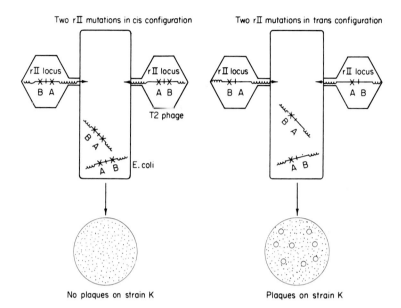

Two rII mutations in cis configuration

Two rII mutations in trans configuration

No plaques on strain K

Plaques on strain K

FIG. 6. Schematic representation of complementation in bacteriophage T4. In the left-hand drawing, a bacterial cell is infected with two viruses, each with a mutation in both the A and B cistrons of the *rII* gene. Viral replication does not occur, and no plaques are formed. In the right-hand drawing, the cell is infected with one phage bearing a mutation in the A cistron of *rII* (x in the drawing), and a second phage with a mutation in the B cistron. The wild-type A and B cistrons, even though on separate chromosomes, complement each other and allow virus multiplication as shown by plaque formation.

correct) explanation for such behavior is to suppose that the A and B cistrons direct the formation of two different but interacting polypeptide chains, which are required for normal bacterial lysis during phage infection.

If the cistron is considered to be the smallest functional genetic unit, a gene may be defined as the collection of cistrons which specifies an entire protein molecule. Evidently if an enzyme is made up of several different polypeptide chains, its gene is composed of the several cistrons responsible for the individual chains. This terminology becomes somewhat unsatisfactory when, as in the case of human hemoglobin, the protein consists of nonidentical subunits, but the genetic representations of the subunits are not close together on the chromosome (i.e., not linked). Should these be called two independent cistrons or two independent genes? It would seem that either designation is adequate. Another terminological problem arises, because there are certain stretches of DNA which determine only an RNA molecule, but no corresponding protein (i.e., ribosomal and transfer RNA loci). In these cases, again, cistron seems a perfectly reasonable designation although the "one cistron–one polypeptide chain" correspondence, of course, no longer holds.

c. *Topological Considerations. i. The chromosome.* It has long been accepted, on the basis of both cytogenetics and the analysis of the frequencies of recombination between different mutant organisms, that hereditary characters can be represented as a linear array, and maps, showing the order of genes and their relative distances from one another, have been constructed by both these methods. In the cytogenetic method, a correlation is made between visible alterations in chromosome structure and phenotypic changes in the organism in question. In recombinational analysis, genes are ordered on the reasonable assumption that if two genetic determinants are located at different points on a linear chromosome, the probability of breakage occurring between the markers is directly proportional to the distance between them. Thus, if two nearly identical chromosomes are placed side by side, exactly in register, and they break at the same point, the broken end of one can rejoin with the broken end of the other to produce genetic exchange or recombination. (Usually, reciprocal joining of the other pair of broken ends occurs at the same point.) Therefore, the probability of recombination between any two markers is greater when the physical distance between them on the chromosome is larger rather than smaller, and this probability can be used as a measure of genetic distance or linkage.

Interestingly enough Jacob and Wollman (*53*) have established that the entire genetic map of *E. coli* behaves as though it were a closed circle, since there are always two linkage distances between a pair of loci, depending on which way the circle is traversed (*91*) (Fig. 7). Cairns (*16*) has provided physical evidence to substantiate these genetic findings. His autoradiographs of tritium-labeled DNA demonstrated that the *E. coli* chromosome is a single piece of double-strand DNA, 1100 to 1400 μ long which appears to exist as a circle. Circular forms of DNA have since been demonstrated in bacteriophages ϕX174 (*61*) and lambda (*77*), the animal virus polyoma (*26, 99*), salamander oocytes (*69*), and mitochondria (*63*), showing that this is a common feature of genetic material.

ii. *Clustering of functionally related genes.* With the development of detailed genetic maps in microorganisms, a rather remarkable finding has come to light: there is a tendency for functionally related genes to occur adjacent each to the other in clusters on the chromosome. In *Salmonella typhimurium* ten contiguous genes involved in histidine biosynthesis are located on one segment of the chromosome. Clustering of genes involved in the biosynthesis of arginine, cysteine, leucine, trypto-

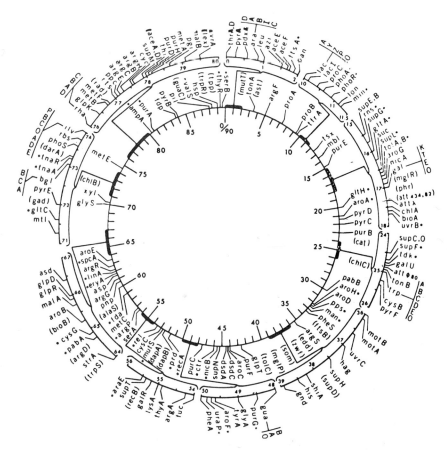

Fig. 7. Linkage map of *Escherichia coli*, drawn to scale (reprinted from Taylor and Trotter, *91*). The inner circle depicts the whole chromosome. The letters refer to the various genes, which are identified in Table 10 of Taylor and Trotter (*91*). The numbers refer to the time (minutes) required for the injection of the male chromosome into a female, starting at "0." Certain portions of the map (e.g., the 0- to 2-minute segment) are shown on arcs of the outer circle with a 4-fold expanded time scale in order to accommodate markers in crowded regions. Data from Taylor and Trotter (*91*).

phan, pantothenate, and other compounds also occurs, such that approximately 70% of known loci in *Salmonella* map in clusters (*23*). A similar phenomenon occurs in *E. coli* (*91*) and to a lesser degree in *B. subtilis* (*25*), yeast (*28*), and *Neurospora crassa* (*43*). In *Pseudomonas aeruginosa* (*50*), however, enough mapping has been done to indicate that there is little or no association of functionally related genes. In higher organisms, as well, some evidence exists for gene clustering. For example, five genes necessary for body segment development in *Drosophila melanogaster* appear to be clustered (*64*), and a series of genes for tail development in mice (*27*) show a similar pattern.

The high frequency of gene clustering in certain microorganisms suggests that such an arrangement has some evolutionary functional advantage. The development of the operon concept, wherein functionally related genes are regulated as a unit (see below), offers a convincing argument for a functional role of gene clusters. In addition, the proximity of genes within a cell implies a proximity of gene products, which might facilitate assembly of subunits into enzymes and enzyme complexes concerned with a given metabolic pathway.

Stahl and Murray (*87*) have suggested that close linkage of functionally related genes may offer an organism a better opportunity to pass on advantageous combinations of genes to its progeny. Their argument states that if the functioning of two or more gene products is dependent upon some mutual interaction of these proteins (or polypeptide chains), it is advantageous for the genes involved to be closely linked, minimizing the probability of their separation during genetic recombination. Whereas the gene combination ab may offer selective advantage to an individual, recombination with the $a'b'$ genome may give rise to $a'b$ and ab' progeny, which do not retain that selective advantage. The more closely linked a and b are, the higher the percentage of progeny which will contain the desirable ab combination. The authors propose that proximity of certain genes may have evolved because of these considerations.

iii. Gene topology. The great sensitivity of the *rII* system of bacteriophage T4, described above, made it possible to detect very rare recombinational events, and thereby to ask whether individual mutations within the *rII* gene could be represented as a linear array. Benzer (*11*) took an ingenious approach to this topological problem by attempting to arrange *rII* mutations in dictionary order. This was done by determining whether or not different pairs of nonreverting *rII* mutants (i.e., deletions) could recombine to form wild-type viruses and marking the results of the crosses by either 1 or 0 (since topology deals with qualitative relationships such as connectedness rather than quantitative measures, the actual frequency of recombination was irrelevant in these crosses). From the results a square matrix was formed in which all the mutants in question were listed in the same order both horizontally and vertically (Fig. 8). At the intersection of any two mutants, either 1 or 0 was inserted indicating whether or not recombination was observed between them.

If the different deletions in the *rII* locus could be arranged in a

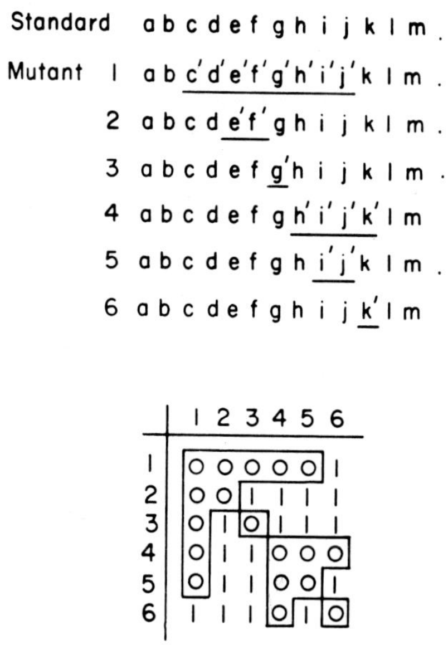
Standard a b c d e f g h i j k l m
Mutant 1 a b c' d' e' f' g' h' i' j' k l m
2 a b c d e' f' g h i j k l m
3 a b c d e f g' h i j k l m
4 a b c d e f g h' i' j' k' l m
5 a b c d e f g h i' j' k l m
6 a b c d e f g h i j k' l m
1 2 3 4 5 6
1 0 0 0 0 0 1
2 0 0 1 1 1 1
3 0 1 0 1 1 1
4 0 1 1 0 0 0
5 0 1 1 0 0 1
6 1 1 1 0 1 0

FIG. 8. An example of a recombination matrix (reprinted from Benzer, *11*). The upper portion represents a linear standard structure (wild type) and six mutant structures where the primes indicate mutation, arranged in dictionary order. In the lower drawing these are shown as a matrix. A one indicates that the standard can be produced by recombination, while a zero indicates that it cannot.

unique linear order (dictionary order), with the positions of large deletions being assigned according to where they began in the gene, then the recombination matrix would have a unique property when the mutations were listed down and across in this order. Obviously, the diagonal, which represents the intersection of any mutant with itself, will be 0. Starting at the diagonal, however, and moving either across a row or down a column, in a dictionary-order matrix, once a 1 has been encountered, then all further elements in that row or column must also be 1. If this sort of array can be generated for any set of elements, it implies that they can be arranged in a unique one-dimensional sequence. One hundred forty-five nonreverting *rII* mutants were examined in this manner. It was found that their recombination matrix had dictionary-order characteristics, indicating that mutations in this single gene form a unique linear sequence. Although this type of analysis does not rigidly exclude certain other special geometrical configurations, their occurrence seems extremely unlikely. Obviously the linear DNA molecule is ideally suited to specify the linear order of genes and of individual sites within a gene.

iv. Colinearity of the gene and polypeptide chain. Soon after the promulgation of the Watson–Crick DNA model it was generally assumed that the linear order of DNA bases somehow directed the linear order of amino acids in a polypeptide chain. This sequence hypothesis has now been proven to be correct by direct experimental observations. Using *amber* mutants* of the head protein of bacteriophage T4D (see Section II,A,1) Sarabhai *et al.* (*79*) examined *amber*-terminated peptide fragments and found that the points of termination correlated exactly with the linear order of mutations producing these fragments. In their work on the relationship between the A gene and the A protein of tryptophan synthetase of *E. coli,* Yanofsky and his co-workers (*105, 108*) isolated the altered α-protein from 16 missense mutants and determined the position of the amino acid substitution in each of these mutant proteins. The linear order of the altered amino acid residues corresponded exactly with the order of the mutations which gave rise to them, thus establishing the colinearity of the gene (cistron) and its protein.

d. Mutations. We shall now consider the chemical basis of the following types of permanent changes in a genetic determinant; i.e., mutations—*transitions, transversions, frame-shifts, recombinational mutations,* and *deletions.*

i. Transitions and transversions. Watson and Crick (*95*) have presented the most reasonable hypothesis to date to account for spontaneous mutations. Very infrequently one of the DNA bases involved in specific pairing may assume a rare tautomeric configuration; for example, the enol instead of the keto form of thymine. The enol tautomer of thymine can form its hydrogen bonds with guanine rather than with its normal partner, adenine. Similarly, if cytosine tautomerizes to the amino form, it would bond to adenine rather than guanine (Fig. 9). Obviously these tautomeric shifts upset the normal rules of base-pairing and if such tautomerization occurs during the synthesis of DNA an incorrect mutant base is inserted into the newly formed strand.

Transitions are mutations in which one purine (A or G) at a given site is replaced by the other (G or A) or else one pyrimidine (U or C) is replaced by the other (C or U). Transitions occur spontaneously, but may also be induced by chemical mutagens, such as the base analog 2-aminopurine or by reagents which chemically attack DNA by alkylation or deamination (e.g., nitrous acid). Single transitions appear genetically as point mutations which, under certain circumstances, may revert to wild type if the original base-pair reforms by back mutation.

A given purine can also be replaced by a pyrimidine in the equivalent

* *Amber* mutants are named after a graduate student, Bernstein, who was working with them at Cal Tech. Amber in German is Bernstein.

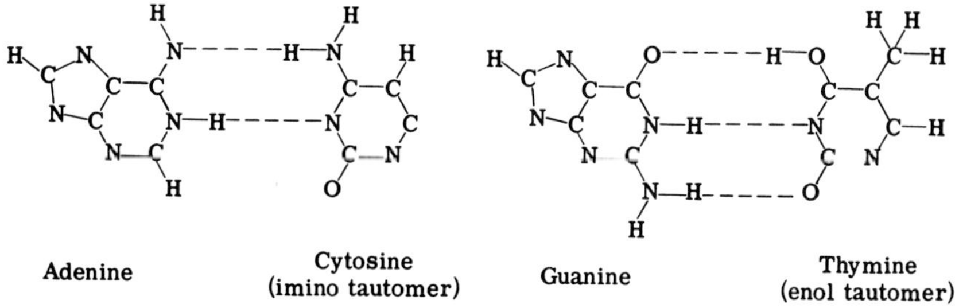

| Adenine | Cytosine (imino tautomer) | Guanine | Thymine (enol tautomer) |

FIG. 9. Possible abnormal base-pairing, which would lead to mutation.

position. In this case, of course, the complementary pyrimidine on the opposite strand is changed to the required purine during the next round of DNA replication. This type of mutation, called a transversion, has been found in the A cistron of the tryptophan synthetase gene in *E. coli*, especially when mutations are induced by a mutator gene present in the organism (*106*). In addition, a comparison of the amino acid sequences of homologous proteins from different species suggests that a considerable number of evolutionary changes have been the result of transversions (see Jukes, *59*).

ii. Frame-shifts. It has been suggested that certain mutagens, for example the acridines, can intercalate between two normally contiguous bases of a DNA strand causing the insertion or deletion of one or more bases during DNA synthesis; this process produces frame-shift mutations (*21*). Since each amino acid is coded by three consecutive bases in the DNA, the result of either the insertion or deletion of one or two bases is to force the translation mechanism, which decodes the genetic message three sequential bases at a time, to operate out of phase. A mutant polypeptide is formed in which an entire sequence of amino acids is abnormal, starting at the point coded by the frame-shift. The abnormal sequence may terminate abruptly if the frame-shift leads to the premature reading of a nonsense codon. However, the original base insertion or deletion may be compensated for by a second deletion or insertion further down the gene, allowing the resumption of the original reading frame and the normal completion of the original protein chain. A beautiful verification of this mechanism of mutation (as well as an incidental demonstration of the chemical direction of messenger RNA translation and a confirmation of the genetic code assignment to given amino acids) was based on the analysis of the amino acid sequence of phage T4 lysozyme in wild-type and acridine-induced mutants (*89*).

The sequences of comparable peptides were examined in the wild-type enzyme and in the proteins produced by several mutant phages con-

taining acridine-induced frame-shifts in the lysozyme gene of T4
(Fig. 10). The sequence in question in the wild-type enzyme is: Thr-
Lys-Ser-Pro-Ser-Leu-Asn-Ala. As shown in Fig. 10 this change can be
explained if, and only if, an adenine or a guanine is deleted from the
second codon of the messenger for the wild-type octapeptide and rein-
serted into the penultimate triplet.

A similar analysis (71) has shown that a two-base addition to the
third codon and the insertion of a single base into the last triplet
induced by acridine orange gives rise to the amino acid sequence: Thr-
Lys-Ser-Val-His-His-Leu-Met-Ala. This amino acid sequence can be
accounted for if the two bases GU or UG are inserted into the third
triplet coding for the wild-type sequence, and if G or A is added to
the last triplet (Fig. 10).

iii. Recombinational pseudomutations. When the genomes of two parent organisms differ from each other in a given triplet, genetic exchange
within the triplet may produce a mutant protein with an amino acid
not present in either of the parents (*103*). For example, AUA (*ile*) and

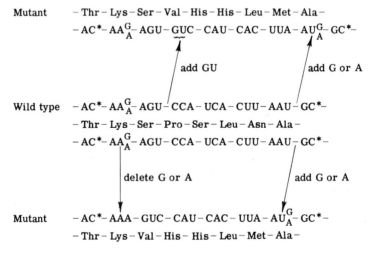

Fig. 10. Results of frame-shift mutagenesis by acridine on the composition of a
portion of the messenger RNA of the lysozyme gene of bacteriophage T4 and the
resulting change in the amino acid sequence of the polypeptide fragment coded
by it. In the center of the drawing, there is a representation of the wild-type
sequence and immediately above and below it the most probable messenger codon
assigned to each amino acid residue. An asterisk indicates no restriction on the base
in the third position of the codons in which it occurs. The arrows symbolize mutagenesis by acridine and the amino acid and base sequences on the top and bottom
of the drawing show two resultant mutant messengers and polypeptides. (From
Streisinger *et al., 89,* and Okada *et al., 71.*)

AGU (*ser*) can give rise to the recombinational mutants AUU (*ile*) and AGA (*arg*).

iv. Deletions. Rarely, long stretches of DNA may be lost giving rise to deletions. These mutants do not revert to wild type, and, in certain instances, recombinational analysis has shown that the genes on either side of a deletion are closer together than in the wild-type organism. In certain deletions in bacteriophage, chemical analysis has also shown that affected viruses contain less DNA than normal (*98*).

v. Physiology of mutations, missense, and nonsense. As explained above, the exchange of one base for another is called a point mutation, which may have one of several effects on the protein normally coded for by that gene. (1) The amino acid sequence may remain unchanged especially if the base change is in the third position of a codon, because the genetic code is degenerate, i.e., several codons specify the same amino acid (see Chapter VI). (2) The wild-type amino acid residue may be replaced by a different one (missense). Such a change might, rarely, produce a mutant organism, which is favored by the environment over the original wild type; or the growth of the missense mutant might not differ significantly from that of the wild type. In either case the mutation leads to acceptable missense. More likely a missense mutation will cause the production of a partly or completely inactive protein, in which case the mutation will be disadvantageous or even lethal (unacceptable missense). It has been suggested, on the basis of the chemical similarity of the amino acids designated by related triplets (differing from one another by a single base), that the code may have evolved in such a way as to minimize the probability of unacceptable missense mutations (see Chapter VI). (3) Another possible consequence of single-base mutation changes is the conversion of normal triplets to those which cause growing polypeptide chains to terminate prematurely, i.e., nonsense mutations. When written in messenger RNA language, the chain-terminating triplets are UAG (*amber*) (*12, 96*), UAA (*ochre*) (*12*), and UGA (*13*). It has been found that when a normal triplet in the messenger of the coat protein gene of bacteriophage T4 is converted to UAG, the growing peptide chain terminates at that point, and the partially completed protein is released from the ribosomes (*79*). Therefore it has been proposed that the normal stop signal for peptide chain growth, i.e., the triplet marking the end of a cistron, is one of these three nonsense codons. Indirect evidence suggests, but does not prove, that the natural stop may be the *ochre* triplet UAA (*12*).

There must also exist genetic signals which mark the beginning of gene transcription. These may be identical to the promoters detected in genetic experiments (*57*). The chemical nature of the transcription starting signals is not yet known, but the RNA polymerase (which catalyzes gene transcription) is known to attach only to a limited number of sites on the DNA molecule, and, presumably, these regions are related to the signal for the beginning of transcription (see Chapter II). It is also found that, when DNA is used as a template for *in vitro* RNA synthesis, the resulting messenger usually begins with a purine at the 5'-hydroxyl end (see Chapter II), which also suggests the existence of specific signals on the DNA for the initiation of RNA synthesis. On logical grounds, one would assume that a stop signal for transcription is also required, which perhaps is somehow related to the translation stop.

II. INTERACTIONS BETWEEN GENETIC SITES

A. Alterations in Mutant Gene Function

The restoration of a biological activity lost by mutation can occur in one of a number of different ways (see Fig. 11): (1) *Reversion:* a return to the wild-type base sequence by a second mutational event, or the alteration of a mutated base to one differing from the wild-type residue, but resulting in a triplet which codes for the wild-type amino acid (due to the ambiguity of the code). (2) *Pseudoreversion:* the mutated base is changed to one which results in a codon for a nonwild type, but acceptable missense amino acid. (3) *Suppression:* a mutational event restoring biological activity, but which involves a DNA base or base sequence other than that originally mutated. (4) *Phenotypic curing* (not depicted in Fig. 11): environmental conditions not causing changes in the DNA base sequence of the mutant, but which temporarily alter transcription and/or translation of the DNA information in such a way as to result in functional protein from a mutated gene.

1. *Suppression*

This subject may be discussed under two general categories. If the suppressor mutation occurs in the same cistron as the original mutation it is termed an *intracistronic suppressor;* if it occurs elsewhere it is called an *extracistronic suppressor.*

a. Intracistronic Suppression. Yanofsky *et al.* (*107*) have demonstrated that, among tryptophan-independent strains derived from a mutation involving a single amino acid change in the α-protein of *E. coli* tryptophan synthetase, there are found a number of strains which synthesize proteins with different amino acids substituted for the orig-

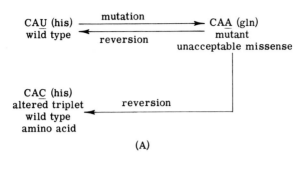

(A)

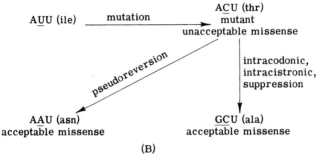

(B)

Fig. 11. Triplets are given as RNA codons, rather than DNA sequences. (A) In this example reversion occurs by further mutating the third base of the altered triplet to the original U, or to a C, both of which resultant triplets code for histidine (underlined letters). (B) Pseudoreversion involves the further alteration of the originally mutated base (middle position in this example) to give a codon for a nonwild type, but acceptable amino acid. Alteration of one of the other two bases in the codon is known as intracodonic intracistronic suppression (underlined letters).

inally altered residue. Using the codon assignments (derived from *in vitro* experiments; see Chapter VI) for the amino acids found at the mutated locus, Yanofsky *et al.* concluded that a nonacceptable missense triplet may be converted to an acceptable missense triplet by a single base change at any of the three positions of the triplet.

If the originally mutated base is changed in such a conversion, then the process is reversion (wild-type amino acid residue restored) or pseudoreversion (acceptable missense amino acid residue restored) as defined above. However, if one of the other two bases in the mutated codon is changed, so that the triplet codes for the wild type or another acceptable amino acid, this is referred to as intracodonic intracistronic suppression (Fig. 11).

The tryptophan synthetase α-protein also provides an excellent example of extracodonic intracistronic suppression, in which the original mutation is retained, but a second amino acid substitution occurs within the same protein, 36 residues removed from the first mutation (*104*). These second site revertants result in a protein which can now assume an enzymically active tertiary structure, prohibited by the first amino acid substitution.

A third type of intracistronic suppression, which may be either intra- or extracodonic, is possible for frame-shift mutations. Since the insertion or deletion of one or two bases into the DNA disrupts the correct reading frame of all triplets following it, such mutations obviously alter the entire amino acid sequence distal to the mutation. In the case of a single base deletion, a second mutation (the suppressor) in the same cistron, consisting of a base insertion, restores the proper reading frame to all but the region between mutations. If the second mutation is near enough to the first, the resultant protein will differ from the wild type by only a few missense amino acids and may be functional. Such a case of intra- cistronic suppression in bacteriophage lysozyme has been discussed earlier in this chapter (see Fig. 10).

b. Extracistronic Suppression. Whereas intracistronic suppression in- volves the further alteration of the DNA sequence of the originally mutated cistron, extracistronic suppression results from mutations in other cistrons.

An extracistronic suppressor may leave the structure of the mutant protein unchanged, but alter the nature of the environment so that the abnormal protein can function. While examining extracts of a mutant strain of *Neurospora crassa* lacking tryptophan synthetase, Suskind and Kurek (*90*) found that this enzymic activity could be restored by puri- fying the protein from an inorganic inhibitor whose action was mimicked by zinc. Extracts from a suppressed strain of this mutant were found to be active in an unpurified state, but they were still inactivated by the addition of zinc, suggesting that the suppressor might be an altered zinc permease, which prevented the accumulation of inhibitory concentrations of the metal ion.

More frequently, extracistronic suppression involves the alteration of the amino acid sequence of a mutant protein to give a functionally active molecule. Recent work on the action of extracistronic suppressors has been concerned primarily with those which result in the formation of new species of amino acyl tRNA's. These suppressor tRNA's insert an acceptable amino acid into a site designated in the mutant messenger RNA by nonsense or unacceptable missense triplets (see Chapter III).

It has long been noted in bacteria that certain suppressor mutations are capable of suppressing a large number, or a class, of mutations, which may occur almost anywhere on the chromosome or even in bacteriophage infecting a suppressor-containing host. One such class of suppressible mutations are the nonsense mutations *amber* and *ochre* which result in a polypeptide chain termination at the point of mutation.

By an elegant analysis of amino acid substitutions in revertants of an *amber* mutant of *E. coli* alkaline phosphatase, Weigert and Garen (*96*) were able to identify the triplet UAG as the *amber* codon, while Brenner and associates (*12*) simultaneously came to the same conclusion, based primarily on mutagenic data obtained with T4 bacteriophage mutants. In addition, the latter group established the *ochre* triplet as UAA. More recently Brenner *et al.* (*13*) have determined that UGA is a nonsense triplet in the *rII* region of T4 bacteriophage.

In *E. coli* there are a number of loci for *amber* and *ochre* suppressors, each of which, thus far studied, results in the substitution of a different amino acid for the nonsense codon. One of the most extensively studied is the *Su-1* (*Su-A*) suppressor gene, which results in the insertion of serine in *amber* mutants (*97*). In an *in vitro* protein-synthesizing system, utilizing mRNA from an *amber* mutant of bacteriophage R17, Capecchi and Gussin (*17*) demonstrated the role of tRNA in the action of the *Su-1* suppressor. They found that the coat protein of the bacteriophage was not synthesized unless serine-accepting tRNA from the suppressor strain of *E. coli* was added. Since they added purified tRNA, stripped of amino acids, from the *Su-1* strain, it was evident that the nonsuppressed (*su⁻*) strain could charge the tRNA (also see Goodman *et al.*, *40*).

Another type of suppression involving the altered reading of a triplet by means of a new species of tRNA has been described by Carbon and associates (*18*) in the tryptophan synthetase of *E. coli*. In this case a triplet coding for glycine (GGA) was mutated to the triplet corresponding to arginine (AGA), which resulted in an enzymically inactive protein. A strain containing an extracistronic suppressor was then isolated, which was capable of synthesizing small amounts of active A protein with the normal glycine residue instead of arginine. By examining the *in vitro* coding properties of synthetic polynucleotides containing AGA triplets, Carbon *et al.* were able to demonstrate the incorporation of glycine into polypeptides, which was dependent on tRNA isolated from the suppressor strain in question. No glycine incorporation occurred using tRNA from strains not carrying the suppressor under study, indicating that the suppression involved the synthesis of an altered tRNA that accepted glycine and recognized the AGA triplet.

A similar case of missense suppression in the tryptophan synthetase system was reported by Gupta and Khorana (*45*) who demonstrated the appearance of a glycyl-tRNA, which recognizes the UGU triplet (ordinarily coding for cysteine), in a suppressor-bearing strain substituting glycine for cysteine to give an active protein.

Although most suppressor mutants thus far studied have been found to elaborate an altered tRNA, it is conceivable that suppression could result from mutations involving other components of the complicated machinery of transcription and translation. For instance, a mutation in the RNA polymerase, causing that enzyme to make infrequent errors in the transcription of DNA, might result in the occasional synthesis of a messenger RNA in which an acceptable codon was inserted in place of an unacceptable or nonsense codon. On the translation level, defective amino acyl synthetases, peptide synthetases, or ribosomes might occasionally misread a mutant codon in a messenger RNA and insert an acceptable amino acid in its place.

Apirion (4) has described such a situation in *E. coli*, where a temperature-sensitive revertant of a tryptophan auxotroph was found to have temperature-sensitive 50 S ribosomal subunits. Genetic analysis revealed the existence of an extracistronic mutation, which is presumably responsible for the altered ribosomal structure. Although Apirion was unable to demonstrate any increase in the mistake level during the *in vitro* translation of synthetic messengers with these altered ribosomes, the assumption is that mistakes do occur *in vivo*, resulting in the suppression of the tryptophan auxotroph.

2. Phenotypic Curing

This phenomenon involves the temporary modification of the amino acid sequence of an inactive protein to give a functional molecule and does not affect the DNA of the mutant, being, therefore, nonheritable. A mutation can be cured by alteration of either gene transcription or messenger translation.

Champe and Benzer (*19*) demonstrated that the RNA base analog 5-fluorouracil phenotypically cured certain *rII* mutants of T4 bacteriophage. This effect in T4 appears to be specific for transition mutations resulting in AT pairs. Since 5-fluorouracil is known to substitute extensively for uracil in RNA, Champe and Benzer postulated that the base analog is incorporated as uracil during transcription, but is occasionally mistakenly recognized as cytosine during translation. This effect of 5-fluorouracil would result in the mutant T being finally interpreted as a C, with the resultant insertion of the wild-type amino acid residue.

In vitro coding data, however, indicate that 5-fluorouracil is faithfully translated as U and not as C (*44*). This has led Heidelberger (*48*) to suggest that the mistake is made on the transcriptional level, but the *in vitro* transcription of synthetic polydeoxyribonucleotides (*14*) shows that 5-fluorouracil is transcribed as U and is mistaken for C no more often than U itself. Hence there is no *in vitro* evidence to date which convincingly differentiates between transcription or translation as the site of action of 5-fluorouracil in phenotypic suppression.

Streptomycin is also able to change the reading of the code, which results in the phenotypic curing of many mutations (*41*). Mutations which can be cured by streptomycin are known as conditionally strepto-mycin dependent or CSD mutants, since, in the case of auxotrophs, their growth is conditionally dependent on the presence of streptomycin (or related antibiotics) in the absence of the growth factor involved. Because of the growth-inhibiting effect of streptomycin, CSD mutants were first noted in streptomycin-resistant strains, and it was originally thought that their curing was dependent upon the mutation to streptomycin resistance. It has since been demonstrated, however, that curing of CSD mutants does not depend on streptomycin resistance, provided that low concentrations of the antibiotic are used. The effect is most easily demon-strated by placing a crystal of streptomycin on a lawn of CSD-mutant bacteria. The zone of growth inhibition immediately around the crystal is surrounded by a ring of growth, where the streptomycin concentration is high enough to cure the mutation but low enough to allow growth of the streptomycin-sensitive organism.

Using this technique, Whitfield, Martin, and Ames (*100*) demonstrated that, among mutants in the *C* gene of the histidine operon in *Salmonella typhimurium*, phenotypic curing occurred in 19 of 20 *amber* mutants, 5 of 7 *ochre* mutants, and in 20% of missense mutants, while none occurred in frame-shift mutants.

Apirion and Schlessinger (*5*) have shown that the potential for curing in a CSD mutant is maximal in streptomycin-sensitive strains, and that mutations to streptomycin resistance almost invariably hinder or elim-inate phenotypic curing. This loss of phenotypic curing is most pro-nounced in streptomycin-resistant mutants which are mapped at a locus involved in the structure of the 30 S ribosomal particle. It seems that streptomycin "cures" by causing ribosomes to misread the code, occa-sionally allowing the incorporation of an acceptable amino acid in response to a mutant nonsense or unacceptable missense triplet. Apirion and Schlessinger argue that the ability to misread the code in the pres-ence of streptomycin is optimal for wild-type ribosomes, and that muta-tions to streptomycin resistance at a ribosomal-determining locus usually diminish or eliminate the ability of ribosomes to misread the code, lead-ing to a loss of curability.

Davies, Jones, and Khorana (*22*) have demonstrated a fairly specific effect of streptomycin on the *in vitro* reading of synthetic polyribonucleotides. They found that streptomycin results in the misreading of pyrimidines, so that a given pyrimidine at the 5'-terminal end of a triplet may be misread as the other pyrimidine (i.e., C read as U and U read as C), and that a pyrimidine in the middle of a triplet may be misread as either a pyrimidine or a purine. Such misreadings are markedly influenced by the nature of adjacent bases, and purines are not misread at all.

B. Coordinate Expression of Functionally Related Genes

1. *The Operon*

In Section I,B,2, we drew attention to the frequent contiguity of functionally related genes on the chromosome of certain species of bacteria. Specific examples which have been studied are the groups of genes controlling the catabolism of arabinose (*ara*) or lactose (*lac*) in *E. coli,* the biosynthesis of histidine (*his*) in *S. typhimurium,* and tryptophan synthesis (*trp*) in both organisms. In many instances, linked genes which control related functions are activated or inactivated more or less simultaneously. A gene cluster of this kind is called an operon (*54, 56*), the concept of which has been germinal in modern biological thinking. For the sake of organization, our discussion of coordinated gene function cannot follow the historical development of the subject, and so we shall discuss only certain basic characteristics of operon activity: the polygenic messenger RNA, polarity, and induction and repression.

2. *Strand Selection in DNA Transcription*

The linear structure of the Watson–Crick DNA model correlates well with the linear nature of genes and chromosomes. However, a given length of double-strand DNA offers two sequences of bases which might serve as a template for RNA synthesis, assuming that each strand can be traversed in only one direction (i.e., $3' \rightarrow 5'$) by the RNA polymerase (see Chapter II). While most studies on the *in vitro* transcription of DNA indicate that both strands are copied, it appears that *in vivo* and, in some instances, *in vitro*, only one strand serves as a template (see Chapter II for details). That both strands in a given segment might be used to code for two different proteins seems unlikely in light of the genetic evidence, which indicates that a single base change (in both strands because of base-pairing) results in only one altered protein.

Polarity data (see below), studies on the initial kinetics of enzyme induction (*1*), and direct measurements on the growth of messenger RNA molecules (*51*) show that the transcription of DNA begins at the operator end of an operon. Since the RNA polymerase copies the DNA strand only in the $3' \to 5'$ direction, the fact that certain operons are oriented in opposite directions on the chromosomes* means that transcription must proceed in different directions in different regions of the chromosome; furthermore, since the two strands are antiparallel, the polymerase must transcribe one strand in some operons and the opposite strand in other operons.

3. Polygenic Messenger RNA

A gene is active when messenger RNA is transcribed from one of its two strands, the sense strand. When an entire operon is activated, the question arises of whether its genetic information is transferred into a single long strand of RNA (polygenic messenger), or whether each individual gene is transcribed into a separate messenger RNA molecule. The original statement of the operon model (56) left this question open.

To estimate the length of a given messenger molecule it must be isolated from the total population of RNA molecules in the cell. An ingenious technique which has proved invaluable in this regard is that of hybridization of the messenger with the specific DNA from which it was transcribed (see Chapter II). Specific DNA–RNA hybrid formation occurs because the sequence of bases in the RNA is determined by Watson–Crick base-pairing with the sense strand of the DNA template, allowing the transcribed RNA molecule to form a stable double-strand structure with this strand of DNA. The problem of messenger isolation thus becomes one of isolating the DNA of a single operon from that of the rest of the genome. An elegant method for purifying the required genes has been to isolate a bacteriophage particle into the chromosome of which only the desired bacterial genes have been inserted. When the virus replicates, the DNA of the desired operon, integrated into such a "specialized high-frequency transducing" phage, also replicates, allowing it to be prepared in quantity. The corresponding messenger RNA molecules isolated from extracts of normal uninfected bacteria can be purified by hybridization. An experiment of this sort has been carried out by Imamoto et al. (52). It shows that all five genes of the tryp operon of E. coli are transcribed into a single long strand of messenger RNA. Earlier experiments on the his and lac operons also led to the conclusion that the messenger is polygenic, although on the basis of less direct evidence. The tryp experiments, just referred to, also show that the transcription of the messenger starts at one end of the operon, the

---

* The histidine and tryptophan operons are oriented "clockwise," whereas the leucine operon is oriented "counterclockwise" on the chromosome of Salmonella typhimurium.

operator end (see below), and proceeds away from it, rather than occurring on all the genes of the operon simultaneously.

The point on the DNA where the transcription is thought to begin is called the promotor (in the *lac* system). Its existence has, to date, only been detected by genetic experiments.

4. *Polarity*

Polar mutations are defined as follows: In an operon which consists of the adjacent genes *A, B, C,* and *D,* with *A* at the operator end, mutations in *A* which lower the rate of synthesis of the products from *B, C,* and *D* are said to be *polar.* A polar mutation in *B* decreases the function of *C* and *D,* but (except in very special cases) not that of *A.* In general, polar mutations inhibit the expression of genes of the same operon, but only when those genes are located on the operator-distal side of the affected gene, i.e., on the side away from the operator region.

Polarity is so characteristic of operons that the presence of a mutation in one gene which decreases the function of adjacent genes has been used as evidence that the group functions as an operon. The direction of polarity establishes the operator end of an operon since, in almost every cast studied, only the operator-distal genes are affected by polar mutations.

The mechanism of polarity is not yet understood, but, nevertheless, many interesting features of the phenomenon have been described. One of these concerns the types of mutations which cause polar effects. When polarity was originally discovered, in the *lac* operon, it was noted that only certain mutations in *z,* the gene closest to the operator, decreased the function of the other genes. A closer investigation in both the *lac* (*70*) and *his* (*65*) systems has revealed that only mutations leading to the chain-terminating nonsense triplets cause polar effects in the operator-distal genes of the operon. In addition to single-base changes, polarity may arise as a secondary consequence of frame-shift mutations (*66*), since, by the addition or deletion of a base, the reading frame can be shifted so that it encounters a nonsense triplet normally composed, in the wild type, of two adjacent sense triplets. For example, if in the wild-type sequence, – – UUA, AUA, GAA – –, the A in the first triplet is deleted, the sequence becomes – –UUA, *UAG,* AA– – –, which contains the polar *amber* triplet UAG.

In our earlier discussion of suppression, we noted that the function of a gene which has been inactivated by a nonsense mutation can be restored by a suppressor mutation allowing the nonsense triplet to be translated as an amino acid. The polar effect of nonsense mutations can also be corrected by suppression (*65*). Since suppression occurs when a nonsense triplet is translated as a codon for an amino acid, and, since polar effects are corrected by the same mechanism, it seems reasonable to suppose that the effects of polarity are entirely due to the polypeptide chain termination which occurs when the RNA translation mechanism encounters a nonsense triplet in the messenger.

This conclusion is strengthened by the observation that nonsense mutations in the coat protein gene of the RNA bacteriophage f2 produce polar effects on other viral genes when the synthesis of the proteins in question is measured in an *in vitro* system where no messenger RNA synthesis takes place (*109*). Furthermore, a systematic study of messenger RNA production in polar mutants of the *trp* system of *E. coli* showed that, at least in a certain class of such mutants, the weak polars, messenger production is normal (*52*). On the basis of such evidence, a translation mechanism has been proposed (*2, 88*) which accounts for polarity and which takes into account the polygenic messenger and the fact that polar effects disappear when the original polar mutation is suppressed. This mechanism states that ribosomes start only at the operator end of a polygenic messenger. If a nonsense mutation occurs somewhere along the messenger, then a certain percentage of the attached ribosomes will become detached at the nonsense codon, causing a decrease in the number of ribosomes traversing the section of messenger distal to the mutation. This would evidently produce a decrease in the rate of production of the proteins coded for by the genes beyond the nonsense mutation.

Several complications, not, by any means, completely resolved at this writing, arise from a study of the strong polar mutations occurring in several operons. One such complication is the interesting position effect noted in the *z* gene of the *lac* system of *E. coli* (*70*). The intensity of the polar effect of an *amber* or *ochre* mutation in this gene is inversely proportional to the distance of the mutation from the operator itself. Thus, mutations at the beginning of the gene—close to the operator—are very strongly polar while those farther away from the operator are less so. A similar position effect has also been found in the *G* gene of the *his* operon of *S. typhimurium* (*65*), which also lies next to the operator region of that operon. Interestingly enough, at least in the *his* operon, the position effect of nonsense mutations in genes farther from the operator is much less striking and, in at least one gene, the polar effect is independent of its intragenic location.

A second complicating feature of polarity revolves around the rate of production of messenger RNA from operons containing a polar mutation. Early studies on the *lac* system (although somewhat misinterpreted at the time) had shown that strongly polar mutants produced little or no messenger corresponding to the entire *lac* operon *(6)*. A systematic study of polar mutations of the *trp* system has also shown that in strongly polar mutants of the *E* gene (the gene nearest the operator) messenger production is reduced by 85 to 90% *(52)*. Strongly polar mutations in genes farther from the operator may also curtail messenger formation, especially of that messenger from genes operator-distal to the mutation. When polar *trp* mutations are suppressed, messenger RNA production, as well as translation, is also restored to normal. Since polarity undoubtedly reflects a primary defect in the process of messenger translation, the latter must in some way influence messenger production.

The fact that mutations with the strongest polar effects occur near the operator region of the first gene of an operon, and have the strongest effects on messenger formation, also hints at the possibility that the translation of this first gene may somehow be especially important in the transcription of the messenger from the entire operon *(1)*. These considerations may also help explain why pronounced position effects only occur in the first genes of operons. Polarity leads us to a model of operon function where ribosomes attach to the operator end of a polygenic message; and where translation of the message is somehow required for the transcription of the remaining genes of the operon.

5. *Induction and Repression*

a. Introduction. This subject has been a central theme of molecular biology for many years. In fact, attempts to understand enzyme induction and repression have led to many of the concepts presented in this and other chapters of this volume.

The physiological importance of induction is immediately apparent. To a growing culture of microorganisms, a relatively low concentration of an inducer, such as arabinose or lactose, is added. Within several minutes, the intracellular concentration of enzymes required for the metabolism of these sugars is increased by 10- to 100- or even 1000-fold. In enzyme repression the added material inhibits the synthesis of one or a group of enzymes usually involved in the biosynthesis of the added corepressor.

Although induction and repression have been studied much more extensively in bacteria than in other cells, similar regulatory phenomena occur in other microorganisms such as yeast and *Neurospora*, as well as in plant and animal cells. The following discussion will be limited entirely to bacterial systems, because, at the present time, not enough data are available to warrant the assumption that the mechanisms proposed for nonnuclear microorganisms will apply strictly to higher organisms. It must be admitted that analogies of this sort have been very useful in the past, since the citric acid cycle, glycolysis, the genetic code, ATP as an energy source, and many other molecular aspects of biological systems are almost invariant from the lowest to the highest forms of living things. The case of genetic regulation may prove to be different, however, since the hypotheses used to explain bacterial enzyme induction involve the organization of the DNA, and it is precisely this feature which differs significantly between bacteria and the higher organisms. Bacterial cells do not appear to have a distinct nucleus and the DNA is not associated with basic proteins, whereas the DNA of eukaryotic cells such as yeast, fungi, and animals is complexed with cationic proteins to form the structurally organized chromosome (77).

In bacteria, protein synthesis is thought to occur, at least in part, on a complex of messenger RNA, ribosomes, transfer RNA, etc., together with the DNA from which the messenger is being transcribed (15). In eukaryotes, however, such a complex could not play a role in the formation of proteins since most of the DNA is confined to the nucleus, where little protein synthesis occurs, while peptide bond formation takes place in the cytoplasm. An additional complexity is found in animal cells where there are at least two distinct classes of protein-forming centers—those attached to membranes and those apparently free in the cytoplasm. The latter sites are responsible for the synthesis of proteins destined to remain in the cell; the membrane sites produce proteins that will leave the cell (82).

Finally, a single cell of higher organisms contains, in addition to the nucleus, one or more minor genetic systems such as mitochondria, chloroplasts, etc. The nuclear and extranuclear genomes undoubtedly function in a coordinated way, but the mechanisms of this coordination are not yet understood. Evidently, then, regulation in nonbacterial cells involves elements either entirely absent in bacteria or else present in primitive form in bacterial cells. Models derived from the study of bacterial systems can therefore only suggest possible basic mechanisms which might occur in eukaryotic cells; but it is also likely that the greater complexity of the higher organisms has produced mechanisms unique to themselves. (For further discussion of this point, see Tomkins and Ames, 92.)

b. *Messenger RNA and Enzyme Induction.* It was the consideration of the mechanism of inducer action in bacteria which led to the idea of messenger RNA. The inducer can rapidly initiate enzyme synthesis; its withdrawal can, just as rapidly, stop enzyme formation; and a corepressor can equally rapidly repress the formation of the enzymes under its control.

These observations suggested the existence of a very labile intermediate in protein synthesis lying somewhere between the genes and the polypeptides. This intermediate is now known to be messenger RNA. DNA–RNA hybridization experiments done with several operons have shown quite clearly that when the operon is induced (or derepressed), the corresponding messenger rapidly accumulates in the cell, and when the inducer is removed (or when the corepressor of a repressible operon is added), these messenger molecules disappear (6, 47). Labeling experiments done with the *tryp* system of *E. coli* (52) (presented in much greater detail in Chapter II) show that the increases and decreases of messenger concentration are due to changes in its rate of synthesis, rather than to its degradation. It is certainly not excluded that, in some systems, selective protection of the messenger from degradation also influences its concentration. Therefore the induction of an operon involves the stimulation of the rate of its transcription into messenger RNA; when the inducer is removed from the culture, messenger synthesis is slowed, and the normal, rapid rate of messenger degradation quickly lowers the induced level to the uninduced basal concentration. Analogously, the messenger from repressible operons is made at a rapid derepressed rate in the absence of the corepressor. The latter inhibits messenger synthesis causing the derepressed level to fall to the repressed value. Thus, the major factor controlling the rate of enzyme synthesis in the bacterial systems examined so far is the concentration of the messenger RNA which codes for the enzymes. The mechanism by which messenger synthesis is controlled is the subject of the succeeding parts of this chapter.

c. *Regulatory Genes and Structural Genes.* One of the most important discoveries about inducible and repressible systems in bacteria has been that certain regulatory genes control the expression of other structural genes. Similar observations had been made earlier in plants (67), but there it was not possible to correlate the action of regulatory genes with enzyme induction.

In the *lac* system of *E. coli*, a regulatory gene *i* was identified which controls the expression of the *lac* structural genes, *z*, *y*, and *a* (56). When the i^+ allele is present, inducers of the *lac* operon, e.g., lactose, can stimulate the synthesis of the *lac* enzymes. When the i^- allele is present, the *lac* enzymes are made at a high rate even in the absence of the inducer. A third allele, i^s (102), has also been described in which the *lac* enzymes are synthesized at a rather slow rate which cannot be increased by the inducer.

It has been found that the structure of β-galactosidase, the product of the *z* gene, is independent of the configuration of the *i* gene. Conversely, a mutation in the z gene does not influence the induction of its product (58). Therefore, in this case, structural and regulatory genes are distinct entities. So far as is known, the i gene is the principal regulatory gene for the lac operon. Other genetic elements, to be discussed presently, such as the operator and the promotor also influence lac operon activity, but they do not produce corresponding gene products and are not, therefore, strictly speaking, genes.

Many other operons in $E.$ $coli$ such as trp and ara also appear to be controlled by a single regulatory gene. Other systems, however, such as his in $S.$ $typhimurium$ or the alkaline phosphatase gene of $E.$ $coli$ are controlled by several genes (3). The multiplicity of regulatory genes may be due to the fact that the exogenous inducer or corepressor must be metabolically transformed into the biologically active compound, and the methods used to isolate regulatory mutants do not usually discriminate between mutations in such accessory regulatory genes, responsible for inducer modification, and true regulatory genes. The latter are of the greatest interest, because their products presumably interact directly with the machinery of DNA transcription or of protein synthesis.

For example, the repressible his operon of $S.$ $typhimurium$ is controlled by the supply of exogenous histidine; but recent experiments have shown that the true corepressor is not the free amino acid but histidyl-tRNA (80). Therefore, mutations affecting the regulation of the operon are found in the genes controlling the histidine-activating enzyme and the corresponding transfer RNA (78, 83).

One technique which is especially useful in avoiding complications with secondary regulatory genes is the use of inducers which are not metabolized by the organism under study (gratuitous inducers), but, unfortunately, this is not always possible.

In the lac system several gratuitous inducers are known. Using these, it has been found that the i gene still regulates the function of the operon, indicating that i is a true regulatory gene. The mechanism of action of the i gene has been under investigation for some time. One of the most elegantly conceived experiments in molecular biology, performed by Pardee, Jacob, and Monod (72), involved a diploid bacterium containing two different alleles of the i gene, one the wild type i^+ and the other, the i^- mutant. Such an organism requires exogenous inducer to synthesize β-galactosidase. Therefore, in genetic terminology, i^+ is dominant to i^-. Since the i^+ gene inhibits lac operon function (in the absence of inducer), whereas i^- does not, these investigators proposed that the i^+, but not i^-, elaborates an inhibitory product or repressor. In the i^+/i^- organism used, the two i alleles were on separate chromosomes. Furthermore, it was found that i^+ is dominant to i^- even when the structural genes of the *lac* operon are on the chromosomal segment containing the i^- allele. Therefore, the i^+ dominance is exerted over genes in the trans position, i.e., i^+ is trans dominant to i^-. Trans dominance implies that the dominant allele elaborates a product (in this case, repressor) which can diffuse through the cytoplasm to influence the activity of genes on a separate chromosome. Inducers are thought to act by inhibiting repressor function (see below).

In the *ara* system of *E. coli*, dominance tests of the regulatory C gene have suggested that the product of the wild-type allele, C^+ is required for operon function (*81*) and that the inducer arabinose somehow facilitates this action. When a regulatory gene produces a repressor, the system is under negative control; when the product of a regulatory gene stimulates an operon, as in *ara*, the system is under positive control.

The alkaline phosphatase system of *E. coli* shows both types of regulation, and it has been suggested that one of the regulatory genes R_1 elaborates a gene activator which is converted, under the influence of a second regulatory gene R_2 to a repressor (*33*). Superficially, therefore, it appears as if either of several mechanisms may control gene activity, depending on the system. However, it may be that the general features of gene regulation are constant and this apparent diversity may only be the result of insufficient data.

Striking progress has recently been made on the difficult problem of the chemical nature of the products of the regulatory genes and on their mechanism of action. Genetic experiments show that nonsense mutations in regulatory genes can be suppressed by second mutations known to correct the function of mutant enzymes inactivated by nonsense mutations. On this basis it has been proposed that regulatory gene products are proteins (*34*) [although previously it had been suggested that they might be polynucleotides with no corresponding polypeptide (*73*)].

Several years ago, the product of one of the regulatory genes, R_2 of the *E. coli* phosphatase system, was isolated in homogeneous form and its amino acid composition was determined (*35*). Surprisingly, the synthesis of this protein is repressed by inorganic phosphate just as is the enzyme it serves to regulate.

More recently, the regulatory model proposed by Jacob and Monod has received striking confirmation from the experiments of Gilbert and Muller-Hill on the chemical nature of the i gene product (*38, 39*). These workers have purified and partially characterized a protein, obtained from i^+ but not i^- strains of *E. coli*, which tightly binds the gratuitous inducer isopropylthiogalactoside. The molecular weight of the *lac* repressor is estimated to be 169,000, and it is composed of four subunits. Like the R_2 gene product in the phosphatase system, the amino acid composition of the i gene product is that of a typical protein. As predicted from genetic experiments, the synthesis of the *lac* repressor is not controlled by inducers of the *lac* operon. These brilliant experiments have laid the groundwork for an understanding of the molecular basis of induction and repression.

d. Site of Regulatory Gene Action. In the operon model, the repressor is a protein which prevents gene transcription by binding directly to the DNA at the operator region of an operon, where messenger synthesis is initiated (*55*).

The evidence for this proposed model rested in part on the isolation of an interesting group of constitutive mutants of the *lac* operon, which are dominant to the wild type, but only when the mutation is cis to the structural genes it regulates. These mutants were termed o^c, for operator constitutive. Cis dominance shows itself in the following way. In a haploid o^c organism, the *lac* operon is almost fully induced in the absence of the inducer. (In some o^c's, full induction can be obtained only if the inducer is present.) In the diploid organism, $i^+o^+z^-/i^-o^cz^+$ (that is a bacterium containing the wild-type i gene, wild-type operator, and defective z gene on one chromosome and a defective i gene, o^c, and normal z gene on the other), the synthesis of β-galactosidase (the product of the z allele) is not inhibited by the repressor manufactured by the trans i^+ allele. In the diploid $i^+o^cz^-/i^-o^+z^+$, however, β-galactosidase synthesis is repressed unless inducer is added. Therefore, the o^c allele renders the operon insensitive to repressor action only when it is on the same chromosome as the structural genes it controls. In other words, the o^c allele is cis dominant to the o^+ gene.

The most reasonable explanation for the cis dominance of o^c mutations is that the o region does not code for a diffusible product, but represents a region on the DNA which is not transcribed into messenger RNA. Jacob and Monod (*55*) proposed that the operator is the site of the repressor's direct attachment to the DNA. Therefore, it need not be transcribed into messenger RNA.

The experiments of Gilbert and Muller-Hill (*39*) have again confirmed this hypothesis by showing that the partially purified *lac* repressor binds very tightly to the DNA of the *lac* region (the latter having been separated from the rest of the *E. coli* genome by its inclusion in bacteriophage ϕ80) only when the operator is in the (+) configuration. Repressor binding to o^c DNA is much weaker. Finally, as predicted, isopropylthiogalactoside, the gratuitous inducer used for purification of the repressor, weakens the bond between the repressor and the o^+ DNA.

At the time of this writing, therefore, it appears as if the model proposed to explain the regulation of the *lac* operon has been verified to

the smallest detail. A similar model, again proposed by Jacob and
Monod (*55*) to account for the behavior of bacteriophage λ, has also
been verified by chemical experiments similar to those described for the
lac system (*75, 76*). It seems, therefore, that gene regulation occurs
primarily by attachment of the regulatory gene product to the DNA.
Since the *trp* genes of *E. coli* can also be incorporated into the phage
φ80, it may be only a short time until this regulatory scheme is tested
in a repressible biosynthetic system. Until this latter aim is accomplished, it is difficult to be sure of the generality of the Jacob–Monod
mechanism, but for the present it seems the most reasonable hypothesis
to account for genetic regulation in bacteria.

<h2 style="text-align:center">REFERENCES</h2>

1. D. H. Alpers and G. M. Tomkins. The order of induction and deinduction of
the enzymes of the lactose operon in *E. coli. Proc. Natl. Acad. Sci. U. S.* **53,**
797 (1965).
2. B. N. Ames and P. E. Hartman. The histidine operon. *Cold Spring Harbor
Symp. Quant. Biol.* **28,** 349 (1963).
3. B. N. Ames and R. G. Martin. Biochemical aspects of genetics. The operon.
Ann. Rev. Biochem. **33,** 235 (1964).
4. D. Apirion. Altered ribosomes in a suppressor strain of *E. coli. J. Mol. Biol.*
16, 285 (1966).
5. D. Apirion and D. Schlessinger. The loss of phenotypic suppression in streptomycin-resistant mutants. *Proc. Natl. Acad. Sci. U. S.* **58,** 206 (1967).
6. G. S. Attardi, S. Naono, J. Rouviere, F. Jacob, and F. N. Gros. Production of
messenger RNA and regulation of protein synthesis. *Cold Spring Harb. Symp.
Quant. Biol.* **28,** 363 (1963).
7. O. T. Avery, C. M. MacLeod, and M. McCarty. Studies on the chemical nature
of the substance inducing transformation of pneumococcal types. *J. Exptl.
Med.* **79,** 137 (1944).
8. G. W. Beadle and E. L. Tatum. Genetic control of biochemical reactions in
Neurospora. Proc. Natl. Acad. Sci. U. S. **27,** 499 (1941).
9. M. Beer and C. R. Zobel. Electron stains. II. Electron microscopic studies on
the visibility of stained DNA molecules. *J. Mol. Biol.* **3,** 717 (1961).
10. S. Benzer. Fine structure of a genetic region in bacteriophage. *Proc. Natl.
Acad. Sci. U. S.* **41,** 344 (1955).
11. S. Benzer. On the topology of the genetic fine structure. *Proc. Nat. Acad. Sci.
U. S.* **45,** 1607 (1959).
12. S. Brenner, A. O. W. Stretton, and S. Kaplan. Genetic code: The 'nonsense'
triplets for chain termination and their suppression. *Nature* **206,** 994 (1965).
13. S. Brenner, L. Barnett, E. R. Katz, and F. H. C. Crick. UGA: A third nonsense triplet in the genetic code. *Nature* **213,** 449 (1967).
14. H. Bujard and C. Heidelberger. Fluorinated pyrimidines. XXVII. Attempts to
determine transcription errors during the formation of fluorouracil-containing
messenger ribonucleic acid. *Biochemistry* **5,** 3339 (1966).
15. R. Byrne, J. G. Levin, H. A. Bladen, and M. W. Nirenberg. The *in vitro*
formation of a DNA-ribosome complex. *Proc. Natl. Acad. Sci. U. S.* **52,** 140
(1964).

16. J. Cairns. The bacterial chromosome and its manner of replication as seen by autoradiography. *J. Mol. Biol.* **6,** 208 (1963).

17. M. R. Capecchi and G. N. Gussin. Suppression *in vitro:* Identification of serine-sRNA as a "nonsense" suppressor. *Science* **149,** 417 (1965).

18. J. Carbon, P. Berg, and C. Yanofsky. Studies of missense suppression of the tryptophan synthetase A-protein mutant A₃₆. *Proc. Natl. Acad. Sci. U. S.* **56,** 764 (1966).

19. S. P. Champe and S. Benzer. Reversal of mutant phenotypes by 5-fluoracil. An approach to nucleotide sequences in m-RNA. *Proc. Natl. Acad. Sci. U. S.* **48,** 532 (1962).

20. E. Chargaff. Structure and function of nucleic acids as cell constituents. *Federation Proc.* **10,** 654 (1951).

21. F. H. C. Crick, L. Barnett, S. Brenner, and R. J. Watts-Tobin. General nature of the genetic code for proteins. *Nature* **192,** 1227 (1961).

22. J. Davies, D. S. Jones, and H. G. Khorana. A further stduy of misreading of codons induced by streptomycin and neomycin using ribopolynucleotides containing two nucleotides in alternating sequence as templates. *J. Mol. Biol.* **18,** 48 (1966).

23. M. Demerec. Clustering of functionally related genes in *Salmonella typhimurium. Proc. Natl. Acad. Sci. U. S.* **51,** 1057 (1964).

24. G. A. DiMayorca, B. E. Eddy, S. E. Stewart, W. S. Hunter, C. Friend, and A. Bendich. Isolation of infectious DNA from SE polyoma-infected tissue cultures. *Proc. Natl. Acad. Sci. U. S.* **45,** 1805 (1959).

25. D. Dubnau, C. Goldthwaite, I. Smith, and J. Marmur. Genetic mapping in *B. subtilis. J. Mol. Biol.* **27,** 163 (1967).

26. R. Dulbecco and M. Vogt. Evidence for a ring structure of polyoma virus DNA. *Proc. Natl. Acad. Sci. U. S.* **50,** 236 (1963).

27. L. C. Dunn and E. Caspari. A case of neighboring loci with similar effects. *Genetics* **30,** 543 (1945).

28. G. R. Fink. A cluster of genes controlling three enzymes in histidine biosynthesis in *Saccharomyces cerevisiae. Genetics* **53,** 445 (1966).

29. H. Fraenkel-Conrat and R. C. Williams. Reconstitution of active tobacco mosaic virus from its inactive protein and nucleic acid components. *Proc. Natl. Acad. Sci. U. S.* **41,** 690 (1955).

30. H. Fraenkel-Conrat. The role of the nucleic acid in the reconstitution of active tobacco mosaic virus. *J. Am. Chem. Soc.* **78,** 882 (1956).

31. H. Fraenkel-Conrat and B. Singer. Virus reconstitution. II. Combination of protein and nucleic acid from different strains. *Biochim. Biophys. Acta* **24,** 540 (1957).

32. R. E. Franklin and R. G. Gosling. Molecular configuration in sodium thymonucleate. *Nature* **171,** 740 (1953).

33. A. Garen and H. Echols. Properties of two regulating genes for alkaline phosphatase. *J. Bacteriol.* **83,** 297 (1962).

34. A. Garen and S. Garen. Genetic evidence on the nature of the repressor for alkaline phosphatase in *E. coli. J. Mol. Biol.* **6,** 433 (1963).

35. A. Garen and O. Nozumi. Isolation of protein specified by a regulator gene. *J. Mol. Biol.* **8,** 841 (1964).

36. M. Gellert. Formation of covalent circles of lambda DNA by *E. coli* extracts. *Proc. Natl. Acad. Sci. U. S.* **57,** 148 (1967).

37. P. Gerber. An infectious DNA derived from vacuolating virus (SV$_{40}$). *Virology* **16**, 96 (1962).
38. W. Gilbert and B. Muller-Hill. Isolation of the *Lac* repressor. *Proc. Natl. Acad. Sci. U. S.* **56**, 1891 (1966).
39. W. Gilbert and B. Muller-Hill. The *Lac* operator is DNA. *Proc. Natl. Acad. Sci. U. S.* **58**, 2415 (1967).
40. H. M. Goodman, J. Abelson, A. Landy, S. Brenner, and J. O. Smith. Amber suppression: A nucleotide change in the anticodon of a tyrosine transfer RNA. *Nature* **217**, 1019 (1968).
41. L. Gorini and E. Kataja. Phenotypic repair by streptomycin of defective genotypes in *E. coli. Proc. Natl. Acad. Sci. U. S.* **51**, 487 (1964).
42. M. Goulian, A. Kornberg, and R. L. Sinsheimer. Enzymatic synthesis of DNA. XXIV. Synthesis of infectious phage ϕX174 DNA. *Proc. Natl. Acad. Sci. U. S.* **58**, 2321 (1967).
43. S. R. Gross and A. Fein. Linkage and function in *Neurospora. Genetics* **45**, 885 (1960).
44. M. Grunberg-Manago and A. M. Michelson. *Polynucleotide analogues.* IV. Polyfluorouridylic acid and copolymers containing fluorouridylic acid. *Biochim. Biophys. Acta* **87**, 593 (1964).
45. N. K. Gupta and H. G. Khorana. Missense suppression of the tryptophan synthetase A-protein mutant A78. *Proc. Natl. Acad. Sci. U. S.* **56**, 772 (1966).
46. G. D. Guthrie and R. L. Sinsheimer. Observations on the infection of bacterial protoplasts with deoxyribonucleic acid of bacteriophage ϕX174. *Biochim. Biophys. Acta* **72**, 290 (1963).
47. M. Hayashi, S. Spiegelman, W. Franklin, and S. E. Luvin. Separation of the RNA message transcribed in response to a specific inducer. *Proc. Natl. Acad. Sci. U. S.* **49**, 729 (1963).
48. C. Heidelberger. Introduction. *Progr. Nucleic Acid Res. Mol. Biol.* **4**, 1 (1965).
49. A. D. Hershey and M. C. Chase. Independent functions of viral protein and nucleic acid in growth of bacteriophage. *J. Gen. Physiol.* **36**, 39 (1952).
50. B. W. Holloway, L. Hodgins, and B. Fargie. Unlinked loci affecting related biosynthetic steps in *Pseudomonas aeruginosa. Nature* **199**, 926 (1965).
51. F. Imamoto, N. Morikawa, and K. Sato. On the transcription of the tryptophan operon in *E. coli. J. Mol. Biol.* **13**, 169 (1965).
52. F. Imamoto, J. Ito, and C. Yanofsky. Polarity in the tryptophan operon of *E. coli. Cold Spring Harbor Symp. Quant. Biol.* **31**, 235 (1966).
53. F. Jacob and E. L. Wollman. Genetic and physical determinations of chromosomal segments in *E. coli. Symp. Soc. Exptl. Biol.* **12**, 75 (1958).
54. F. Jacob, D. Perrin, C. Sanchez, and J. Monod. L'opéron: Groupe de genes à expression coordonnée par un operateur. *Compt. Rend.* **250**, 1727 (1960).
55. F. Jacob and J. Monod. Genetic regulatory mechanisms in the synthesis of proteins. *J. Mol. Biol.* **3**, 318 (1961).
56. F. Jacob and J. Monod. On the regulation of gene activity. *Cold Spring Harbor Symp. Quant. Biol.* **26**, 193 (1961).
57. F. Jacob, A. Ullman, and J. Monod. Génétique biochimique—le promoteur élément génétique necéssaire a l'expression d'un opéron. *Compt. Rend.* **258**, 3125 (1964).
58. F. Jacob and J. Monod. Genetic mapping of the elements of the lactose region in *E. coli. Biochem. Biophys. Res. Commun.* **18**, 693 (1965).

59. T. H. Jukes. "Molecules and Evolution." Columbia Univ. Press, New York, 1966.
60. A. D. Kaiser and D. S. Hogness. The transformation of *E. coli* with deoxy-ribonucleic acid isolated from bacteriophage λdg, *J. Mol. Biol.* **2**, 392 (1960).
60a. P. Karlson. "Introduction to Modern Biochemistry." Academic Press, New York, 1965.
61. A. K. Kleinschmidt, A. Burton, and R. L. Sinsheimer. Electron microscopy of the replicative form of the DNA of bacteriophage φX174. *Science* **142**, 961 (1963).
62. A. Kornberg. Biologic synthesis of DNA. *Science* **131**, 1503 (1960).
63. A. M. Kroon, P. Borst, E. F. J. Van Bruggen, and G. J. C. M. Ruttenberg. Mitochondrial DNA from sheep heart. *Proc. Natl. Acad. Sci. U. S.* **56**, 1836 (1966).
64. E. B. Lewis. Genes and developmental pathways. *Am. Zoologist* **3**, 33 (1963).
65. R. G. Martin, H. J. Whitfield, D. B. Berkowitz, and M. J. Voll. A molecular model of the phenomenon of polarity. *Cold Spring Harbor Symp. Quant. Biol.* **31**, 215 (1966).
66. R. G. Martin. Frameshift mutants in the histidine operon of *Salmonella typhimurium. J. Mol. Biol.* **26**, 311 (1967).
67. B. McClintock. Controlling elements and the gene. *Cold Spring Harbor Symp. Quant. Biol.* **21**, 197 (1956).
68. F. Miescher. Über die Chemische Zusammensetzung der Eiterzellen. *Hoppe-Seyler's Med. Chem. Unters.* p. 441 (1871).
69. O. L. Miller, Jr. Structure and composition of peripheral nucleoli of salaamander oocytes. *Natl. Cancer Inst. Monograph* **23**, 53 (1966).
70. W. A. Newton, J. R. Beckwith, D. Zipser, and S. Brenner. Nonsense mutants and polarity in the *Lac* operon of *E. coli. J. Mol. Biol.* **14**, 290 (1965).
71. Y. Okada, E. Terzaghi, G. Streisinger, J. Emrich, M. Inouye, and A. Tsugita. A frame-shift mutation involving the addition of two base pairs in the lysozyme gene of phage T4. *Proc. Natl. Acad. Sci. U. S.* **56**, 1692 (1966).
72. A. Pardee, F. Jacob, and J. Monod. The genetic control and cytoplasmic expression of "inducibility" in the synthesis of β-galactosidase by *E. coli. J. Mol. Biol.* **1**, 165 (1959).
73. A. Pardee and L. S. Prestidge. On the nature of the repressor of β-galactosidase synthesis in *E. coli. Biochim. Biophys. Acta* **36**, 545 (1959).
74. G. Pontecorvo. "Trends in Genetic Analysis." Columbia Univ. Press, New York, 1958.
75. M. Ptashne. Isolation of the λ phage repressor. *Proc. Natl. Acad. Sci. U. S.* **57**, 306 (1967).
76. M. Ptashne. Specific binding of the phage repressor to λ DNA. *Nature* **214**, 232 (1967).
77. H. Ris and B. L. Chandler. The utrastructure of genetic systems in prokaryotes and eukaryotes. *Cold Spring Harbor Symp. Quant. Biol.* **28**, 1 (1963).
78. J. R. Roth and B. N. Ames. Histidine regulatory mutants in *Salmonella typhimurium*. II. Histidine regulatory mutants having altered histidyl-tRNA synthetase. *J. Mol. Biol.* **22**, 325 (1966).
79. A. S. Sarabhai, A. O. W. Stretton, S. Brenner, and A. Bolle. Co-linearity of the gene with the polypeptide chain. *Nature* **201**, 13 (1964).
80. S. Schlessinger and B. Magasanik. Effect of α-methylhistidine on the control of histidine synthesis. *J. Mol. Biol.* **9**, 670 (1964).

81. D. E. Sheppard and E. Englesberg. Further evidence for positive control of the L-arabinose system by gene *ara* C. *J. Mol. Biol.* **25,** 443 (1967).

82. P. Siekevitz and G. E. Palade. A cytochemical study on the pancreas of the guinea pig. *J. Biophys. Biochem. Cytol.* **7,** 619 (1960).

83. D. F. Silbert, G. K. Fink, and B. N. Ames. Histidine regulatory mutants in *Salmonella typhimurium.* III. A class of regulatory mutants deficient in tRNA for histidine. *J. Mol. Biol.* **22,** 335 (1966).

84. R. L. Sinsheimer. A single-stranded deoxyribonucleic acid from bacteriophage φX174. *J. Mol. Biol.* **1,** 43 (1959).

85. S. Spiegelman, I. Haruna, I. B. Holland, G. Beaudreau, and D. Mills. The synthesis of a self-propagating and infectious nucelic acid with a purified enzyme. *Proc. Natl. Acad. Sci. U. S.* **54,** 919 (1965).

86. J. Spizizen. Transformation of biochemically deficient strains of *Bacillus subtilis* by deoxyribonucleate. *Proc. Natl. Acad. Sci. U. S.* **44,** 1072 (1958).

87. F. W. Stahl and N. E. Murray. The evolution of gene clusters and genetic circularity in microorganisms. *Genetics* **53,** 569 (1966).

88. G. S. Stent. The operon: On its third anniversary. *Science* **144,** 816 (1964).

89. G. Streisinger, Y. Okada, J. Emrich, J. Newton, A. Tsugita, E. Terzaghi, and M. Inouye. Frameshift mutations and the genetic code. *Cold Spring Harbor Symp. Quant. Biol.* **31,** 77 (1966).

90. S. R. Suskind and L. I. Kurek. On a mechanism of suppressor gene regulation of tryptophan synthetase activity in *Neurospora crassa. Proc. Natl. Acad. Sci. U. S.* **45,** 193 (1959).

91. A. L. Taylor and C. D. Trotter. Revised linkage map of *Escherichia coli. Bacteriol. Rev.* **31,** 332 (1967).

92. G. M. Tomkins and B. A. Ames. The operon concept in bacteria and higher organisms. *Natl. Cancer Inst. Monograph* **27,** 221 (1967).

93. A. Tsugita and H. Fraenkel-Conrat. *In* "Molecular Genetics" (S. H. Taylor, ed.), Part 1, p. 477. Academic Press, New York, 1963.

94. J. D. Watson and F. H. C. Crick. Molecular structure of nucleic acids. *Nature* **171,** 737 (1953).

95. J. D. Watson and F. H. C. Crick. The structure of DNA. *Cold Spring Harbor Symp. Quant. Biol.* **18,** 123 (1953).

96. M. G. Weigert and A. Garen. Base composition of nonsense codons in *E. coli.* Evidence from amino acid substitutions at a tryptophan site in alkaline phosphatase. *Nature* **206,** 992 (1965).

97. M. G. Weigert and A. Garen. Amino acid substitutions resulting from suppression of nonsense mutations. I. Serine insertion by the *Su*-1 suppressor gene. *J. Mol. Biol.* **12,** 448 (1965).

98. J. Weigle, M. Maselson, and K. Paigen. Density alterations associated with transducing ability in the bacteriophage lambda. *J. Mol. Biol.* **1,** 379 (1959).

99. R. Weil and J. Vinograd. The cylic helix and cylic coil forms of polyoma viral DNA. *Proc. Natl. Acad. Sci. U. S.* **50,** 730 (1963).

100. H. J. Whitfield, Jr., R. G. Martin, and B. N. Ames. Classification of aminotransferase (C gene) mutants in the histidine operon. *J. Mol. Biol.* **12,** 335 (1966).

101. M. H. F. Wilkins, A. R. Stokes, and H. R. Wilson. Molecular structure of deoxypentose nucleic acids. *Nature* **171,** 738 (1953).

102. C. Willson, D. Perrin, M. Cohn, F. Jacob, and J. Monod. Non-inducible mutants of the regulator gene in the "lactose" system of *Escherichia coli.* *J. Mol. Biol.* **8,** 582 (1964).

103. C. Yanofsky. Amino acid replacements associated with mutation and recombination in the A gene and their relationship to *in vitro* coding data. *Cold Spring Harbor Symp. Quant. Biol.* **28,** 581 (1963).

104. C. Yanofsky, V. Horn, and D. Thorpe. Protein structure relationships revealed by mutational analysis. *Science* **146,** 1593 (1964).

105. C. Yanofsky, B. C. Carlton, J. R. Guest, D. R. Helinski, and U. Henning. On the colinearity of gene structure and protein structure. *Proc. Natl. Acad. Sci. U. S.* **51,** 266 (1964).

106. C. Yanofsky, E. C. Cox, and Y. Horn. The unusual mutagenic specificity of an *E. coli* mutator gene. *Proc. Natl. Acad. Sci. U. S.* **55,** 274 (1966).

107. C. Yanofsky, J. Ito, and V. Horn. Amino acid replacements and the genetic code. *Cold Spring Harbor Symp. Quant. Biol.* **31,** 151 (1966).

108. C. Yanofsky, G. R. Drapeau, J. R. Guest, and B. C. Carlton. The complete amino acid sequence of the tryptophan synthetase A protein (α subunit) and its colinear relationship with the genetic map of the A gene. *Proc. Natl. Acad. Sci. U. S.* **57,** 296 (1967).

109. N. D. Zinder, D. L. Englehardt, and R. E. Webster. Punctuation in the genetic code. *Cold Spring Harbor Symp. Quant. Biol.* **31,** 251 (1966).

II

Synthesis of DNA and Its Transcription as RNA

E. Peter Geiduschek

I. INTRODUCTION

In this chapter on RNA and DNA synthesis, I shall describe and analyze a number of aspects of the two reactions:

$$n \{4 \text{ dNTP}\}^* + \text{DNA (Template)} \rightarrow \text{DNA (Template product)} + 4\,n\text{PP}_i$$

and

$$n \{4 \text{ NTP}\}^* \xrightarrow{\text{(DNA template)}} \text{RNA} + 4\,n\text{PP}_i$$

Both reactions involve, as their main propagation step, the formation of 3′,5′-internucleotide linkages in sequences specified by templates.

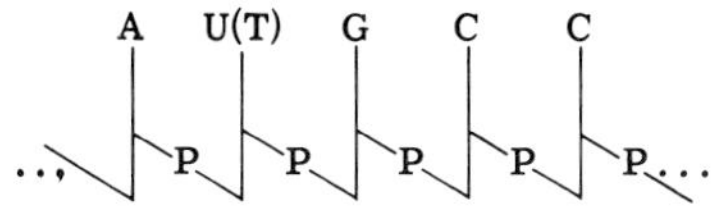

The special character of these reactions, from the chemical point of view, derives from the fact that they are polymerizations that occur on templates. To understand them, it is not only necessary to know the mechanism of propagation but those of initiation and termination† and the relation of product to template, as well.

---

* Deoxyribonucleotide triphosphates. In certain viruses one of the common deoxy-nucleotides is entirely replaced by a virus-characteristic component whose precursor is the corresponding nucleoside triphosphate, e.g., the T-even bacteriophage DNA's contain 5-hydroxymethylcytosine instead of cytosine and certain *B. subtilis* phage DNA's contain 5-hydroxymethyluracil or uracil instead of thymine.

† For DNA synthesis *in vivo* it is not clear whether a distinctive termination step can be specified.

The specificity of the nucleic acid syntheses on templates is guided by two principles: (1) the complementary base-pairs of Watson and Crick and (2) minor nucleic acid constituents such as the methylated, thiolated, and glucosylated nucleotides inserted at the macromolecular level. No exceptions to those two rules are known.

From the biological point of view, the most striking aspect of this entire subject, insofar as we understand it, is, of course, that the nucleic acids are suited to their diverse biological roles in so extraordinarily simple a manner. Beyond this, it is clear that replication and genome

Deoxyribonucleotide triphosphates:

dNTP:

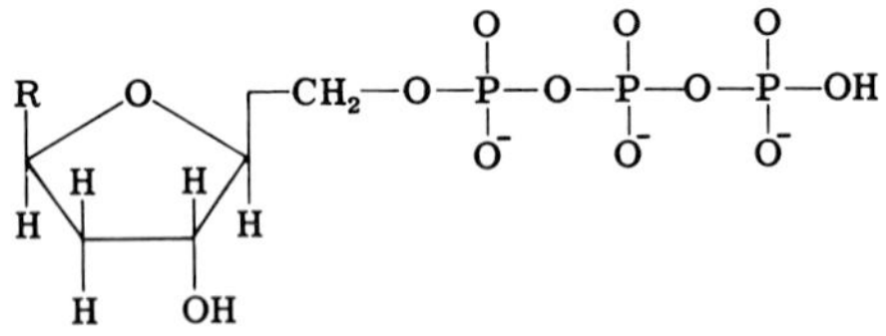

dATP: R =

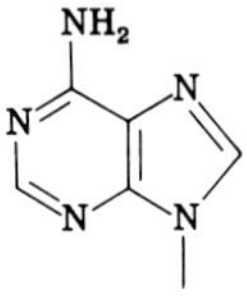

dCTP: R =

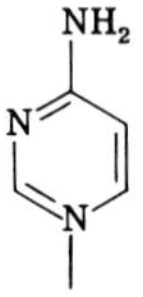

dGTP: R =

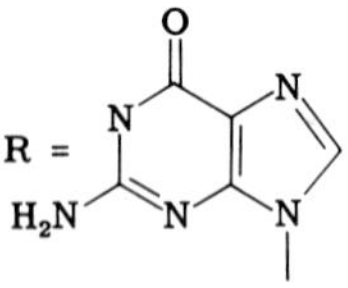

TTP: R =

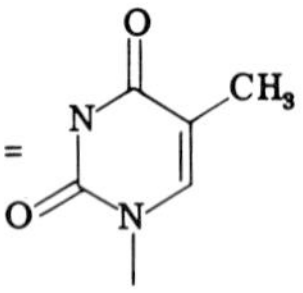

Ribonucleotide triphosphate:

NTP:

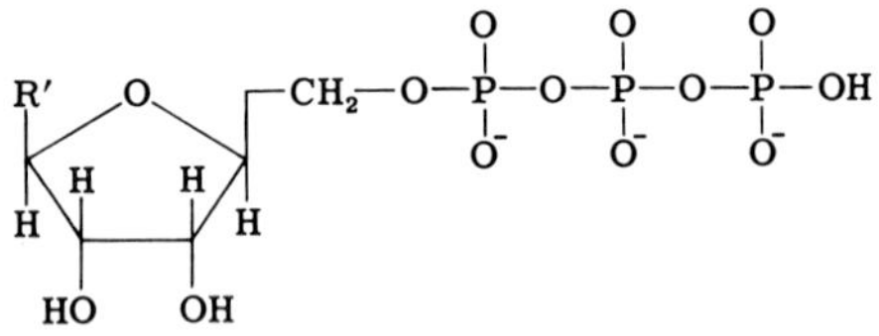

ATP, GTP, CTP; R' corresponds to R above

UTP: R' =

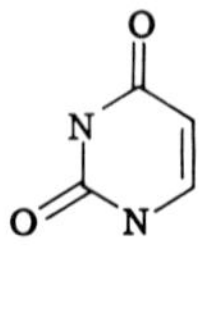

expression are highly organized and regulated activities. Since the main thrust of much current work in biochemistry and developmental biology is to understand this aspect of nucleic acid synthesis, a major part of the chapter that follows is devoted to it. An attempt has been made to specify, where possible, known or postulatable components and mechanisms of this organization and regulation. Currently, this involves a certain amount of guesswork. Some attempt has been made to label speculation as such.

I have, to a certain extent, chosen the perspective that I thought suited to the subject of protein biosynthesis—RNA in the near foreground and DNA in the middle background. Thus, certain topics have received more attention than might otherwise have been accorded them (e.g., stability of mRNA), or less (e.g., transformation, recombination); others have been omitted altogether.

II. DNA SYNTHESIS *IN VIVO*

A. CONTINUITY OF BACTERIAL AND VIRAL CHROMOSOMES

One of the key achievements of classic genetics was the recognition of the relationship between genetic linkage and chromosome structure. It has been a major recent achievement of molecular genetics to show that the same relationship exists at the level of DNA, using bacterial and viral chromosomes. The chain of developments is a long one, including the identification of DNA as the bacterial transforming principle and as the invasive principle of bacteriophage. Some of the most concrete and direct evidence is relatively recent and comes from experiments involving the genomes of the bacteriophage lambda* (λ) and its host, *Escherichia coli*. Both of these are contained in single linkage groups. The DNA of phage λ is a single, continuous structure of molecular weight 33×10^6, and the individual polynucleotide strands of many λ DNA molecules isolated from the mature virus can be shown to be themselves continuous (*76, 345*). Some variants of λ have a deletion covering a region of the chromosome necessary for the establishment of lysogeny. The DNA from such phages (e.g., $\lambda b_2 b_5 c$) is smaller; markers located on either side of the deleted region are genetically linked more closely in $\lambda b_2 b_5 c$ than in wild-type λ. The λ DNA can be fragmented by shearing, and genes located on

---

* Bacteriophage λ has the characteristic form of a regular polyhedral protein coat or head, measuring approximately 55 mμ across, which encloses the viral DNA and to which a thin tail (140 mμ long) is attached.

either side of the linkage map have been located in different halves of the viral DNA (*156, 157*).

Over smaller distances and within restricted regions of chromosomes, the correspondence between genetic and physical proximity becomes more variable due to fluctuations in genetic recombination frequency, which can be quantitatively very impressive (*184, 259*).

The *E. coli* genome is also contained in a single, much larger, circular linkage group and Cairns (*51, 52*) was able to demonstrate the existence of a continuous DNA-containing structure 1100 μ long. A DNA molecule of this size and in the B form (i.e., with base-pairs stacked perpendicularly to the helix axis and 3.4 Å apart) would have a molecular weight of 2.2×10^9. This is only a little less than the DNA content per cell of very slowly growing *E. coli* (Table I). Is the DNA in this structure itself continuous, or are there non-DNA linkers or interruptions? The technical difficulties which must be surmounted in providing answers to these questions are prodigious: (1) Fragility of very long DNA molecules places great restrictions on the methods that may be used to dissociate DNA from protein and other cell constituents. Using extraction methods which are, at least mechanically, particularly gentle, Massie and Zimm (*241*) have been able to isolate DNA with a molecular weight of approximately 2.5×10^8 from *E. coli* and *Bacillus subtilis*, but they could prepare no larger subunits of the chromosomes. (2) DNA from all sources contains a small amount of peptides even after rigorous purification. In fact, it has been suggested that even the simple chromosomes of bacteria and viruses contain peptide linkers in their DNA chains (*22, 241*). How-

TABLE I

DNA CONTENT AND REPLICATION IN *E. coli* 15 THY-MET-ARG-TRY GROWING AT DIFFERENT RATES[a]

Carbon source	Doubling time	DNA content (μg/10⁷ cells)	Percent increase DNA-aa[b]
Glucose	40	0.138	42
Succinate	70–75	0.103	21
Aspartate	120	0.070	18
Proline	180	0.051	8
Acetate	270	0.055	—
Resting *E. coli* chromosome[c]	—	0.045	—

[a] From (*214*) and (*215*).

[b] Residual DNA synthesis after shift from steady growth to amino acid starvation in the presence of thymine.

[c] Assuming 2.2×10^9 as the molecular weight of the completed *E. coli* chromosome (*51*).

ever, unequivocal evidence that such linkers exist has not been provided.

We shall postulate tentatively that DNA-associated peptides have some other origin, and that the DNA of a single replicating unit forms a unitary double strand structure. This is the assumption adopted for the subsequent discussion of replication in bacteria. There is evidence that the chromosomes of higher plants and animals contain many simultaneously replicating units (53, 146, 163a, 284, 352–354).

Little information is available on the continuity of the single strands of bacterial chromosomes, although this has important significance for the mechanics and topology of replication. In this connection, two problems can be distinguished. (1) What is the state of integrity of previously replicated DNA strands? At present, the possibility that several single-strand breaks exist in a bacterial chromosome, at any instant, cannot be excluded. The technical obstacles to answering this question by direct macromolecular analysis are very considerable because of the conflicting requirements of DNA purification and of the fragility of very long polynucleotide chains. (2) What is the covalent linkage of newly replicated DNA at the time of synthesis? A recent experiment bearing on this question is discussed in Section II,E of this chapter.

B. Semiconservative Replication

Part of the extraordinary effect that the Watson–Crick (364) hypothesis had on the development of molecular genetics derived from the DNA structure itself; in a way that is without parallel, the structure suggested function—the semiconservative mechanism of replication. To distinguish between the hypothesis and possible alternative modes of replication (77) Meselson, Stahl, and Vinograd (246) developed the method of equilibrium centrifugation in a density gradient and Meselson and Stahl (245) then performed their $^{15}N \rightarrow {}^{14}N$ isotopic transfer experiment of DNA synthesis. The familiar result of this experiment is that only three density classes of DNA could be distinguished: all ^{15}N, all ^{14}N, and hybrid. This is consistent with the conservation of a subunit of the intact DNA-containing half the ^{15}N during replication, as the semiconservative replication hypothesis would require. A number of subsidiary experiments culminating in Cairns' measurement (52) of the length of E. coli DNA established beyond reasonable doubt that the conserved unit is a single DNA polynucleotide chain. Experiments modeled on the Meselson–Stahl experiment have established the generality of semiconservative replication in many other organisms (61, 82, 155, 191, 208, 245, 325, 346).

The Meselson–Stahl experiment also provides evidence for the orderliness of replication in a population of exponentially growing cells, since all heavy *E. coli* DNA is converted to hybrid DNA before any light DNA can be found. The observation implies that replication is subject to the following three conditions: (1) Replication is sequential in each chromosome. (2) The replication sequence is repeated in succeeding generations. (3) The cell population has a relatively uniform replication rate. The appreciation of this aspect of the Meselson–Stahl experiment has led to extensive progress in understanding the control of replication in bacteria. Before discussing this subject, however, we shall first say something about the morphological basis of ordered replication.

C. The Morphological Basis of Ordered Replication

Replication of the chromosome of *E. coli* is unidirectional. Under ideal circumstances—which correspond to those of the Meselson–Stahl experiment—there is only one growing point per chromosome. These are the conclusions that can be drawn from experiments by Cairns (*51, 52*) and Bonhoeffer and Gierer (*37*).

Bonhoeffer and Gierer explored the following implication of the Meselson–Stahl experiment: the length of time that it takes to make a DNA molecule hybrid over length L, can be measured by density-shift labeling experiments and is simply the inverse of the propagation rate of replication. The ratio of the nucleotide incorporation rate per nucleus to the propagation rate, in the proper units, then gives the number of growing points per chromosome. In this way, it was possible to show that there were no more than two and most probably only one growing point per *E. coli* chromosome. Cairns' experiment involved an examination of autoradiographs of intact *E. coli* chromosomes (Fig. 1) and led to the interpretation of the progress of replication in terms of the model shown in Fig. 2.

Replication proceeds along one growing point from an origin. Partially replicated *E. coli* chromosomes are closed; the replicated and unreplicated parts are fused to each other at two points—the growing point and the origin—which may be distinguished, the one from the other, by the autoradiographic grain density pattern. If there is only one growing point on the replicating *E. coli* chromosome, its rate of progression must be very fast. A replication cycle of 40 minutes corresponds to polymerization of 1400 base-pairs per second, or 8000 turns of the DNA-B helix per minute.

An intriguing, and as yet unsolved, problem posed by the nature of the replicating mechanism is: How does the replicating chromosome un-

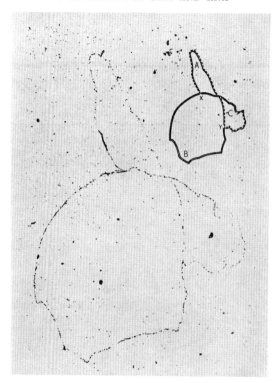

Fig. 1. Autoradiograph of a chromosome of *E. coli* K12 Hfr labeled with tritiated thymidine for two generations and extracted with lysozyme. Exposure time 2 months. The scale shows $100\,\mu$. Inset: to guide the eye, the same structure is shown diagrammatically and divided into three sections (A, B, and C) that arise at the two forks (X and Y) (*52*).

wind the parental, and wind up the daughter, DNA helices? For a closed circular double-strand DNA molecule to keep winding in one sense, it must have at least one swivel—a place at which continuous rotation in one sense around covalent bonds is possible. A single-strand break in the double helix would serve as such a swivel; so would a single-strand non-DNA linker, inserted into the chromosome. If it contains only one swivel, then the components of an entire chromosome and all appendages—including RNA polymerase, nascent polysomes, and growing polypeptide chains—have to rotate together. We have just reviewed the relevant facts for bacterial chromosomes; the existence of several points of free rotation within one chromosome is within the realm of possibility. However, circular DNA molecules without swivels exist. These are the doubly closed circular "type I" DNA's of polyoma, papilloma, and SV40 viruses, of the replicative form of phage ϕX174, of λ, and of mitochondria (*69*,

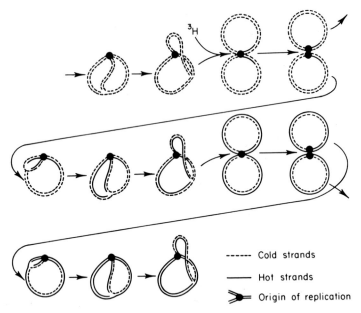

FIG. 2. A diagrammatic representation of the replication cycle which is postulated to have produced the autoradiograph shown in Fig. 1. Each round of replication (long arrow) begins at the same place and proceeds in the same direction (*52*).

180, 202, 203, 266, 309, 326, 359, 360). Perhaps they are, in fact, non-replicative forms and cannot be replicated until at least one point of free rotation is introduced.

D. CONTROL OF REPLICATION

Replication in a given strain of bacteria starts within a characteristic initiation region of the chromosome which is genetically determined. The initiation of a round of replication requires protein synthesis. The evidence for these two statements, which will now be described briefly, comes from experiments of the past 7 years with *E. coli* and *B. subtilis* (*216*). Only some of the simpler results will be presented.

When an *E. coli* auxotroph is deprived of its required amino acid, protein synthesis stops. If the strain is stringent, the rate of RNA synthesis immediately decreases by a factor of approximately 15, but DNA synthesis continues for some time. Certain strains of *E. coli*, growing in a complete medium with glucose as the carbon source, increase their total DNA content by approximately 40% during this starvation period. When amino acids are restored after such an extended period of starvation,

DNA synthesis lags behind RNA and protein synthesis; ultimately the population of cells resumes steady growth characterized by the normal proportions of protein and nucleic acids.

Maaløe and co-workers postulated the following interpretation of these and other observations (*138, 229, 230, 314*). A logarithmic population of cells contains chromosomes at all stages of their replication cycle. Completion of an already initiated replication cycle does not require the maintenance of protein synthesis. Consequently those cells caught in mid-cycle at the time of starvation complete a round of replication; the residual DNA synthesis corresponds to the amount of DNA needed to complete a set of chromosomes randomly phased in their replication cycle, and containing only one growing point.

As a matter of fact, the amount of residual DNA synthesis during amino acid starvation changes with bacterial growth rate (Table I). Such variations suggested that the number of simultaneously active replicating regions per bacterium and the fraction of the cell cycle occupied with replication might vary with growth rate. Subsequent experiments have put this question on a firm basis (*153*). The *E. coli* B/r replication transit time—the time that it takes for a single growing point to traverse the entire *E. coli* B/r chromosome—is relatively constant at bacterial doubling rates greater than 1 per hour (at 37°C), but the transit rate also slows down at lower growth rates.

In rapidly growing *E. coli* B/r, the interval between divisions is less than half the replication transit time. Under these circumstances each chromosome contains more than one growing point and replication is multifurcate. No cells in a population growing steadily in this way contain unforked, resting chromosomes—not even those cells that are newly generated by septum formation and parent cell separation (*66, 153*). At sufficiently slow growth rates, replication does not occupy the entire cell cycle and there is a nonreplicative period comparable to the so-called G periods of eucaryote cells.

Many of the experiments that placed the initial hypotheses about control of replication on an experimental footing were initiated by Lark, Sueoka, and their collaborators (*215, 347*). If the model outlined above is correct, then amino acid starvation should phase replication. At the reinitiation of replication after such a starvation period, thymine should be incorporated into DNA located near the initiation point. If the initiation locus does not move in succeeding generations, then after each cycle of amino acid starvation and reinitiation, DNA synthesis should occur at the same locus. Lark, Repko, and Hoffman (*217*) showed that this is the situation in *E. coli* 15 Thy- Arg- Met- Try-, and an experiment of theirs is summarized in Fig. 3: after one cycle of amino acid starvation,

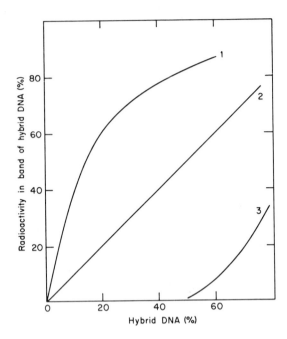

FIG. 3. An experiment which provides evidence that the *E. coli* chromosome has a specific region at which replication is initiated after amino acid starvation. *E. coli* 15 Thy-Arg-Met-Try- are subjected to the following regimen:

(a) growth in complete medium →

(b) amino acids minus 90 minutes to complete replication cycle →

(c) amino acids + ^{3}H-thymine to label beginning 10% of chromosome →

(d) grow 6 generations in complete medium to randomize replication →

(e) amino acids minus 90 minutes to complete replication cycle

They are then transferred to complete medium containing BU instead of thymine, so that all newly replicated DNA is density labeled. The fraction of the DNA that has been replicated after transfer to BU is equal to (% DNA of hybrid density)/(200 − % DNA of hybrid density). The ordinate gives the percent of ^{3}H-label contained in newly synthesized hybrid DNA. Curve 1, result of the above experiment. Curve 2, same, but step (e) is omitted. This control shows that cells were randomized in their replication cycle by step (d). Curve 3, steps (b), (d), and (e) omitted, so that density transfer follows immediately after ^{3}H-labeling. After Lark (*215*).

thymine is incorporated into DNA for 0.1 generations. Growth with unlabeled thymine is then continued for as many as six generations and
followed by another cycle of amino acid starvation. When DNA synthesis is now reinitiated in a medium containing the high density DNA
precursor, 5-bromouracil (BU), the BU is predominantly incorporated
into the chromosome segment that incorporated ^{3}H-thymine during the
previous reinitiation period after amino acid starvation (Curve 1).

Contrast this result with the consequences of a different labeling
regime: Exponentially growing E. coli 15 T$^-$ are briefly labeled with ^{3}H-
thymine and transferred immediately to BU. In this case the BU is
initially incorporated into chromosome segments that were not labeled
with ^{3}H-thymine (Curve 3).

This experiment specifies the retention of the replicative initiation point
over several, but not over several dozen or several hundred, generations.
It therefore does not directly establish whether the origin of replication
is located in the same region of the chromosome in all the cells of the
population. However, experiments to decide this point can be devised,
because the following consequences should be observable if all the cells
of a strain of bacteria initiate their rounds of replication at the same
chromosomal locus. (1) Resting and logarithmically growing populations
should have different distributions of gene abundances, with those genes
closer to the replicative origin being relatively more abundant in the
growing cells. (2) Cells synchronized in their replication cycles by amino
acid starvation should be capable of having the beginnings and ends of
their chromosomes selectively labeled (e.g., by density labeling; cf.
Fig. 3).

In order to make tests of these predictions practicable, some means of
measuring relative gene abundances is required and the best-studied
technique is that of bacterial transformation. We therefore turn first to
B. subtilis where relative transforming activities of different genes were
first measured by Sueoka and collaborators. Of course, the absolute transforming activity of markers depends not only on the relative abundance
of the corresponding DNA segments but also on other factors. The most
easily interpretable results are therefore provided by the ratio of transforming activities of DNA extracted from steadily growing cells and from
units, such as spores, all of whose chromosomes have completed replication. The results of such experiments permit the construction of a
chromosome map (Fig. 4).

Upon germination of B. subtilis spores, genes located near the replicative origin are the first to be duplicated. When B. subtilis spores grown
in ^{15}N–D$_2$O medium are germinated in ^{14}N–H$_2$O, the heavy parental
chromosome is replicated, and the density of the corresponding DNA seg-

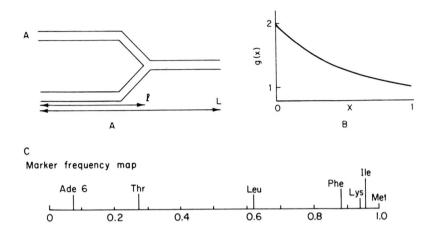

C

Marker frequency map

Density transfer map

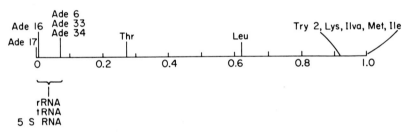

FIG. 4. (A) Relationship of abundance of genes to distance from the replicative origin. Replicating chromosome with 1 replicative fork. $x = l/L$ is the fractional distance along the chromosome from the origin of replication. The chromosome is shown for convenience as a linear structure, but this does not affect the model. (However, attempts to demonstrate physical or genetic circularity of the B. subtilis chromosome have not been successful thus far.)

(B) Expected distribution of abundances of genes in a population of chromosomes randomly phased in replication which proceeds at a constant rate along the structure shown in A. Genes located at the origin are twice as abundant as genes located at the other end. The distribution function is $g(x) = 2^{1-x}$.

(C) Results of marker frequency and density transfer analysis of B. subtilis gene order, presented as chromosome map. (After 279, 333.)

ments is changed to that of hybrid DNA (containing one strand of $D^{15}N$ and one strand of $H^{14}N$–DNA). The order of transfer of genetic markers from heavy to hybrid density during replication (Fig. 5) permits the construction of a chromosome map of replication sequence (279). These two ways of constructing a B. subtilis chromosome map give, for the most

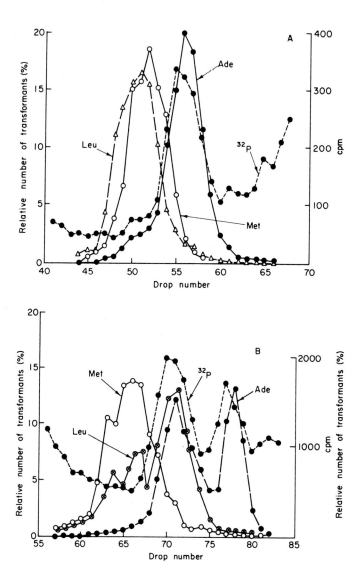

FIG. 5. The ordered replication of *B. subtilis* genes (271); *B. subtilis* spores are prepared in $D_2O-{}^{15}N$ medium and germinated in ${}^{32}P-H_2O-{}^{14}N$ medium. At two successive stages of germination, samples are withdrawn, the DNA is prepared and centrifuged in CsCl density gradients. Samples are collected from the gradient and assayed for the transforming activity of several markers (see Fig. 4 for location on chromosome map) as well as incorporated ${}^{32}P$. Three density classes of DNA are clearly distinguishable—heavy, hybrid and light. The ${}^{32}P$-label is mainly confined to the latter two since it is incorporated into newly synthesized DNA. A shows the adenine marker being replicated before the leucine and methionine markers. B shows the effects of multifurcate replication. The DNA has been prepared later than A; *ade* is replicated a second time before *met* is replicated for the first time.

part, consistent results [Fig. 4(C)]. The important conclusions which may be drawn from these experiments and others like them are: (1) Replication of the *B. subtilis* chromosome is unidirectional; (2) at intermediate steady rates of growth there is only one growing point per chromosome, and replication is continuous throughout the cell cycle; (3) a unique initiation point for replication can be distinguished—in two strains of *B. subtilis* it is shown to be an identical inherited property. Up to this point, the evidence about replication which has been presented establishes (1) the genetic linkage and physical continuity of the *E. coli* chromosome but not the order of replication and (2) the order of replication of the *B. subtilis* chromosome but not the genetic linkage.

However, analysis of the order of replication of *E. coli* genes has been undertaken recently (*1, 24, 376*). One method, which we shall not describe in detail, involves density labeling of DNA regions replicating near the beginning and end of the replication cycle (cf. Fig. 3, for example), infecting with the transducing phage P1, and analyzing the transducing ability of density-labeled progeny phage for different bacterial genes. The results of such an analysis locate the starting point for DNA replication after amino acid starvation for several strains of *E. coli* and, with somewhat less assurance, the direction of replication (Fig. 6). It is interesting, for historical reasons, to note that the strains of *E. coli* chosen for this analysis include those whose genetic transfer during conjugation has widely different origins and differing (clockwise or anticlockwise) sense (*152*). The replicative origins and directions during normal vegetative growth are not determined by the origins and direction of genetic transfer during conjugation.

What determines the initiation of chromosome replication and the location of the replicative origin? From a biochemical point of view this is an unsolved problem, but the formal replicon theory of Jacob, Brenner, and Cuzin (*176*) has been important in focusing attention on the essential properties of ordered replication. An independently replicating unit, such as an *E. coli* chromosome, is called a replicon. The replicon is postulated to possess two essential control elements: a site at which replication is initiated (which is unfortunately called the replicator) and a product of one of its genes which starts replication at the replication site (and perhaps continues replication). This gene product is called the initiator and it is postulated to be chromosome specific: In a cell containing two different replicons (such as a chromosome and a *Flac* episome) the initiator of one replicon does not function at the replicator of the other. Thus, if bacteria containing an episome with an initiator that is inactivated at higher temperatures, are shifted to nonpermissive temperatures, they should continue to divide and to replicate only their chro-

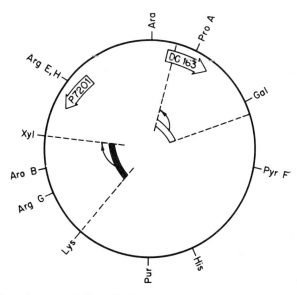

Fig. 6. Genetic map of *E. coli,* showing the location of regions in which the origins of replication after amino acid starvation are located and the direction of replication (*376*): closed rectangles, strains (F⁻)DG68, (F⁻)DG75, and (Hfr) P7201; open rectangles, strain (Hfr)DG163. The open arrows point to the origin of transfer of genes during conjugation of Hfr strains and indicate the direction of transfer. The loci defining position on the chromosome are those of genetic markers concerned with arginine (Arg), xylose (Xyl), lysine (Lys), purine (Pur), histidine (His), pyrimidine (Pyr), galactose (Gal), proline (Pro), arabinose (Ara), and aromatic amino acids metabolism (Aro).

mosomes, thereby diluting out the episomes in the population. Strains of *E. coli* carrying temperature-sensitive *Flac* episomes, which behave in this way, are known. Episomes can also be integrated into the bacterial chromosome [see reviews by Jacob and Wollman (*179*) and Driskell-Zamenhof (*84*)]. If the chromosomal replicon bearing an integrated episome is to have a unique replication cycle, the function of one of the initiators must be suppressed. The question of which one remains active during normal replication is only now being resolved, the current results having been stated above and presented in Fig. 6.

An *E. coli* cell growing at intermediate growth rates and carrying an episome contains several concurrently replicating units. There is evidence that the distribution of these units among the two daughter cells is orderly so that the chromosome and episome synthesized in a progenitor cell are distributed together (rather than independently) in the progeny (*72*). The replicon theory proposes that DNA attaches to a cellular site, postulated to be the plasma membrane, and that the movement and partition of this into progeny cells assures such ordered segregation. The mode of attachment must, in addition, be such that segregation is orderly over several generations, even for bacterial chromosomes with several replicating forks. When replicating chromosomes recombine, structures of considerable complexity can result. Bacteriophage T4 provides a striking illustration of these possibilities. Recombination and replication–initiation are both frequent (*338, 370, 371*) and the resulting replicating structures appear to be tightly interconnected and intricate (*163*).

Thus two regions of the bacterial chromosome might conceivably have special physical properties by virtue of their special location and function: the growing point and the origin of replication. The former, associated with the replicative enzyme system, could have special conformational properties and both could be bound to cytoplasmic membrane (*310*). The exploration of these possibilities has provided some suggestive evidence: Newly replicated DNA is found to be membrane-associated and retains this association in certain methods of extraction (*85, 106, 139*). In other extraction methods, it exhibits distinctive denatured DNA-like properties (*200, 274*). While neither result specifies the properties of newly replicated DNA directly, both attest to its distinguishability. Another aspect of this question is discussed in Section II,E.

Finally, a few words about more complex relationships in DNA replication. When cells contain plasmids* whose replication is not coordinately controlled with chromosomal replication, they may undergo drastic changes of DNA content and composition. The most completely studied systems of this kind involve the induction of lysogenic viruses which triggers a sequence of steps leading to the autonomous and rapid replication of the viral episome (*152, 179*). When a normal lysogenic phage is induced, viral DNA synthesis is followed by the action of other viral genes specifying proteins of the viral coat so that the induced cells ultimately produce mature progeny virus. However, when "maturation defective" lysogens are induced (these are lysogens which lack the function of a gene-controlling expression of part of the viral genome), replication continues in the absence of progeny maturation (*83, 89, 186*). The most ubiquitous plasmids are those corresponding to mitochondrial and chloroplast DNA (*158*). The consequences of noncoordinate replication of plasmid and chromosome have been explored by Chiang (*59*) and Chiang and Sueoka (*60*) in an alga, *Chlamydomonas reinhardtii*, exploiting the fact that the plasmid DNA's

---

* Plasmids are extrachromosomal replicating elements. Those plasmids which can also exist as integrated parts of the chromosome are called episomes.

of this organism have characteristic buoyant densities that permit them to be distinguished from the major DNA component. Chiang has shown that the absolute abundance of chloroplast satellite DNA, and its abundance relative to chromosomal DNA undergo drastic changes characteristic of different stages of the life cycle of *Chlamydomonas reinhardtii;* chloroplast DNA is most abundant in vegetable cells. At one stage of zygospore maturation, another satellite DNA becomes abundant. Chiang (*59*) postulates that this might be mitochondrial in origin. During maturation of sea urchin and frog eggs, great quantities of a cytoplasmic DNA, probably of mitochondrial origin, accumulate. This appears to be another example of the noncoordinate replication of an extrachromosomal DNA.

An entirely separate question, currently actively pursued by many, is: What part of the proteins required for mitochondrial or chloroplast function is coded by plasmid genes and what part by nuclear genes?

E. NEWLY REPLICATED DNA

The topology of parental DNA during replication poses an as yet unsolved mechanical problem which was stated in Section II,C. What of the newly replicated DNA? A recent experiment by Okazaki *et al.* (*273, 274*) provides important insights into its organization. They have examined the single-strand lengths of newly replicated DNA in *E. coli* (15 T⁻ and B) and T4-infected *E. coli* B and find that newly replicated, briefly pulse-labeled DNA is isolated in very much shorter (sedimentation constant about 10 S) segments than the bulk of the DNA. The subsequent integration of these smaller segments into larger pieces can be followed quite readily and the integration does not require thymine. One must conclude either that the newly replicated DNA is selectively very sensitive to a variety of isolation and denaturation methods or that replication is discontinuous, yielding short DNA segments which are subsequently joined together (Fig. 7). More recently it has been shown that newly synthesized *short* DNA segments have both chemical polarities (*285, 348a*). This suggests that if any of the models of Fig. 7 is valid, then it is model C.

F. REPAIR OF DNA

Replication is quantitatively by far the most important aspect of DNA synthesis in growing cells. However, enzyme-mediated reactions involving DNA synthesis are also required for recombination and repair. Repair occurs in a variety of cells after inactivation by ultraviolet light, X-rays, decay of incorporated ^{32}P, and reaction with alkylating agents such as the nitrogen and sulfur mustards, mitomycin C, methyl- and ethyl methane sulfonate, N-nitro-N'-nitrosoguanidine, and others. The discussion offered here is necessarily very brief and for a more detailed summary of the available information, the reader is referred to reviews by Howard-Flanders and Boyce (160) and Strauss (344). Of the repair processes, that involving ultraviolet lesions is the best studied and understood. Two entirely distinct repair mechanisms of ultraviolet damage exist: (1) photoreactivation of the lesions which is less closely related to DNA synthesis *in vivo* and *in vitro* and (2) dark repair to which we confine our discussion.

Among products of DNA irradiation at 280 mμ are thymine dimers (T̂T) (28, 29, 361). These arise predominantly or exclusively from the reaction of adjoining thymine residues on the same polynucleotide chain. The other pyrimidine dimers, ĈT and ĈC, are also formed in ultraviolet irradiated DNA. In *E. coli* recovering in the dark from such irradiation, T̂T-containing tri- and tetranucleotides appear in the acid-soluble fraction as the result of excision from the DNA. This excision is generally accompanied by much more extensive solubilization of DNA than

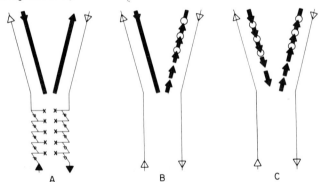

Fig. 7. Three possible relationships of newly replicated and parental template DNA in the replicating region. The direction of replication is from top to bottom. ————, parental DNA; the chemical polarities are indicated at the bottom and symbolized ($\rightarrow$) by arrows pointing from the 3'-OH to the 5'-OH that are joined by an internucleotide linkage. This is the direction of growth of DNA chains *in vitro*. ➡ newly synthesized DNA; ○ repair-linkage of newly synthesized DNA.

(A) All newly synthesized DNA is continuous from the replicative origin. Two chemical polarities of chain elongation are required. (B) Half the newly replicated DNA is synthesized in short segments. (C) All newly synthesized DNA is initially in short pieces. (B and C) All DNA could be synthesized with the same chemical polarity. A repair system is required to form the intact chromosome.

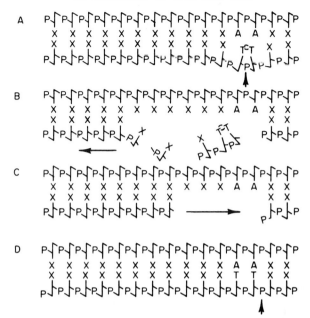

FIG. 8. A possible sequence of steps in the repair of ultraviolet-irradiated DNA containing a pyrimidine dimer (*162*). (A) An ultraviolet irradiated DNA twin helix containing a thymine dimer (arrow). (B) A single-strand fragment of unknown length containing the dimer is released from the DNA by enzymes. This excision is accompanied by DNA breakdown with the release of 5′-nucleotides containing thymine. (C) Resynthesis by a DNA repair polymerase. (D) The reconstruction of the DNA twin helix is completed as the last phosphodiester link (arrow) is joined. The available experimental data do not establish the sequence of steps B and C.

can be accounted for by $\widehat{TT}$ and other pyrimidine dimer nucleotides. Presumably the excised portions are contiguous with thymine and other dimers and may be several hundred nucleotides in length (*319, 321*). Such cells, recovering from ultraviolet irradiation can be shown to synthesize DNA dispersively, i.e., in small segments at many loci (*282*).* Presumably this is the repair synthesis that fills up the single-strand holes created by excision and degradation and involves the resynthesis of one strand of DNA against the template of its complement. Thus a possible sequence of enzymic reactions involved in the dark repair of ultraviolet damage is: (1) excision of pyrimidine dimers; (2) solubilization which, it is reasonable to suppose, should involve an exonuclease [certain ultraviolet-sensitive mutants of *E. coli* are defective in steps (1) and (2)]; (3) resynthesis; and (4) rejoining (Fig. 8).

I do not discuss recombination at all. However, it is important to

* A similar dispersive DNA synthesis can be observed in *B. subtilis* recovering from DNA alkylation by methyl methane sulfonate (*289*).

point out that many single *E. coli* mutants defective in recombination are also defective in repair of ultraviolet lesions; these pathways therefore may use, or be affected by, the products of common enzymes. This is a theme to which I return briefly in Section III of this chapter.

G. Viral Replication

A number of the larger bacteriophages produce part or all of their own replicative, recombination, and repair machinery. For example, the *E. coli* bacteriophage T4 chromosome contains the genes of an entire pathway for DNA synthesis (*63*). Among the products of these genes is a new DNA polymerase which can be distinguished from that of the host (*9*). As a result of the genetic studies of Epstein and collaborators (*93*), the gene responsible for this polymerase can be identified as T4 gene 43 (*80*).

Certain mutants in gene 43 cannot synthesize DNA at high temperatures, and the DNA polymerase isolated from some of these mutants can also be shown to be thermolabile. Among temperature-sensitive mutants in gene 43 (*336*), some have the properties of a temperature-sensitive mutator gene. At temperatures not sufficiently high to block viral development completely, these gene 43 mutants generate mutations in other T4 genes at abnormally high frequencies. Clearly, therefore, the T4 DNA polymerase which is the product of this gene is not only required for DNA synthesis but is also involved directly in DNA polymerization. This does not imply that the gene 43 protein is the sole DNA replicative enzyme; it could be a part of the replicative apparatus and its function may include repairlike processes.

Another T4 gene whose action is required for sustained viral DNA synthesis is gene 30 which has been identified as coding for a polynucleotide-joining enzyme, or ligase, which catalyzes the formation of phosphodiester bonds on a template; the enzyme therefore repairs single-strand breaks in double-strand DNA (*97, 365*). In host cells infected with T4 gene 30-defective mutants, viral DNA replication commences but stops very soon (*32*). The "defective" newly synthesized DNA is made in small segments which are not joined together (*274*). Much of the newly synthesized DNA is ultimately degraded (*159*).

There is an obvious superficial resemblance between these facts about the role of the T4 ligase and the model of replication presented in Fig. 7. It must also be admitted unfortunately that the resemblance and superficiality are interdependent: (a) the product of T4 gene 30 is not the only one determining a ligase-defective phenotype of viral replication (*270*); (b) the ligase function of T4 gene 30 is exerted on parental DNA and on other than newly replicated DNA (*295, 303*); and (c) the function of the T4 gene 30 ligase that makes it essential to sustained replication may be its antagonism of a nuclease (*207*). Further work on the viral DNA polymerase and ligase will be required to define the enzymic relationships of replication, repair, and recombination.

III. DNA SYNTHESIS *IN VITRO*

A. DNA POLYMERASE

Polymerization of DNA, dependent on the presence of four nucleoside triphosphates, has now been observed with enzymes purified from several sources. The most extensively studied DNA polymerases are those isolated from *E. coli* and calf thymus but, as summarized in Table II, highly purified enzymes have been isolated from other sources. Some of the salient features of the enzymic synthesis of DNA are:

(1) The enzymic polymerization of DNA (to be more precise—of DNA containing all 4 nucleotides) requires a template.

(2) The nucleotide composition of the product corresponds to that of the template; high molecular weight polydeoxyribonucleotide is synthesized.

(3) Nearest neighbor frequencies of the synthesized product conform to the requirements of the replication of antiparallel chains base-paired according to the Watson-Crick hypothesis. For example, in a double helical AT and GC base-paired DNA, the sequence ApG occurs on one polynucleotide strand every time the sequence CpT occurred on its complementary strand. The nearest neighbor distributions of three *in vitro* synthesized DNA's are listed in Table III in terms of these pairs of complementary dinucleotide sequences, which are seen to be replicated with equal frequency *in vitro*.

(4) 3'-OH terminated polynucleotide chains are the most active templates for DNA polymerases. 3'-OH ends are generated by endonucleases such as pancreatic DNase I and by sonic disintegration of DNA (*291*), and these treatments activate DNA templates. The *E. coli* enzyme binds to helical DNA only at molecular ends and at internal chain interruptions ("nicks"; *91*). In some instances, the 3'-OH template

E. P. GEIDUSCHEK

TABLE II
Properties of Some DNA Polymerases

	Calf thymus	E. coli	T2, 4	T5	B. subtilis	M. lysodeikticus
pH optima	7.0	7.4	8–9	8.1	8.2[a]	7.4–8[b]
Mg^{2+} optima, mM	4–8	7	6	10	≥ 1.5	3–5[b]
Purification[c]	150-fold	1500	600	400	600	1500
Template requirement	S	N,S[d]	S	S	N,S	N,S
Sedimentation coefficient		5.6				
Molecular weight	—	1.09×10^5	1.12×10^5	—	—	(ca.) 8×10^{4e}
Maximum extent of synthesis: Moles nucleotide product/ Mole nucleotide template	1	<1	>1	>1	<1	<1
Reference	(33, 383)	(18, 91, 297)[f]	(8a, 9, 126)	(27, 278)	(275)	(390)

[a] Varies with buffer ion.

[b] Varies with buffer and nature of template.

[c] The relative response of enzyme to different templates, e.g., endonuclease-activated calf thymus DNA and poly d(A-T), changes as purification proceeds. The figures cited are for those templates giving the most impressive result.

[d] Template activities are comparable at 37°C; at 20°C S > N. S = single strand; N = double strand.

[e] Approximate value from gel filtration.

[f] See also Jovin, Englund, and Bertsch referred to in (278).

TABLE III
NEAREST NEIGHBOR FREQUENCY OF *in Vitro* SYNTHESIZED DNA AND RNA

Sequence[a]	T2 DNA[b]	T2 RNA[c]	λ+ DNA[b]	λ+ RNA[d]	M. lysodeikticus DNA[b]	M. lysodeikticus RNA[e]
ApA ŪpŪ	0.119 0.108	0.111 0.106	0.069 0.073	0.076 0.086	0.019 0.017	0.019 0.018
CpA ŪpG	0.061 0.064	0.061 0.063	0.070 0.069	0.072 0.079	0.052 0.054	0.053 0.056
GpA ŪpC	0.057 0.061	0.059 0.057	0.060 0.064	0.060 0.058	0.065 0.063	0.066 0.065
CpU ApG	0.054 0.055	0.054 0.057	0.056 0.053	0.053 0.055	0.050 0.049	0.047 0.048
GpU ApC	0.052 0.049	0.051 0.048	0.054 0.054	0.058 0.051	0.056 0.057	0.057 0.053
GpG CpC	0.034 0.031	0.036 0.034	0.062 0.063	0.052 0.050	0.112 0.113	0.112 0.111
ŪpA	0.082	0.089	0.047	0.051	0.011	0.009
ApŪ	0.103	0.104	0.070	0.076	0.022	0.026
CpG	0.031	0.030	0.064	0.053	0.139	0.140
GpC	0.038	0.040	0.072	0.069	0.121	0.121
A	0.326	0.320	0.247	0.259	0.147	0.146
Ū	0.315	0.315	0.252	0.272	0.145	0.148
G	0.181	0.186	0.249	0.240	0.354	0.356
C	0.177	0.179	0.252	0.229	0.354	0.351

[a] Ū = Uracil for RNA, thymine for DNA.
[b] E. coli DNA polymerase (185).
[c] M. lysodeikticus RNA polymerase (369).
[d] E. coli RNA polymerase (232).
[e] E. coli RNA polymerase [data of (331) which differs by less than 0.007 for all frequencies from data of (164)].

chain end has been shown to serve as a primer* so that the synthesized DNA is joined to the template (*33*, *34*, *292*).* For T4 DNA polymerase the primer requirement appears to be absolute (*126*). However, *E. coli* DNA polymerase catalyzes DNA polymerization on a circular, single-strand DNA template without 3′-OH ends. This appeared, at first sight, to be a fundamental difference between the T4 and *E. coli* DNA polymerases but emerges, on detailed study, as nothing of the kind. Although the circular DNA template clearly is not a primer, yet a primer is evidently required and present.†

The *E. coli* DNA polymerase has now been extensively studied in pure form. It is probably composed of a single polypeptide chain and has a molecular weight of 1.1×10^5 and a sedimentation coefficient, $s_{20,w}$ of 5.6 and therefore must be a relatively compact protein (*91*). From the known degree of purification of this enzyme, its abundance in *E. coli* had been estimated as approximately 140 molecules per growing chromosome. This is, at best, an approximate estimate because of uncertainties in specifying the degree of purification (Table II). The enzyme catalyzes the polymerization of DNA chains on single stranded ϕX174 DNA templates at almost 10^3 nucleotides/minute (*91*).

A major preoccupation in the study of DNA polymerization *in vitro* has been the relationship of template and primer to product. In their template requirements, the enzymes fall into two classes: at 30° or 37°C, the *E. coli*, *B. subtilis*, and *Micrococcus lysodeikticus* enzymes all utilize native and denatured DNA templates with comparable efficiency, whereas calf thymus, T2, and T5 polymerases require denatured DNA templates. The former group of enzymes permit manyfold replication of primer. On the other hand, calf thymus polymerase only replicates its template once; evidently synthesis stops because the template is converted to an inactive ordered conformation and must be denatured to function once more (*33*). (We shall see below that this distinction between the

---

* The words primer and template have distinct meanings. A primer stimulates synthesis of the product by its involvement in initiation of polymerization. The product is covalently linked to the primer. A template determines the composition of the product and directs sequence throughout the polymerization but does not have to be covalently linked to the product. A number of polynucleotide syntheses show a primer requirement or stimulation without having a template requirement; among these are polynucleotide syntheses with polynucleotide phosphorylase and with DNA and RNA terminal addition enzymes. (See review of these topics by Michelson, *248*.)

† The kinetics of DNA synthesis on purified circular single-strand DNA templates are complicated and variable. They are marked by considerable lags and synthesis is greatly stimulated—especially in the presence of DNA ligase—by oligodeoxynucleotides (*123*, *124*).

enzymes is not an absolute one.) The denatured DNA template is also the primer of the calf thymus and T4 DNA polymerase, so that the synthesized product is covalently attached to the template; primer and product cannot be detached by denaturation, but can be separated by sonic degradation followed by denaturation.

It is with the *E. coli* DNA polymerase that the detailed and elegant studies of Kornberg and collaborators have unfolded the full complexity of DNA polymerization *in vitro* and the problems of correlating *in vivo* and *in vitro* replication (*124, 125, 172, 253, 254, 292, 296, 317*). We shall consider three reactions: replication of the native DNA template, repair synthesis on partly single-strand DNA molecules, and replication of single-strand circular DNA.

The replication of density-labeled, double-strand helical DNA by the *E. coli* DNA polymerase *in vitro* does not show the simple parent–progeny relationship of the Meselson–Stahl experiments. The buoyant density of DNA in the reacting system changes gradually as synthesis proceeds and exhibits a considerable heterogeneity. Part of this complication, of course, arises from the relatively slow rate at which polymerization proceeds *in vitro* on templates that are small relative to the intact chromosome. Thus, *in vitro*, even molecules whose molecular weights are only a few million are, predominantly, incompletely replicated at any instant, whereas the opposite is the case *in vivo* (see Section II of this chapter).

However, this is not the sole explanation of the observed complication. Replication of the initially linear template in fact leads to the production of branched product so that entirely double-strand molecules can contain unequal proportions of template and product. Some of the branches are large enough to be visualized in the electron microscope by the Kleinschmidt method (*204*), but these are not the analogs of replicative forks of the growing *E. coli* chromosome (cf. Section II). The product and the template can be separated by denaturation. While the template strands behave like normal, single-strand DNA, the product has unique properties; its denaturation is readily reversible so that it cannot be converted to an exonuclease I susceptible (i.e., disordered) form. These properties suggest the model for the replication of double-strand DNA by *E. coli* polymerase shown in Fig. 9.

Evidently, although the *E. coli* DNA polymerase can utilize native DNA as its template, it interacts with the secondary structure in an erratic way, leading to the synthesis of an aberrant product. Such secondary structure interactions are temperature dependent, and, at lower temperatures, the enzyme ceases to function on native DNA templates.

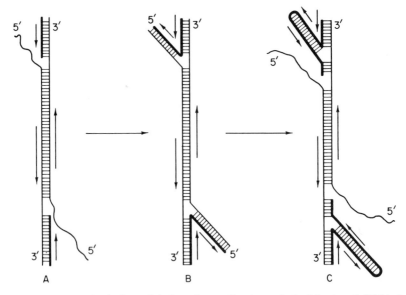

Fig. 9. A hypothetical model for the replication of double-strand DNA by
E. coli DNA polymerase *in vitro* (*317*). The DNA template is indicated by thin
lines, the synthesized product by heavy lines. Hydrogen-bonded purine–pyrimidine
pairs are indicated by horizontal lines.

Diagram A shows the unipolar synthesis of DNA from the 3′-OH ends of
the template. Erratic progress in copying the template, perhaps as in diagram B,
leads (diagram C) to the production of branched molecules containing self-
complementary renaturable hairpin chains of newly synthesized DNA

A distinctly different relationship unfolds when DNA repair–resynthesis is studied. The *in vitro* resynthesis experiments involve the
following steps (Fig. 10); double-strand DNA is a substrate for the
E. coli exonuclease III, which degrades polynucleotide chains only from
their 3′-OH ends yielding nucleoside 5′-phosphate and 3′-OH terminated
polynucleotide chains (*293, 294*). The material produced by exonuclease
III degradation of native DNA is the template and primer for a DNA
synthesis which is more rapid at 37°C than that obtained with native
DNA templates. Thus the ensuing DNA polymerization has two kinetically distinguishable parts. The first process to occur is the resynthesis in
which double-strand DNA is built up again; when this step is complete,
polymerization slows [Fig. 10(B)]. The product of the repair synthesis
is covalently attached to the primer strand of the template. The connection involves, as expected, the polynucleotide synthesized at the
very beginning of the *in vitro* reaction. The repair polynucleotide product
is linear, it has the secondary structure properties of single-strand DNA

after denaturation, and it shows no evidence of looped or branched structure [Fig. 10(C)]. However, if synthesis is continued beyond the repair stage, the repaired template soon functions as a native DNA template: separable polynucleotide chains, which have the hairpin-branched structure characterized by reversible denaturation are formed. At lower temperatures, as we have mentioned, the second part of the reaction can be almost entirely suppressed.

The synthesis of DNA on a single-strand circular DNA template

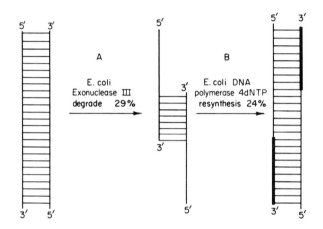

Fig. 10A

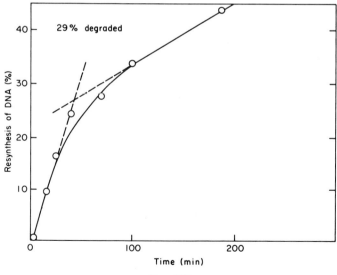

Fig. 10B

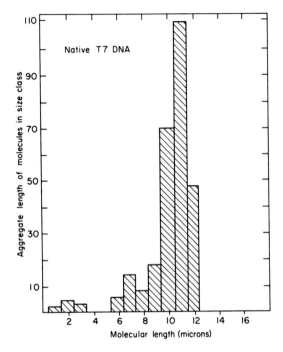

Native T7 DNA

FIG. 10C-I

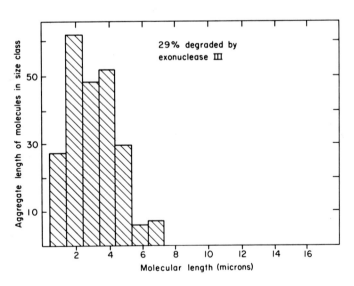

29% degraded by
exonuclease III

FIG. 10C-II

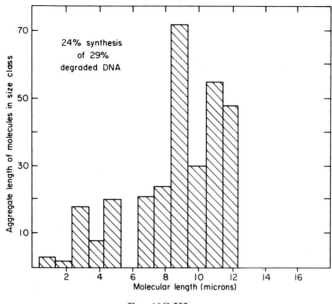

FIG. 10C-III

FIG. 10. Repair–resynthesis of DNA by *E. coli* DNA polymerase (*294*, *297*). (A) Model of the reaction. The heavy lines represent newly synthesized product. Step A indicates the polarity of exonuclease III. Step B shows the covalent linkage of product to primer and the linear extension of the molecule, when resynthesis is not more extensive than the degradation of Step A.

(B) Kinetics of resynthesis of T7 DNA (*294*). The T7 DNA has been incubated with *E. coli* exonuclease III for a period sufficient to remove 29% of the DNA as nucleotides. The exonuclease has then been inactivated and the DNA is used as template primer for synthesis with *E. coli* DNA polymerase. The rate of deoxyribonucleotide incorporation changes relatively sharply when the quantity of nucleotide incorporated is almost equal to the quantity of nucleotide removed by exonuclease.

(C-I to -III) Length distribution of DNA molecules corresponding to Model A. The ordinate shows the number of molecules of length l_i multiplied by l_i; this is the aggregate length of all molecules in each size class. Neither molecules with branches nor molecules of length greater than the length of a whole T7 DNA molecule are found after 24% resynthesis of the degraded DNA.

provides further insight into DNA polymerization *in vitro* (*123–125, 254*). The *E. coli* bacteriophages M13 and φX174 are the sources of such DNA templates. M13 DNA is circular, single strand, and has a molecular weight of 2×10^6. As far as is known, the DNA rings contain no free 3′-OH terminals, phosphomonoester groups, or internucleotide linkages other than phosphodiester bonds. Synthesis on single-strand M13 DNA as template yields circular helical DNA which can be denatured to yield the original circular single-strand template molecules and linear product

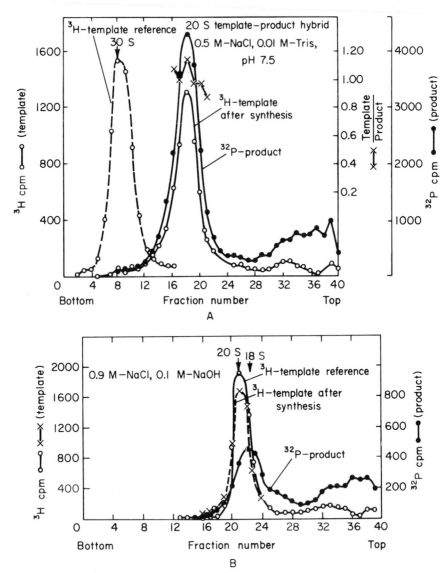

Fig. 11. Sedimentation analysis of template and product after DNA synthesis on a template of single-strand circular M13 DNA (254).

(A) Sedimentation through a 5–20% sucrose gradient in 0.5 M NaCl at pH 7.5 after a reaction yielding an average of approximately 1 μmole ³²P-labeled DNA synthesized per μmole ³H-labeled DNA template. ○---○, template reference: unreacted single-strand circular M13 DNA (³H labeled); ○——○, ³H-template after synthesis; ●——●, ³²P-product after synthesis; ✕——✕, μmole template/μmole product.

(B) Sedimentation through an alkaline 5–20% sucrose gradient (0.9 M NaCl, 0.01 M NaOH): ○——○, ³H-template reference; unreacted single-strand circular M13 DNA; ✕---✕, ³H-template after synthesis; ●——●, ³²P-labeled product. The sedimentation constant for the leading ³²P zone (18 S) is that expected for linear DNA having the same length as the template. The circular template has not been cleaved by endonuclease during the enzymic synthesis.

molecules. Many of the latter correspond in size to the length of the template (Fig. 11). They are terminated at the 5′ (initiated) end by nucleoside 5′-monophosphate and their synthesis appears not to be initiated at a unique sequence, since all nucleotides (A, T, G, and C) are found at the starting end. This contrasts strongly with what happens in RNA transcription on a DNA template with RNA polymerase *in vitro* (Section V). The differences probably can be explained in this way (*123*): (1) the synthesis is oligonucleotide-primed; only an initiating nucleoside triphosphate would provide a nucleoside 5′-triphosphate end for the *in vitro* synthesized polymer. (2) The primer oligonucleotides have no specificity (of course) and can interact with the template at many loci to constitute the required primer–template complex. (3) *E. coli* DNA polymerase has an, until recently unsuspected, associated 5′-exonuclease activity. After extensive synthesis, the survival of primer in the synthesized product is barely detectable. The survival of the nucleotide that is initially added *in vitro* is also uncertain.

B. Comparison of *in Vitro* and *in Vivo* DNA Synthesis

When DNA synthesis *in vitro* is examined as a representation of replication, a number of points of difference are immediately apparent: the unidirectional replication of a double-strand template (Fig. 2) would involve two chemical polarities of the parental DNA template, since the complementary chains of DNA are antiparallel. However, only one polarity of replication—nucleotide addition to the 3′-OH terminal of a growing chain—is represented by DNA polymerase (Fig. 9). The other polarity can be represented by the following hypotheses (Fig. 7): (1) polymerization of deoxyribonucleotide 5′-triphosphates onto polynucleotide 5′-triphosphates generates a 5′-triphosphate end for the chain at each step; this *a priori* plausible mechanism has not been demonstrated *in vitro*; (2) replication involves only the polarity of synthesis represented by DNA polymerase. Along one direction, replication proceeds from the 3′-OH end of the newly synthesized primer strand. The other strand is replicated in short segments in the opposite direction; as each segment is completed, it is joined to the preceding segment. In this hypothesis, then, the problem of unidirectional replication is solved by having one of the strands synthesized in the opposite sense to the unique polarity of propagation. (3) Both strands are replicated in short segments, which are subsequently connected by polynucleotide joining enzymes or ligases. This is the picture suggested by recent experiments described in Section II,E.

There are also quantitative discrepancies. DNA polymerization *in vitro* is about two orders of magnitude slower than replication *in vivo*. Assuming that *in vitro* synthesis is continuous (as with RNA synthesis *in vitro*, see Section IV) then the highest observed propagation rate *in vitro* with purified enzyme is 10 nucleotides per second per synthetic site. The maximum *in vivo* replication rate of the *E. coli* chromosome is of the order of 1500 nucleotide pairs per second. To what extent the rapidity of *in vivo* replication is made possible by a special conformation of the template, by mechanochemical coupling of polymerization to chromosome movement, by the concurrent synthesis of several contiguous DNA segments, or by other factors, is simply not known.

What, then, is the role of DNA polymerase in replication *in vivo*? It could be a polymerizing enzyme of DNA replication proceeding according to hypotheses (2) or (3) above. In that case the replicating machinery must, in addition, contain other enzymes for synthesizing primer oligonucleotide and for assembling the newly synthesized chromosome from its synthesized pieces. Obviously, the template must also be presented to the replicating machine *in vivo* in a manner that avoids the *in vitro* teratosynthesis. The replicator swivel that has been postulated by Cairns (*52*) to drive chromosomal rotation during replication could serve to unfold the template in the replicative fork; alternatively, this could be a function of the replicating machinery itself if, as seems plausible, the entire chromosome does not rotate together with the replicative fork.

The more than 100 DNA polymerase molecules per replicating chromosome in rapidly growing *E. coli* cannot all be simultaneously active in the replicating region; the experiment of Bonhoeffer and Gierer (*37*) referred to in Section II,C excludes this possibility. DNA polymerase could also be responsible for DNA resynthesis during intracellular repair. Certainly, the experiment described in Fig. 10 is a plausible *in vitro* representation of the degradation and resynthesis that follows pyrimidine dimer excision, and it is reasonable to suppose that an enzymelike exonuclease III and DNA polymerase could be responsible for the postulated steps (2) and (3) of ultraviolet repair (*160*) (Figs. 7 and 8). These two possibilities for DNA polymerase function *in vivo* are not mutually exclusive.

DNA polymerase, DNA ligase, DNA exonucleases, and other enzymes may be common to all the pathways of DNA polymerization—replication, repair, and recombination. This is a particularly attractive possibility sustained as yet more by suggestion (Sections II,E and II,G) than by solid evidence. Of course, the genetic and physiological evidence discussed in Section II of this chapter shows that the replicative system must also contain those components that endow it with its characteristic orderliness of initiation and regulation. Discovering the way in which these enzymes are adapted to their tasks in the cell is a major current problem.

IV. RNA SYNTHESIS *IN VIVO*

The description of RNA synthesis which follows is almost completely confined to what has been found out about bacteria and their viruses. From the point of view of a comparison of *in vitro* and *in vivo* transcription and of basic mechanisms, this is useful but comes at the expense of omitting description of phenomena restricted to eucaryotes. Some of these omissions are covered in Chapter V.

Discussion will be restricted to (1) the rates of synthesis of various classes of RNA in bacteria, (2) the synthesis and lifetime of messenger, (3) transcription of operons, and (4) RNA synthesis and virus development.

A. RNA SYNTHESIS IN BACTERIA

We begin by reviewing some of the information on the nucleic acid content of bacteria growing at different rates. The availability of these data has been important for formulating current ideas on the integration of nucleic acid and protein metabolism and also helps to provide some idea of the rates of transcription of bacterial genes (*231, 305*). Table IV

TABLE IV

NUCLEIC ACID CONTENT OF *S. typhimurium* GROWING AT 37°C AT DIFFERENT RATES (*231, 306*)

1 Growth rate (doublings/ hour), μ	2 Cell dry weight (μg/10^7 cells)	3 DNA content (% dry weight)	4 Genome equivalents/ cell	5 RNA/DNA	6[a] tRNA/ Total RNA
2.4	7.7	3.0	4.5	10.3	0.11
1.2	3.2	3.5	2.4	6.3	0.11
0.6	2.1	3.7	1.7	4.9	0.14
0.2	1.6	4.0	1.4	3.0	—

[a] Data taken from (*306*) for *S. typhimurium* 5120 at 30°C adjusted to growth rate data at 37°C by multiplying μ (30°C) by 1.4. There is considerable variation among different measurements of tRNA content in slowly growing bacteria. This is attributed (*201, 306*) to problems in extracting RNA from slowly growing cells, and in distinguishing tRNA from other low molecular weight polynucleotides.

TABLE V
NUCLEIC ACID CONTENT AND AVERAGE POLYMERIZATION RATE IN RAPIDLY GROWING E. coli B (231, 305, 306)

RNA/DNA = 10.0 = 22 × 10^9 amu RNA per chromosome equivalent of DNA
Growth rate = 1.8 doublings per hour at 30°C
First-order growth rate constant = 3.5 × 10^{-4}/second

	Transfer	5 S	Ribosomal 23 S	Ribosomal 16 S	Messenger
1. RNA type					
2. Molecular weight	2.65 × 10^4[a]	4 × 10^4[b]	1.66 × 10^6[c]		—
3. Relative abundance (Σ = 1.0)[d]	0.13	0.03[e]	0.81		0.03[f]
4. Molecules/genome equivalent	1.1 × 10^5	1.1 × 10^4	1.1 × 10^4		—
5. Transcription rate at 30°C nucleotides/second/genome equivalent[a]	3 × 10^3	4.6 × 10^2	1.3 × 10^4	6.5 × 10^3	—
6. Estimated number of chromosomal transcription sites/genome equivalent	45[h]	12[k]	5[l] / 4[m]	5[l] / 4[m]	—
7. Transcription rate at 30°C nucleotides/second/cistron	70	55[k]	2.6 × 10^3[l] / 3.3 × 10^3[m]	1.3 × 10^3[l] / 1.1 × 10^3[m]	—
8. Molecules/second/cistron	0.9	0.5[k]	0.8–1.0	0.7–0.8	—

[a] After (226).
[b] After (308).

[c] After (*213*).

[d] After (*306*).

[e] Corresponding to 1 molecule 5 S RNA per 70 S ribosome.

[f] After (*313*).

[g] For synthetic rates at 37°C these figures can be multiplied by approximately 1.4.

[h] The degeneracy of transfer RNA is further discussed in Chapter III. There are at least 32 distinct anticodons in different tRNA molecules (*71*). By direct hybridization, Goodman and Rich (*122*) and Giacomoni and Spiegelman (*114*) found about 20 complementary sites per genome equivalent. Zehavi-Willner and Comb (*387*) have repeated these measurements, using the more efficient hybridization technique of Gillespie and Spiegelman (*117*), and find a more than twofold higher capacity of *E. coli* DNA to bind tRNA. The value listed is probably not off by more than 25%.

[k] After (*387*). In *B. subtilis* the number of chromosomal sites is evidently smaller (see Table VI). There are only 2 major species of *E. coli* MRE600 5 S RNA (*46a*). *Either* duplicated genes for 5 S RNA have retained identical sequences *or* some genes are transcribed in greater yield than the average listed in this column.

[l] Data on DNA complementary to 23 and 16 S ribosomal RNA from (*12*). Assume distribution as in *B. subtilis* (*272*), i.e., equal numbers of cistrons for each ribosomal type.

[m] Same as above footnote, but data on *B. megaterium* was used to provide an estimate of the relative proportions of 23 and 16 S cistrons (*381*).

gives a convenient summary. At the highest growth rate, nucleic acids constitute about one-third of the dry weight of the cell and ribosomal RNA is the major component. The stable RNA complement of the cell also includes the transfer RNA's and 5 S RNA. The latter comprises a class of stable RNA molecules that was originally described in bacteria (*307*) and has subsequently been found to be a ubiquitous cell constituent (*387*). The 5 S RNA of *E. coli* has a chain length of 120 nucleotides and is only found on ribosomes in the proportion of one 5 S RNA molecule per 70 S ribosome. It contributes approximately 3% of the total RNA content.

It is known that the ribosomal content of bacteria is proportional to their growth rate and that the DNA content (Table IV columns 3 and 4; see also Table I) and number of nuclei is also dependent on growth rate. The tRNA content of cells growing at different rates is shown in Table IV column 6. Table V shows the relative abundance of ribosomal, transfer, and 5 S RNA is rapidly growing *E. coli*. The number of molecules of stable RNA's per genome equivalent are listed on line 4; the genome equivalent is taken to be the DNA content of the mature *E. coli* chromosome—3.3×10^6 nucleotide pairs (molecular weight of the sodium salt 2.2×10^9, length $1100 \, \mu$) (*52*). The average transcription rates of the stable RNA's are listed on line 5. The remaining part of the RNA, estimated at about 3% of the total, is messenger, the RNA coding for the amino acid sequence of proteins (*45, 130, 131, 178*). As described further on, in more detail, bacterial messenger is unstable, so that its concentration in a steadily growing population results from the balance between synthesis, growth, and degradation.

Systematic measurements of the messenger content of bacteria growing at different rates are not abundant, but Kjeldgaard (*201*), in a recent review, cites evidence that the messenger and ribosome contents of *E. coli* are proportional over a large range of growth rates—an important result to which we return below.

We wish to infer something about the rates at which various RNA molecules are synthesized on their templates. In addition to the above information, we therefore need to know the number of cistrons for the synthesis of ribosomal and transfer RNA. The DNA loci complementary to these RNA's have been identified by DNA–RNA hybridization (*12, 114, 117, 122, 379–381, 387*), and the quantities of DNA complementary to 16 S ribosomal, 23 S ribosomal, transfer, and 5 S RNA in *E. coli*, *B. megaterium*, and *B. subtilis* are listed in Table VI. Thus, potential sites for rRNA and tRNA transcription are located on the chromosome and there is every reason to believe that these are the

TABLE VI

COMPLEMENTARITY OF BACTERIAL DNA TO STABLE RNA

Organism	tRNA	5 S RNA	Ribosomal RNA			References
			Total	16 S	23 S	
E. coli[a]	0.05	0.02	—	—	—	Zehavi-Willner and Comb (387)
	0.025	—	—	—	—	Goodman and Rich (122)
	0.02	—	—	—	—	Giacomoni and Spiegelman (114)
	—	—	0.39	—	—	Attardi et al. (12)
B. megaterium	—	—	0.31	0.14	0.18	Yankofsky and Spiegelman (381); Gillespie and Spiegelman (117)
	—	—	0.38[b]	0.13[b]	0.25[b]	Oishi and Sueoka (272)
B. subtilis[a]	0.027[c]	0.0043[c]	—	0.13[c]	0.26[c]	Smith et al. (333)

[a] Genome size: *E. coli* 2.2×10^9 amu (52), *B. subtilis* 3.5×10^9 amu. [Average of the values—3.0 and 3.9 $\times 10^9$ amu—given by Dennis and Wake (79) and Eberle and Lark (86).]

[b] DNA from exponential cultures; in view of the chromosomal loci assigned to the ribosomal RNA cistrons (Fig. 4), agreement between B and C is not as good as the numbers seem, at first, to indicate.

[c] DNA from stationary cultures; these complementarities yield unambiguous values for the number of loci for each class of RNA per *B. subtilis* genome: tRNA 35; 5 S RNA 3–4; 16 S rRNA 8–9; 23 S rRNA 8–9.

actual sites of synthesis.* Knowing the size of the genome, one can calculate the number of cistron equivalents per genome equivalent for each RNA species (Table V, line 6). For tRNA, the estimate given by the earlier hybridization experiments seems low and a higher value has been used, based on a more recent determination (Table VI). The estimation of the actual number of ribosomal and transfer RNA cistrons per chromosome is more difficult because the average number of gene

* The kinetics of bacterial rRNA synthesis are such that it is not necessary to postulate the existence of an amplification system involving the formation of auxiliary templates (305). If an auxiliary amplification system for transcription of certain loci existed, one might imagine that this could involve plasmids (Section II) or a double-strand replicative RNA form. A search for double-strand ribosomal RNA and ribosomal RNA complements in *E. coli* has been made, utilizing methods capable of detecting one complementary equivalent per bacterium; none could be found (17).

copies per cell depends on the time in the cell cycle at which these cistrons are replicated as well as on the growth rate and stage of the cell population (see Section II of this chapter). However, if the DNA for these experiments had been prepared from cells growing logarithmically and from strains that have the same replication cycle as those used in determining rates of RNA synthesis, then the average rate of RNA synthesis per ribosomal site would be determined without ambiguity. This is the assumption that we make in calculating average rates of polymerization per chromosomal site for rapidly growing cells (doubling time 33 minutes at 30°C; Table V, lines 7 and 8). The calculated average polymerization rates per template are of the order of 1 molecule of tRNA, 16 S, or 23 S rRNA every 1 or 2 seconds assuming uniform synthesis throughout the cell cycle on each template. The rate of production of all these RNA species varies with growth rate.

Beyond this, however, one would want to know how many chains of each type of RNA are being simultaneously transcribed. This requires additional knowledge of the "transit" time for synthesizing an entire chain of each type of RNA from the 5' to the 3' end (see Section V of this chapter) or, equivalently, the propagation rate of RNA transcription. Moreover knowing how this propagation rate depends on the overall growth rate or the rate of protein synthesis would tell us whether the coupling of RNA synthesis to protein synthesis involves changes in the frequency of initiation of new chains only, or changes of initiation frequency and propagation rate.

This subject is currently being taken up vigorously, and sufficient data are now available to provide preliminary answers. The average rate of elongation of all RNA chains has been measured in *E. coli* at 29° and 37°C, at varying growth rates, and also in T4 infected *E. coli*. The rate of propagation of *E. coli* ribosomal RNA and tryptophan operon (*trp*) messenger RNA chains has also been measured. Propagation rates appear to vary little with overall bacterial growth rates at constant temperature. Variations in the rate of RNA production (Table IV) are therefore mostly accomplished by changes of initiation frequency.

For uninfected *E. coli,* average rates of RNA chain propagation have been estimated to be of the order of 15–30 nucleotides/second at 30°C. For rRNA chains and for *trp* mRNA values of the same order of magnitude have been obtained (*44, 121, 235–237, 374, 391*). Thus the very different rates of production of these two kinds of RNA result from different rates of initiation. Clearly transcription units the size of ribosomal cistrons, producing transcripts at the rate of one every 1 or 2 seconds, support the tandem transcription of many RNA chains. In fact, according to the calculation of Table V, the *E. coli* ribosomal cistrons must be practically close-packed with transcribing RNA polymerase molecules.

It is interesting to estimate the number of RNA polymerase molecules required to sustain rapid growth of *E. coli*. For a growth rate constant of 3.5×10^{-4} sec^{-1} (Table V), RNA synthesis is required at a rate of 2.2×10^4 nucleotides polymerized per second per chromosome equivalent of DNA. With polymerization propagating at the rate of 30 nucleotides/ second, 730 molecules of RNA polymerase would be required for each chromosome equivalent or approximately 2500 molecules per rapidly growing cell. We shall see in the next section that this figure is within the bounds of reasonable estimates for the RNA polymerase content of *E. coli*.

B. PROPERTIES OF THE MESSENGER

Less than 1% of the *E. coli* chromosome codes for transcription of ribosomal, transfer, and 5 S RNA. The rest accounts for all the remaining RNA synthesized in the cell, whose only known function is serving as the messenger, coding for protein synthesis. Under normal bacterial growth conditions, this RNA is heterogeneous and labile, with an average lifetime of a few minutes at 37°C. This instability prevents its accumulation to more than a few percent of the total RNA content of the cell; yet analysis of RNA that is pulse labeled for a brief period, makes it evident that a major part of the synthesis is devoted to messenger. Before discussing messenger synthesis and stability, we shall list some properties of mRNA molecules.

Messenger is heterogeneous in sedimentation rate. This size heterogeneity results from two entirely different factors: (1) transcription of different mRNA classes takes place on chromosomal units of differing size and (2) the time required for mRNA polymerization to proceed the length of a large transcription unit can be comparable with the average messenger lifetime. At any instant, the distribution of chain lengths for products of a single transcription unit consequently arises from the presence of completed, incompletely synthesized and partly degraded chains. Thus, even when messenger RNA sedimentation is examined with due regard for possible artifacts arising from degradation during extraction and isolation or from aggregation with ribosomal RNA (*12, 151, 257, 311*), the messenger of a single operon is heterogeneous.

The range of sizes reported for different messengers conforms to these expectations. On the one extreme, a class of messenger in immature duck erythrocytes exhibits sedimentation constants higher than 60–80 S (*13,*

316). The *E. coli* tryptophan operon messenger is heterogeneous as we shall see below, but the largest molecules that can be observed under some circumstances are in the range of 33–35 S (*168, 169*). The other extreme would be represented by monocistronic messenger of small proteins. In practice, part of the rapidly turning-over RNA in, for example, T4-infected *E. coli*, sediments quite slowly. (However, the smallest of these molecules may be nascent or partially degraded messenger fragments, and some of them are not messenger and do not turn over rapidly.)

Whenever messenger is the transcription product of many different parts of the genome—as it is in bacteria—it assumes an average nucleotide content relatively close to that of the respective DNA, even though the transcription product is synthesized asymmetrically, i.e., complementary to only 1 DNA strand at any locus (*132, 148, 357*) and different genes are transcribed at different rates—thus the prematurely archaic term D-RNA.

Further elucidation of the properties of messenger RNA has come from studies on RNA–DNA hybridization. That these hybridization techniques are capable of the specificity required for detailed analysis of messenger species, was originally inferred from the low extent of cross reaction between products of nonhomologous DNA templates with identical average base composition (*136, 239*). A critical test was provided by Attardi and co-workers (*12*), who, under suitable conditions, could attach segments of ribosomal RNA containing several hundred nucleotides to homologous DNA in pancreatic RNase-resistant conformation. Initially, such hybridization analyses were performed with the aid of CsCl density gradient equilibrium centrifugation, but more convenient, inexpensive, and precise methods became available (*35, 117, 267, 268*). From such studies has come much information on frequencies of ribosomal and transfer RNA cistrons, messenger content of cells, and the dynamics of formation of messenger corresponding to single operons, as well as the time relationships of RNA synthesis during viral development.

We turn next to a consideration of messenger synthesis, stability, and content in steadily growing bacteria. Basically, the following kinds of experiment are used to measure these interrelated quantities.

(1) In a very brief pulse of radioactive labeling, the fraction of label entering different classes of RNA is proportional to their relative rates of synthesis. Stable and unstable messenger RNA can be distinguished in hybridization experiments because of their different relative abundance and because of their very different specific activity after a short labeling pulse (*35, 243*).

(2) Unstable RNA is degraded to ribonucleoside phosphates which mix with the RNA precursor pool and are efficiently reutilized for RNA synthesis. In a cell containing unstable RNA, the rate at which the nucleotides (mainly triphosphates with smaller amounts of diphosphates and only very small quantities of other phosphorylated compounds) are labeled after addition of exogenous labeled guanine depends not only on the rate of nucleotide synthesis and the size of the nucleotide pools but also on the rate of messenger breakdown and on the relative proportions of nucleotides contained in unstable RNA (130, 131, 313).

(3) Actinomycin D (AMD) is one of a number of agents which block RNA synthesis (see review by Reich and Goldberg, 288). Analysis of the extent and rate of degradation of label incorporated into RNA to acid-soluble fragments, when AMD is added to cells that have been labeled for varying periods of time, in principle permits the determination of messenger content, stability, and relative rates of synthesis of stable and unstable RNA species (225). It is also possible to quantitate polysome degradation after AMD addition and this defines another aspect of messenger breakdown kinetics (315).

(4) The functional lifetime of a specific messenger can be determined from the residual kinetics of enzyme synthesis after transcription has been blocked, either by removing an inducer or by adding inhibitors of RNA synthesis (16, 30, 141, 188, 189, 193–196, 219, 260, 261).

Each of these methods is attended by particular complications and by the need to justify special assumptions involved in calculating parameters of messenger metabolism from the data. Since there has been much apparent disagreement between conclusions derived from studies using different methods, detailed discussions of problems of interpretation of these measurements are available (23, 313, 327). At the risk of oversimplifying, only those difficulties that are most closely related to the results cited will be considered below.

When method (1) is used to measure the relative rates of messenger RNA and stable RNA synthesis, care must be taken that the quantities of denatured DNA available for hybridization are in excess of all messenger species, even those which are very abundant (243). When DNA is used in great excess in the hybridization assays, then it may be necessary to dilute out the label contained in ribosomal RNA (238). Neglecting the second of these two precautions will generally be less serious, and errors from the above two sources will tend to cancel. For pulses of finite duration, correction for the lifetime of messenger must also be made. These corrections are inaccurate, as we shall see below, but for short pulses (e.g., 1 minute) they are also small. Measurements of this kind have been made by Bolton and McCarthy (35, 36).

They found that 33% of *Pseudomonas vulgaris* RNA pulse-labeled for 1 minute hybridized to excess DNA on an agar column; efficiency of the columns used, as determined by hybridization of T2 DNA messenger was approximately 70%.* 28% of *E. coli* RNA pulse-labeled for 1 minute hybridized to an excess of its DNA; efficiency of the columns used, as determined by eluting and rehybridizing the hybridized messenger, was 80%.† Correcting for the efficiency of hybridization and for an average messenger decay time of 4 minutes, the synthetic rate for messenger becomes 55 and 40% of the total rate of RNA synthesis in these two cases, respectively. The data for *E. coli* are listed in Table VII.

Method (2) has not been fully exploited until recently. For their experiments, Salser *et al.* (*313*) have used guanine requiring mutants of *B. subtilis* and *E. coli* that are also blocked in purine interconversion. For exponential cells grown so that no pool expansion occurs when the label is added, the determination of the specific activity of GTP, as a function of the time after addition of labeled guanine to the medium, permits calculation of the rates of unstable nucleic acid synthesis and breakdown, the rate of stable nucleic acid synthesis, and the messenger content, if the growth rate of the cells, the rate of guanine uptake, and the size of the GTP pool are also known and if a mechanism of messenger degradation can be assumed; for the experiments cited here, decay of messages was assumed to be kinetically first order.

The results of such experiments with *E. coli* and *B. subtilis* give the messenger content, the relative rate of messenger synthesis, and the average decay time (or lifetime) as listed in Table VII. The most striking result is that the distribution of RNA synthetic effort of *E. coli* and *B. subtilis* is definitely very different. *E. coli* synthesize unstable and stable RNA's at comparable rates, and, with a 4-minute average decay time at 30°C,‡ contain about 3% mRNA. *B. subtilis*, on the

---

* This correction for the estimated efficiency of hybridization is, to some extent, arbitrary. If the efficiency corresponds to a 70% probability that a messenger molecule hybridizes to one of many available DNA complements, then the correction is valid. If the less than 100% efficiency of the control arises from partial saturation of T2 DNA sites by the most abundant messenger species, then applying this correction to bacterial messenger involves the tacit ad hoc assumption that DNA saturation of control and experiment are comparable.

† In this case, DNA was definitely in excess. If there is some variation of efficiency of hybridization of different messenger species, then this is an overestimate of the efficiency of the column and leads to an underestimate of the proportion of RNA synthesis devoted to messenger.

‡ The relationship between decay time (τ) and half-life ($t_{1/2}$) of a substance and the first-order rate constant of decay (k) are

$$\tau = \frac{1}{k} = \frac{t_{1/2}}{\ln 2}$$

TABLE VII

PARAMETERS OF MESSENGER METABOLISM IN *E. coli* AND *B. subtilis*

Organism	Method (see text)	Relative rate of unstable RNA synthesis	Relative content of unstable RNA	Average decay time (minutes)	Temperature (°C)	Growth rate (doublings/ hour)	References
E. coli strain							
BB	(1)	0.40	—	—	—	1.0	(243)
R257	(2)	0.45	0.034	4	30	0.9	(313)
K AB 1105	(3)	0.7 ± 0.1	0.015–0.03	2.5	37	0.9	(220)
K 200P	(4) β-Galactosidase	—	—	3.4	30	0.55	(261)
	(4) β-Galactosidase	—	—	2.4	37	—	(261)[a]
	(4) β-Galactosidase	—	—	1.5	37	0.85	(155)
	(4) β-Galactosidase	—	—	2.1	30	—	(195)
B	(4) β-Galactosidase	—	—	2.3	30	—	(189)
K (many strains)	(4) β-Galactosidase	—	—	2.0–3.4	30	0.5–1	(189)
K AB 1105	(3, 4) β-Galactosidase-AMD	—	—	2.5–3.4	30	—	(219)
	(4) Thio-galactoside transacetylase	—	—	1.1	37	—	(195)
K C600	(4) Tryptophanase	—	—	3.5	30	—	(30)
B. subtilis strain							
295	(2)	0.85	0.09	3.6	37	0.50	(313)
295	(2)	0.90	0.09	6.2	30	0.22	(313)
W23	(3)	0.82	0.08	2	33.5	0.85	(225)
W23	(3)	—	0.09	—	37	1.0	(391)
W239	(3, 4) Histidase-AMD	—	—	3.4	37	0.40	(142)

[a] Assuming $Q_{10} = 1.7$.

other hand, devote five-sixths of their RNA synthesis to unstable species, whose average lifetime (6.2 minutes at 30°C) is slightly greater so that mRNA constitutes about 9% of the total RNA at any time. The derived values of messenger lifetime are averages. Evidence for a range of messenger stability in such cells comes especially from the experiments of Schaechter *et al.* (*315*) with *B. megaterium.* More detailed considerations of Salser and co-workers show that, for a distribution of messenger stabilities, the average determined by this method weights the more stable classes more heavily.

Measurement of the decay of radioactivity incorporated into pulse-labeled RNA after AMD addition (method 3) provides a simple way of looking at messenger stability. Analysis of how the fraction of incorporated radioactivity that is unstable to AMD depends on the length of the radioactive pulse yields an estimate of the proportion of synthesis devoted to unstable species (*96, 223, 225*). The ambiguities of this method are: (1) the heterogeneity of messenger lifetimes complicates the comparison of RNA labeled in pulses of different lengths. More accurate kinetics are provided by very brief pulse labeling, but these results predominantly describe the properties of the least stable RNA species and of incompletely synthesized RNA chains. Thus, in the extreme, one has the apparent paradox that in the presence of AMD, part of the label in very briefly pulsed RNA is less stable than are polysomes [e.g., in *B. megaterium* (*315*)]. (2) Nascent RNA molecules whose synthesis is blocked by AMD may have abnormal stability. Incompletely synthesized ribosomal RNA molecules are unstable (*218, 391*) and contribute to an overestimate of the proportion of synthesis devoted to unstable RNA by method (3). (3) AMD might modify the stability of all messenger species, even those that have been completed and have left their templates. A considerable weight of evidence (*219, 220, 260*) indicates that this is not the case. If the effects of AMD on stability of RNA are confined to nascent messenger, then its use to measure β-galactosidase messenger lifetimes by method (4) is valid.

Thus the relative rates of unstable RNA synthesis should be overestimated by the AMD methods (3) and the characteristic lifetime should be underestimated. Comparison of the results for *B. subtilis* using methods (2) and (3) suggests that messenger lifetimes are underestimated by AMD method (3), probably because of messenger heterogeneity and perhaps because of the properties of nascent rRNA chains. The proportion of messenger synthesis in *B. subtilis* is, in any case, so high that it is hard to overestimate it. However, in *E. coli* the situation is less favorable. Leive (*219, 220*) has found it possible to make *E. coli* permeable and susceptible to AMD by a brief treatment with ethylenediaminetetraacetic acid (EDTA), which does not change the rate of protein synthesis. Compared to the pool-labeling kinetics method (2), the use of AMD in *E. coli* leads to an overestimate of the proportion of synthesis devoted to unstable RNA (Table VII).

Method (4) provides a powerful and versatile approach to the kinetic study of transcription and translation. The detailed design of the experiments can be adapted to yield information on the appearance of the capacity to form enzyme and its decay, and on the appearance of enzyme precursor and activity. Use of amino acid analogs provides a means of identifying (in terms of thermal sensitivity) enzyme molecules or subunits made at a specific time. By the proper use of metabolic inhibitors, messenger and protein synthesis can be separated in time, and initiation of translation can be distinguished from completion of already initiated polypeptide chains (*5, 6, 141, 188, 189, 193–196, 261, 280*). For the purpose of this discussion we are interested in the rate of the decay of capacity to form enzyme after removal or dilution of the inducer; this measures the rate of decay of functioning messenger after transcription (actually, initiation of transcription) has been halted.

The β-galactosidase of *E. coli* has been the most extensively studied in this way and many measurements of messenger lifetime have been reported on many different *E. coli* B and K strains (Table VII). At 30°C, half-lives range from 2 to 3.5 minutes. At least a part of this spread corresponds to the limits of reproducibility of the experiments. Leive (*220*), in less extensive experiments, showed that the lifetime of β-galactosidase messenger, estimated from enzyme synthesis after inhibition of RNA synthesis with AMD, falls within this range and that the lifetime of messenger measured with AMD is independent of whether inducer is present or absent. This is of interest because the immediate consequences for RNA synthesis of AMD addition and withdrawal of inducer are quite different; AMD blocks propagation of transcription, while removing the inducer blocks initiation of transcription (see below and Section V of this chapter).

The three other methods that have been described [(1) to (3)] identify messenger as acid-precipitable polynucleotide, while method (4) assays messenger function. Insofar as the results of these two classes of experiment are comparable, they suggest that functional inactivation is concomitant with, or rapidly followed by, degradation and that there is little macromolecular but inactivated messenger in *E. coli* or *B. subtilis*.

We have already remarked that, with β-galactosidase and *B. subtilis* histidase (Table VII), auxiliary experiments distinguish effects on the kinetics of formation of enzymic activity due to messenger decay from those due to protein maturation. Unless such auxiliary evidence is available, studies of the persistence of enzyme or protein synthesis after RNA synthesis is blocked, are no more than suggestive of messenger lifetimes.

Table VII compares these data on messenger synthesis and stability in bacteria under a variety of growth conditions, but systematic data on relative rates of messenger and ribosomal RNA synthesis as a function of growth rate are not yet available. However, we have already mentioned that there is evidence to suggest that messenger content of *E. coli* (as measured by ability to stimulate *in vitro* amino acid incorporation) is proportional to ribosomal content over a wide range of steady growth rates (*201*). If messenger lifetimes proved to be insensitive to growth rate on one hand then one could conclude that messenger and ribosomal RNA synthesis are regulated to protein synthesis in parallel ways. On the other hand, during a transition period of adjustment to amino acid starvation, or when protein synthesis is blocked by chloramphenicol, or in recovery from ribosomal depletion (*102*, *223*, *265*, *356*), the stability of messenger RNA increases greatly. If the relative rate of its synthesis remains the same under such circumstances, then the messenger content of the cells must increase.

While the bacteria provide instances of variations in stability of different messengers in a single cell, the well-known cases of extremely stable messenger come from other systems—hemoglobin messenger in erythrocytes and messenger in sea urchin oocytes (*256*) are two examples. The sensitivity of protein synthesis in *B. subtilis* to actinomycin indicates that these bacteria contain little or no messenger of comparable stability.

The discussion thus far has emphasized the distinction between abundant and stable ribosomal and transfer RNA and the heterogeneous and much less abundant messenger. However, different messenger species differ greatly in their rates of synthesis and this can be demonstrated by hybridization–titration of *E. coli* DNA with pulse-labeled RNA (Fig. 12) (*192*, *243*). Evidently different cistrons may function at very different transcription rates. Some of this variation presumably results from variations in the levels of repression and depression of different genes (Section IV,C); some of it may also result from variations in the intrinsic transcribability of DNA. These variations of transcribability arise from the different frequencies with which different transcription units are initiated. For *E. coli*, one can point to a variation of initiation frequency of more than 3 orders of magnitude (*109*); the range of transcript abundances is probably not greater (*192*). Two factors that determine differences in quantitative function of different genes have now been discussed: variations of transcription rate and variations of messenger lifetime. Together with variations in the efficiency of translation

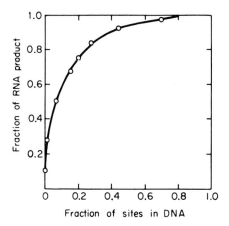

FIG. 12. The variation of abundance of *E. coli* messenger RNA species (*243*). Results of a hybridization titration with ^{32}P pulse-labeled *E. coli* RNA are used to calculate the differential activity of various *E. coli* genes. Abscissa: the fraction of DNA covered with mRNA at saturation is defined as 1.0. Bolton and McCarthy concluded that half the DNA could be so titrated at saturation, evidence that RNA is transcribed off only one coding strand of the DNA. A very small fraction of the genome accounts for an appreciable portion of the unstable RNA product.

of different messages these presumably constitute the basis of the great range of abundances of different proteins.

The nucleotide composition of different portions of bacterial and viral chromosomes tends to be somewhat heterogeneous; so, also, is the messenger of *E. coli* (*36*) and of T4 and λ phages. In the latter two cases, genes of different average nucleotide composition are transcribed at different times so that the composition of the messenger changes during viral development (*197, 329*).

C. TRANSCRIPTION OF AN OPERON

The description of the gross rate of nucleic acid synthesis in bacteria and other cells provides important background information on synthetic capacities, on the integration of macromolecular synthesis, and on the great range of transcriptive capacities of different genes. The experiments to which we now turn provide information on control of initiation of transcription by repressor, on direction and termination of transcription, and on multicistronic messages.

For the sake of simplicity, the emphasis here is on experiments in which quantities and properties of mRNA are measured directly. Attention is therefore mainly confined to the tryptophan (*trp*) and to a lesser extent, the lactose (*lac*) operons in *E. coli*. This is to some extent a distortion of the historical sequence. Many of the major ideas about messenger and about operons result from studies of the *lac* operon (*177, 178*), the detailed analysis of the histidine operon in *S. typhimurium* (*7*), and of other coordinately controlled gene sets contributes importantly to the current state of understanding of this subject (*94*).

After induction (*lac*) or derepression (*trp*), the operon messenger appears in far greater abundance and can be detected by hybridization to homologous DNA. The segregation of the appropriate genes separated from the rest of the bacterial DNA can be achieved on transducing phages; phage λ for the *gal* (*11*), phages P1 and, more recently, λ or 80 for the *lac* genes (*20, 150*) and phage 80 for the *trp* region; F factors have been used in a similar way (*11*). The currently available information comes predominantly from the analysis of RNA synthesis using DNA–RNA hybrids, and from the analysis of kinetics of enzyme synthesis under a variety of conditions in the case of the *lac* operon.

In *E. coli* six enzymes of the tryptophan pathway whose synthesis is coordinately controlled, are located near the lysogenic phage $\phi80$ locus. The DNA segment corresponding to the genes of this pathway can be incorporated into $\phi80$ transducing particles. Different transducing phages may carry different segments of the *trp* operon: $\phi80$ pt$_1$ carries all the *trp* genes except E (anthranilate synthetase), $\phi80$ dt$_0$, a defective phage, and $\phi80$ pt(*A–E*) carry all the *trp* genes and other phages carry other segments of the operon. On the other hand, complete $\phi80$ carries no tryptophan genes (Fig. 13). Accordingly, the quantity of messenger of the entire *trp* operon and of different segments may be measured by hybridization with these viral DNA's (with only minor uncertainties regarding corrections for nonspecific hybridization). For example:

(1)   *trp* Operon messenger $= \begin{pmatrix} \text{Hybridization to} \\ \phi80 \text{ dt}_0 \text{ or } \phi80 \\ \text{pt}(A - E) \end{pmatrix} - (\text{Hybridization to } \phi80 \text{ DNA})$

(2)   *E* Gene messenger $= \begin{pmatrix} \text{Hybridization to} \\ \phi80 \text{ dt}_0 \text{ DNA} \end{pmatrix} - (\text{Hybridization to } \phi80 \text{ pt}_1 \text{ DNA})$

(3)   *D* → *A* Gene messenger $= \begin{pmatrix} \text{Hybridization to} \\ \text{phage } \phi80 \text{ pt}_1 \\ \text{DNA} \end{pmatrix} - (\text{Hybridization to } \phi80 \text{ DNA})$

(4)   *C* → *A* Gene messenger $= \begin{pmatrix} \text{Hybridization to} \\ \text{phage } \phi80 \\ \text{pt}(A-C) \text{ DNA} \end{pmatrix} - (\text{Hybridization to } \phi80 \text{ DNA})$

The exploitation of these possibilities is originally due to Imamoto *et al.* (*168–171*) and has also been pursued by others, notably by Yanofsky

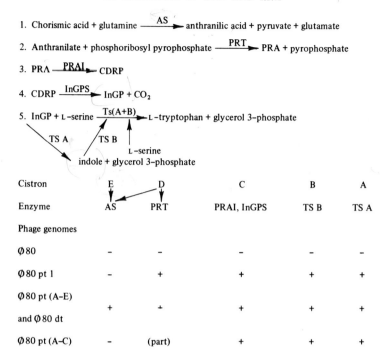

1. Chorismic acid + glutamine $\xrightarrow{\text{AS}}$ anthranilic acid + pyruvate + glutamate

2. Anthranilate + phosphoribosyl pyrophosphate $\xrightarrow{\text{PRT}}$ PRA + pyrophosphate

3. PRA $\xrightarrow{\text{PRAI}}$ CDRP

4. CDRP $\xrightarrow{\text{InGPS}}$ InGP + CO_2

5. InGP + L–serine $\xrightarrow{\text{Ts(A+B)}}$ L–tryptophan + glycerol 3–phosphate

TS A   TS B

L–serine

indole + glycerol 3–phosphate

Cistron	E	D	C	B	A
Enzyme	AS	PRT	PRAI, InGPS	TS B	TS A
Phage genomes					
Ø80	–	–	–	–	–
Ø80 pt 1	–	+	+	+	+
Ø80 pt (A-E) and Ø80 dt	+	+	+	+	+
Ø80 pt (A-C)	–	(part)	+	+	+

FIG. 13. The tryptophan pathway. The cistron designation is according to Yanofsky and Ito (*382*). Abbreviations: AS, anthranilate synthetase; PRT, phosphoribosyl transferase; PRAI, PRA isomerase; InGPs, idoleglycerophosphate synthetase; Ts, tryptophan synthetase; PRA, phosphoribosylanthranilate; CDRP, 1-(*o*-carboxy-phenylamino)-1-deoxyribulose 5′-phosphate.

and his co-workers (*16a, 257a*). The experiments consist of derepressing and rerepressing the *trp* operon of *E. coli* grown to high density. At various times after derepression and rerepression, RNA is pulse-labeled and analyzed by zone centrifugation and DNA–RNA hybridization. In this way, the kinetics of *trp* mRNA synthesis and chain growth are followed.

After derepression, the proportion of incorporated label found in *trp* RNA increases. The transcript segments which first appear are those of the *E* gene followed, in order, by $D \rightarrow A$ segments. The transit time for the entire operon is 5–6 minutes corresponding to 20–24 nucleotides/second at 37°C. Initiation of the derepressed *trp* operon only occurs every 1½–2 minutes at this temperature.

The initiation event at the operator end is independent of the rest of the *trp* genes—it occurs with the same frequency when distal portions of the genetic material are deleted. It is easy to see that with such infrequent initiation and relatively rapid propagation, synchronous de-

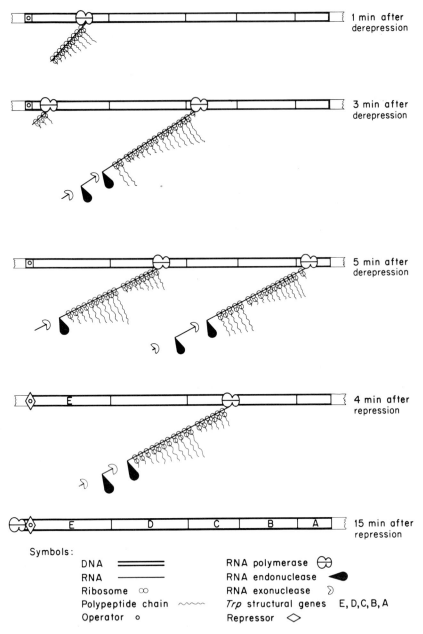

FIG. 14. An interpretation of Tryptophan Operon Transcription. Degradation of message is *speculatively* shown to start before a complete *trp* transcript has been made (see text and contrast Fig. 15). The chemical polarity of destruction is $5' \rightarrow 3$ but the degrading enzymes at work on the nascent RNA chain could be an endonuclease and a $3' \rightarrow 5'$ exonuclease (*109*).

repression of a whole population of cells might yield pulsatile synthesis of different RNA segments and of *trp* enzymes. In fact such pulsatile synthesis has been demonstrated. Upon rerepression, *trp* transcription stops (decreases by a large factor) in a sequence that has the identical polarity and comparable duration: *E* gene transcription ceases at once and *A* gene transcription stops several minutes later with intermediate events polarized $E \rightarrow A$. Thus, repression is the prevention of transcription–initiation; those *trp* polynucleotide chains that are nascent at the moment of rerepression are completed and translated into *trp* enzyme polypeptide chains (*16a, 167, 175, 257a*).

When the kinetics of *trp* RNA and enzyme production after derepression are compared, it becomes clear that translation and transcription are concurrent; the nascent message is already part of a polysome (*175, 257a*). These powerful experiments can be summarized in a simple model (Fig. 14).

Is the entire *trp* message made as one RNA chain? What is its stability? How is it degraded? The answers to these three questions are at this time unclear. The obscurity appears to originate from our imperfect understanding of messenger degradation. Suffice it to say that conditions have been found in which: (a) *trp* RNA segments are degraded about 2 minutes after they are synthesized (at 37°C) with the polarity $(E \rightarrow A)$ of synthesis. Under these conditions, intact *trp* transcripts do not exist since the transit time is longer than the chemical lifetime of message; *trp* *E* and *D* polypeptide chains are made only on nascent, template-attached polysomes (*167a, 258*). The model sketched in Fig. 14 incorporates this mode of degradation. (b) Intact *trp* RNA molecules exist whose size (sedimentation coefficient approximately 33 S) is in the range of $2–2.5 \times 10^6$ amu (Fig. 15). The sum of the molecular weights of the subunits of the enzymes of the tryptophan pathway is estimated* to be approximately 240,000 corresponding to about 2300

---

* The molecular weight of the *D* gene polypeptide is not accurately known. Also, the sedimentation–molecular weight relation for the messenger is based on ribosomal RNA and is not exact.

Each pulse of *trp* transcription ultimately yields, on the average, 100–200 polypeptide chains of each of the *trp* enzymes. Each ribosome symbol therefore stands for several actual ribosomes.

The RNA polymerase σ factor is absent from the picture—none of the frames shows initiation of RNA polymerization. For the sake of concreteness, RNA polymerase without σ is *speculatively* shown in the bottom frame as binding near the *trp* operator but not initiating *trp* RNA synthesis, while *trp* repressor binds to the *trp* operator.

No part of the diagram is to scale.

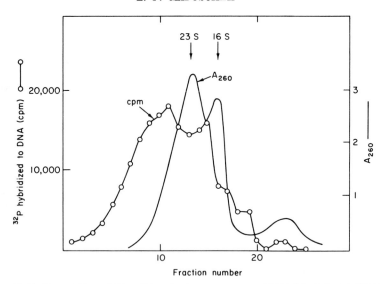

Fig. 15. Sedimentation of *E. coli* tryptophan messenger: RNA labeled with ³²P 7½–10½ minutes after derepressing tryptophan synthesis (in *E. coli* W3102) prepared and sedimented through a 5–30% sucrose gradient containing 0.01 *M* Tris-Cl pH 7.3, 0.05 *M* KCl. Fractions are collected and hybridized to φ80 pt and φ80 DNA (see text) to give an estimate of the sedimentation rate of *trp* messenger (*169*). O———O, total *trp* messenger: difference of ³²P cmp hybridized to φ80 pt₁ DNA and φ80 DNA; ————, absorbance at 260 mμ. Sedimentation is from right to left.

amino acids. The size of *trp* mRNA is therefore appropriate for a multicistronic message coding for the enzymes of the *trp* operon (*169*). Evidence supporting the notion of a multicistronic messenger had previously been presented by Martin (*240*) for the histidine operon although the lack of a sensitive and selective detection method made the task of identification much more difficult.

Studies of the kinetics of induction of the *lac* enzymes β-galactosidase and thiogalactoside transacetylase, which are the first (operator-proximal) and third proteins of the *lac* operon, respectively, also provide evidence that the primary effect of repression is to block the initiation of RNA synthesis. Although the evidence is indirect, the experiments of Kepes suggest that inactivation of β-galactosidase messenger, in effect, also has the same polarity as synthesis; the loss of the ability to initiate the synthesis of new polypeptide chains is rate determining for the functional life time of β-galactosidase messenger. It is not known which enzymes are responsible for the rate-limiting step of messenger inactivation.

The tryptophan operon also provides dramatic evidence of the way in which polarity* is reflected in transcription as well as in translation. The polarity of *amber, ochre,* and phase-shift mutations leading to termination of polypeptide chain synthesis of the *trp* enzymes has been worked out in some detail (*382*); the gradient of polarity in each cistron is related to the distance between the chain-terminating triplet, and the next chain-initiating site. When this distance is short (i.e., for mutants located on the right side of a cistron in Fig. 15) there is no polarity and the distal cistrons of the operon are translated normally; for chain terminating mutants located near the beginning of a cistron there are sharp reductions in the expression of distal genes. Polarity is exhibited in all possible *trp* genes (*E* to *B*). The amount and size of *trp* messenger and the relative proportions of *A* to *C* gene messenger have been measured for a number of these mutants and compared with messenger synthesized in *trp* deletion mutants. In all cases, there is a clear correlation between polarity measured in terms of enzyme activity and messenger survival synthesis; polarity is also manifested as partial or complete absence of message distal to the chain-terminating triplet. The polar mutations do not affect the rate at which RNA chains are initiated after derepression but only specify their premature termination (or degradation) at or near the peptide chain terminating codon with an effectiveness that depends on the length of the untranslatable segment between the chain-terminating and the next translation-initiating signal.

Consistent evidence regarding consequences of polarity for transcription of the *lac* operon has been noted by Contesse, Naono, and Gros (*65*). In fact the original and most extreme effects of this kind were part of the very first experiments on messenger abundance and concerned the *lac* o mutants (Chapter I). Their mode of suppression identifies these mutants as having a chain terminating triplet which is located very close to that end of the *lac* operon at which transcription and the synthesis of the β-galactosidase subunit polypeptide chain are both initiated. When the chain terminating signal is read as such, no *lac* messenger can be detected. The extreme polar effect of polypeptide chain termination very near the operator is to block transcription completely.

D. TRANSCRIPTION DURING BACTERIOPHAGE DEVELOPMENT

For the final topic in the discussion of RNA synthesis *in vivo,* we turn to a description of RNA synthesis during bacteriophage T4 development.

---

* This section should be read in conjunction with Chapter I in which polarity, chain termination, *amber,* and *ochre* mutations are defined and discussed.

Briefly, the facts about T4 development in *E. coli* are these [see the volume by Stent (*338*) for a splendidly written and informative exposition of this subject]: the phage injects its DNA, together with a very small amount of viral protein, into the host cell. Approximately 20 minutes later (at 30°C) the first completed phage appear intracellularly, and cells lyse 30–40 minutes after infection. Normal development in a rich medium yields 100–200 progeny per infected cell. During the period between infection and lysis a complex series of metabolic events occurs. Synthesis of host macromolecules stops very quickly after infection and the expression of viral genes is substituted for that of bacterial genes. Viral genes do not all function at the same time but in a sequence about which much detailed information is available. Among the genes that function at the beginning of viral development are those which specify the enzymes of the pathway of viral DNA synthesis. Approximately 20 of these enzymes are known (*63, 93, 129, 363, 373*); they are some, but by no means all, of the early proteins. Not all the early proteins are subject to synthesis at the same time (*224, 318*).

Although host DNA synthesis shuts off immediately after infection, viral DNA synthesis starts only after several minutes have elapsed and reaches a steady rate 10–12 minutes (at 30°C) after infection. At about this time, synthesis of many, but not all, early proteins ceases (*58*), and a new class of late proteins, which include the structural proteins of the viral coat, start to be made. These late functions are grouped in subassembly pathways (*87, 378*). Genes of early and late functions are segregated on the T4 chromosome in long sequences and genes of a pathway, especially the genes of the pathways of viral coat maturation, tend to be clustered together. Analysis of T4 development owes all of its most recent advances to the development of the proper genetic tools—the conditional lethal *amber* and temperature-sensitive mutations (*93*). From the study of these mutants, and from certain other evidence, it is known that DNA synthesis is required for the expression of late viral genes.

Information regarding the program of viral messenger synthesis which accompanies, and is presumably responsible for, this program of gene expression, is now becoming available. A major experimental tool of this analysis is RNA–DNA hybridization competition, an isotope-dilution method first used by Hall and co-workers (*134, 135*) and Khesin and co-workers (*198, 199*), which is described in Fig. 16(A). The results of an isotope-dilution experiment involving two pairs of labeled and unlabeled RNA samples prepared just before the onset of DNA synthesis (early RNA) and before the appearance of the first intracellular phage (late RNA) are shown in Fig. 16(B) and 16(C). It is evident that RNA synthesized late, i.e., just toward the end of viral development, contains messenger RNA polynucleotide sequences that are absent from (or several hundredfold less abundant in) RNA isolated during the early period. However, late RNA isolated at the end of the viral eclipse period still contains some quantities of most of the RNA species being synthesized during the early period (*32, 134, 135*). More detailed and sensitive analysis which will not be described show that: (1) There is a wide range of abundances of late messenger species, including some that are relatively very abundant. (2) These abundant species are at least 200 times more concentrated at the end of the eclipse period than during the early period; their synthesis is turned on soon after DNA synthesis has started. (3) The concentrations of many early RNA species decrease during the late phase of development; however some early species continue to be synthesized at equal or greater rates late in infection. (4) A sequence of early RNA transcription can be detected during the first few minutes after infection; not all early RNA sequences are transcribed from the very beginning. (5) While early T4 messenger synthesis does not require the expression of viral genes, late messenger is not made unless viral DNA is made, and unless at least one other gene, besides the genes required for DNA synthesis, have functioned (*32, 144, 312*).

Thus the dynamics of transcription during T4 development include the following distinctive control and selection phenomena:

(1) Shut-off of host transcription; initiation of transcription is blocked, as in repression (*188*).

(2) The T4 chromosome also codes for tRNA species (*73, 365a*). Their role in controlling gene expression during phage development is not yet understood; however, the suggestion has been advanced that T4 also exerts control over shut-off of host functions at the level of translation (*145, 161, 348*).

(3) Restriction of early viral transcription to a part of the viral chromosome. This phenomenon has an *in vitro* counterpart that is discussed in the next section of this chapter.

(4) A transient period of early RNA transcription; in part this is analogous to the transient state of the tryptophan operon, which has just been described; different early transcription units may, in addition, be initiated nonsynchronously or propagation may be discontinuous.

(5) Turn-on of late viral transcription.

(6) A depression of the relative rate of transcription of certain early genes at later stages of development, perhaps in a manner analogous to repression.

Enough is known to realize that the phenomenology of transcription applies, with some important variations, to other developing viral systems such as λ and the *B. subtilis* phage SPO1, in part also to vaccinia virus

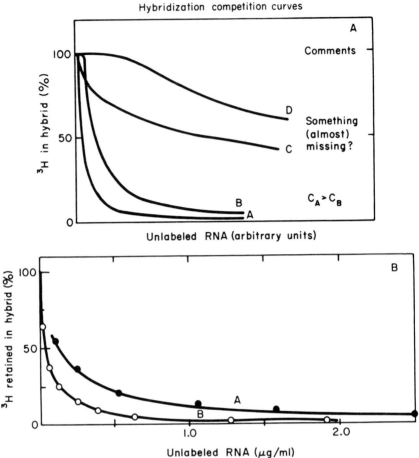

Fig. 16. RNA analysis by hybridization competition: (A) Hypothetical hybridization competition curves. RNA is hybridized to homologous DNA in the presence of varying quantities of unlabeled competing RNA. Ordinate: concentration of unlabeled RNA in arbitrary units. Abscissa: ^{3}H-labeled RNA hybridized to DNA expressed as percentage of label bound when unlabeled, competing RNA is absent. Curve **A**: The labeled RNA contains only 1 labeled species, a, homologous to the DNA. The unlabeled RNA also contains a at the same concentration (relative to total RNA). DNA sites homologous or complementary to a are in excess over the content of labeled a. As unlabeled a-containing RNA is added, the DNA is first saturated and then the probability of binding labeled a molecules to the DNA declines as the proportion of competing unlabeled RNA is increased. If hybridization were ideal (in the experiments to be cited it probably is not), curve **A** would be a rectangular hyberbola, displaced from the origin by the amount of unlabeled a necessary to saturate DNA sites complementary to a. Curve **B**: Same as **A** but a is 3-fold less abundant in the unlabeled RNA than for curve **A**. Curves of type **A** and **B** will also be shown by more complex mixtures of messenger or other labeled RNA species. If the abundances of different RNA species vary, the com-

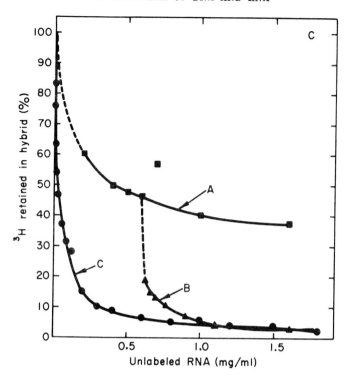

petition curves will not be simple hyberbolas; yet so long as all the labeled RNA species homologous to the test DNA are also present in the unlabeled RNA, all the label can be competed away. Curves **C** and **D**: The labeled RNA contains several labeled species of RNA homologous to the test DNA. Some of these are present in the unlabeled RNA, but others are absent, or present in such low concentrations that they fail to compete away all the hybridized radioactivity effectively. For example, in **C** approximately 30–40% of the label hybridized to DNA is in species that are quite abundant in the unlabeled RNA but 40% of the hybridized label is in species that are rare in, or missing from, the unlabeled RNA. The initial portions of curves **C** and **D** differ because the relative abundances of overlapping RNA species in labeled and unlabeled RNA differ. In actual practice, several complications of hybridization can obscure such distinctions as those between curves **C** and **D** and limit the detail in which these experiments can be interpreted.

(B) Competition of ^{3}H-labeled RNA 0–5 minutes after infection of *E. coli* B with bacteriophage T4 at 30°C. The labeled RNA contains no late T4 messenger (*32*). Curve **A:** ●———●, competition with unlabeled 20-minute RNA; Curve **B:** ○———○, competition with unlabeled 5-minute RNA.

(C) Competition of RNA labeled with ^{3}H uridine from 1–20 minutes after T4 infection at 30°C. The labeled RNA contains late T4 messenger (*32*). Curve **A:** ■———■, competition with unlabeled 5-minute RNA; Curve **B:** ▲———▲, competition with a constant amount (0.6 mg/ml) of unlabeled 5-minute RNA plus increasing amounts of unlabeled 20-minute RNA (the total concentration of unlabeled RNA is indicated on the abscissa); Curve **C:** ●———●, competition with unlabeled 20-minute RNA.

and to early stages of embryonic development (*19, 78, 83, 105, 119, 127, 181, 186, 242, 264, 269, 329, 372*). Some of the mechanisms through which these controls of transcription are asserted are well on the way to being understood. The λ repressor can be regarded as an example of this class. Most of the control mechanisms, however, remain to be elucidated. *In vitro* experiments are beginning to yield some results which are discussed at the end of Section V of this chapter.

V. RNA TRANSCRIPTION *IN VITRO*

A. RNA POLYMERASE

Enzymes catalyzing RNA synthesis on the DNA template have now been isolated and at least partly purified in a great variety of cells and organisms (see review by Elson, *90*, for a listing). The most extensive studies are those on the enzymes from *E. coli* and *M. lysodeikticus* which have contributed importantly to the current understanding of enzyme action and to defining unresolved problems (*56, 104, 165, 234, 262, 341, 343, 366, 368, 388*). The *E. coli* enzyme has been highly purified to yield electrophoretically homogeneous monodisperse protein preparations free of contaminating RNase, DNase, and polynucleotide phosphorylase. When highly purified, the *E. coli* enzyme attains a specific activity of up to 40 μmoles nucleotide polymerized per hour per milligram protein; the *M. lysodeikticus* enzyme has been purified to a comparable extent and represents an approximately 200–400-fold purification from crude bacterial lysates (*367*).

   E. coli RNA polymerase is a polymer of smaller subunits (*47, 174, 389*). Recent resolution of these suggests the following constitution.

Subunit designation	Molecular weight	Subunits per enzyme "monomer"
α	39,000	2
β,β'	155,000; 165,000	1 each
σ	~90,000	≤1
(ω)	(~10,000)	(≤1)

ω May not be an obligate constituent of RNA polymerase; it is missing from some preparations of the enzyme and changes the aggregation properties of the latter. σ Functions cyclically in initiation of RNA synthesis (see below); it is possible that many different σ-like components are present in virus-infected or uninfected cells (*25, 48, 210, 357a*). At low ionic strength *E. coli* RNA polymerase associates to a dimer (molecular weight = $7.2\text{--}8.8 \times 10^6$; sedimentation constant 19–21 S) and higher aggregates (*283, 298, 342, 388, 389*).*

B. ENZYME BINDING TO DNA

At low ionic strength, a stable complex between *E. coli* RNA polymerase and DNA is formed rapidly, even at 0°C where the subsequent RNA synthesis is extremely slow or absent. Although this complex is stable enough to be centrifuged as a zone without dissociation (*99, 187, 298, 299, 340*), complex formation is not irreversible, since bound and free enzyme can undergo isotopic exchange (*340*).

The capacity of DNA to bind polymerase has been determined in a number of ways, including measurement of the amount of enzyme required to give maximal RNA synthesis, sedimentation measurements of DNA–enzyme complexes, ability of enzyme to bind DNA to nitrocellulose filters, and visualization of DNA–enzyme complexes in the electron microscope. The experiments have involved a number of DNA templates including T4, T7, λ, and doubly closed circular polyoma and papilloma DNA's. The quantities of enzyme bound to a number of different DNA templates correspond approximately to 1 enzyme molecule of molecular weight 800,000 to every 2000–2500 Å of DNA (*70, 182, 299, 340*). Since a 2500 Å stretch of DNA codes for a 240-residue polypeptide chain, it was thought that such average spacings of binding sites on DNA might correspond to the average spacing of cistronic transcription units. If this were so, then these binding sites could correspond to *in vivo* utilizable initiation sites for transcription. It does not, at present, look as though these binding experiments identify initiators of *in vivo* transcription units. However, they do establish that binding sites on native DNA for *E. coli* RNA

---

* It is instructive to try to estimate the number of RNA polymerase molecules in an *E. coli* cell. The assumptions are: (1) dry weight: 7.7×10^{-13} gm/cell, (2) 40% protein, (3) RNA polymerase 0.4% of the total protein (180–250-fold purifiable according to Maitra and Hurwitz (*234*); this is a very imprecise way of determining abundance because of the difficulty of quantitatively interpreting enzyme assays in crude extracts), (4) molecular weight of the active unit 3.7×10^5. With these assumptions one calculates that there are 2300 enzyme molecules/cell. A comparably imprecise calculation in Section IV,A suggests that this supply of enzyme molecules would suffice for RNA synthesis during rapid growth.

polymerase are, under certain ionic strength conditions, widely spaced and discrete.*

At different salt concentrations, the capacity, and, presumably, the affinity, of DNA for RNA polymerase can be varied greatly. At low ionic strength the number of RNA polymerase molecules binding to circular polyoma DNA greatly exceeds 1 molecule/2000 Å, but most of the enzyme is bound in a transcriptively inactive form (283, 299). A gradient of affinities between RNA polymerase molecules and λ and T7 DNA has recently been demonstrated by Stead and Jones (337). It seems probable that different templates may show individual and characteristic distributions of stronger and weaker binding sites, but this remains to be investigated in detail. It may also be that the distribution of affinities is partly due to enzyme heterogeneity (283). Unfortunately, therefore, it is not yet possible to correlate enzyme binding sites with sites for the initiation of RNA synthesis *in vivo* or *in vitro*.

The connection between the state of polymerization of RNA polymerase and its binding to DNA has recently become clarified. Clearly the 13 S dissociated form of the enzyme can bind to DNA as such; it may very well function as such (332). The enzyme can also bind to DNA without its σ subunit. However, in view of uncertainty about whether such binding is at or near transcription initiation sites, one should not make too much of this observation.

While denatured DNA binds much greater proportions of enzyme to nucleotide than does native DNA (377), this binding is also reversible.

C. INITIATION OF RNA SYNTHESIS

Once RNA synthesis on the ordered, double-strand DNA template starts, enzyme template interactions are considerably altered. The enzyme becomes much more tightly bound to its template, so that the complex is operationally irreversible (40, 300). This stabilization can be achieved without all four ribonucleoside triphosphates being present and is most responsive to purine nucleotides, since RNA chains are predominantly initiated with purine nucleotides (8, 251). A consequence of this irreversible enzyme–template binding is that the outcome of competition

---

* It is important to stress the distinction between these experiments, which measure the spacing of binding sites on DNA and the spacing of polymerase molecules on the DNA template during active transcription. The latter is determined by the rate of initiation and propagation as well as the spacing of initiation sites.

of different nucleic acid sites for enzyme depends on the order of addition of reactants—the enzyme continues on the native DNA template on which it starts (40, 99, 234, 377).

The reasons for the change of binding concomitant with RNA synthesis are not understood. It may be that there is a structural transition of the enzyme concomitant with the start of RNA synthesis and the release of the σ subunit. Alternatively, it may simply be that RNA synthesis engages a short segment of the coding strand of the DNA with its RNA complement, and therefore frees the antisense strand of DNA to interact more strongly with the enzyme. (The latter alternative is equivalent to saying that the initiation of RNA synthesis further disrupts the secondary structure of the template beyond changes caused by binding.)

As a result of this tighter binding, higher concentrations of salt (e.g., 0.4–0.5 M KCl), which completely prevent the binding of enzyme to DNA (and in the presence of which, RNA synthesis cannot be initiated), do not inhibit the continuation of RNA synthesis once it has been initiated (8, 103, 300).

Many in vitro RNA synthetic systems can be shown to exhibit some degree of selective transcription of the DNA template (62, 111, 112, 127, 149, 198, 228, 264). Since the enzyme–template interaction is essentially irreversible once propagation is under way, this specificity of transcription in vitro should have its origin in the specificity of enzyme binding and initiation. The current emphasis on initiation of RNA transcription and on the biological specificity of the transcript reflects this line of reasoning.

The identification of the nucleotide at the initiated end of an RNA molecule requires a knowledge of the polarity of transcription, which is 5′-P → 3′-OH as the following evidence shows: (1) 3′-dATP is an inhibitor of RNA synthesis in vitro and is incorporated into the terminal nucleotide of the inhibited RNA chain (322). Therefore, normal chain propagation must be by addition of a nucleoside 5′-triphosphate to the 3′-OH which is missing in 3′-dATP. (2) Alkaline hydrolysis of RNA cleaves P–O–5′-C phosphoester bonds and results in the formation of nucleoside 2′,3′-phosphates. Alkaline hydrolysis of newly synthesized RNA chains yields, in addition to these 2′,3′ phosphates, a nucleoside from one end of the RNA chain, and a nucleoside tetraphosphate, presumably a 3′-phosphate 5′-triphosphate, from the other. The appropriate pulse labeling and chase experiments show that the initiated terminal of the RNA chain is the one containing the 5′-triphosphate end and that the 3′-OH end of the molecule is the growing end. This is the same polarity that has been determined directly in vivo (121).

Pursuing this result further, it is clear that RNA synthesis with nucleoside triphosphates labeled with ^{32}P in the γ (distal) position would label only the chain-initiating nucleotide of the synthesized RNA. The nucleotide distribution of chain-initiating ends can be determined by synthesis with all four γ ^{32}P triphosphates in turn, and thus provides a probe of initiation sites for transcription (42, 233). In all cases studied, initiation predominantly places purine nucleotides in the 5'-terminal of the RNA chain. The A/G ratio varies from template to template: For the T2, T5, and SP3 viral DNA templates, initiation is predominantly through ATP, and the A/G ratios for initiation are close to the proportions of A and G in the templates. For other templates, G initiation predominates and is in excess of the G/A proportion in the DNA (Table VIII).

The predominance of initiation with purine nucleotides is a property of native and denatured DNA templates and similar relations exist in syntheses on homopolymer templates. For example, poly $dA + dT$ template yields predominantly poly rA synthesis (55). Thus, the purine-specific initiation of E. coli RNA polymerase appears to be a property of the enzyme that is invariant to the secondary structure of the template.

There is a curious connection between these observations on chain initiation by purine nucleotides and information that has recently come to light about the interaction of denatured DNA with homopolynucleotides. It has been known for some time that DNA's contain long purine and pyrimidine sequences (49, 137, 335). Some of these permit DNA to complex with ribosomal RNA and polynucleotides in a way that does not require annealing but occurs upon heating and rapidly cooling denatured DNA and polynucleotide together (211, 277). Different DNA's are distinguished by the composition of the long sequences responsible for the DNA-polynucleotide interaction. Some DNA's have predominantly G- and C-rich sequences, others predominantly A- (and T-) rich ones. The former interact more strongly with poly IG than with poly A or U. The latter interact more strongly with poly U than with poly IG. The homopolymers can therefore be used to separate complementary DNA chains and this is a matter of considerable practical importance. There is a striking qualitative correspondence between the predominance of A- or G-rich sequences in the DNA as uncovered by homopolynucleotide interactions and the relative frequency of initiation of transcription with A and G nucleotides in native DNA (351) (see Table VIII).

What makes this comparison even more intriguing is, that there are also certain correlations of nucleotide clustering and in vivo transcription which associate pyrimidine clusters with the coding strands

TABLE VIII

INITIATION OF RNA TRANSCRIPTION

DNA	DNA purine content A/(A + G)	Initiation double strand DNA[a]		Initiation denatured DNA[a]		Relative binding[b] $\left(1 - \dfrac{\text{Poly U}}{\text{Poly G}}\right)^{-1}$
		A/(A + G)	purine/total	A/(A + G)	purine/total	
T2	0.65	0.66	0.94	0.42	0.87	0.8
T5	0.61	0.56	0.88	—	—	0.5
SP3	—	0.56	0.81	—	—	—
Cl. perfringens	0.69	0.43	0.88	—	—	0.5
Calf thymus	0.56	0.37	0.80	0.27	0.78	0.33
λ AT rich	0.55	0.30	—	—	—	—
E. coli	0.50	0.26	0.90	0.24	0.91	0.25
λ	0.50	0.28	—	0.16	—	—
λ GC rich	0.45	0.14	—	—	—	—
M. lysodeikticus	0.28	0.13	0.93	—	—	0.15

[a] Data from (232) and (233).

[b] Estimated on the basis of buoyant density shifts of denatured DNA in the presence of equal poly G/DNA and poly U/DNA ratios. The data are internally normalized by setting the value of the ratio for Cl. perfringens DNA to 0.5 (351).

of transcription units in bacteriophage λ, T7, and probably also T4 DNA. Thus, there is good reason for insisting that the pyrimidine clusters must have some signaling or punctuation function for transcription–translation. If they are transcription–initiation signals, long dT:dA and dC:dG sequences might be the enzyme attachment loci. However, the analogy is not quite as simple as it seems. The *in vitro* RNA syntheses on the T4 and λ DNA templates transcribe the viral genes selectively and to widely different extent (see Section V) while the pyrimidine clusters are distributed along these entire DNA templates. Thus, the similarity of the two sets of data which are compared in Table VIII must have a slightly more complex significance. Perhaps binding-initiation signals for RNA polymerase are bipartite and the pyrimidine clusters are a part, but not all, of the signaling sequence.

DNA denaturation greatly increases the number of sites at which polymerase can be bound. The distribution of nucleotides among the 5′-terminals of RNA made on denatured DNA templates still shows the predominantly purine initiation characteristic of the *E. coli* RNA polymerase. However, especially in the case of T2 DNA, there is a pronounced shift toward the initiation of RNA chains at a guanylic acid residue. With *E. coli* DNA as template, the distribution of nucleotides among the termini is almost unchanged. Thus, the interaction of enzyme with template and the nature of the RNA product are drastically affected by secondary structure changes. On the other hand, the terminal nucleotide distributions are relatively little affected. A possible reason has already been mentioned; perhaps the secondary structure-dependent binding site is not transcribed.

RNA polymerase also binds to the ends of DNA molecules (*26*). However, in contrast to the situation in DNA polymerization, synthesis is not predominantly from the ends of the template (*41, 164*).

D. Propagation

The addition of ribonucleotide residues to growing RNA chains is the most studied and, in certain aspects, the best understood part of the RNA polymerization mechanism. Examination of the composition of RNA polymerized on a variety of DNA templates shows the correspondence of overall composition and nearest neighbor frequency of transcript and template (*166, 368*). Data for three templates are collected in Table III. The conclusion was at first drawn that transcription of the template *in vitro* was indiscriminate and uniform. Although, as we show in Section V,I, this proved not to be generally true, the data nevertheless do suggest fidelity of transcription of the template. (The apparent slightly inferior correspondence of λ RNA and DNA nearest neighbor frequencies is a subject to which we return in Section V,I.) More stringent tests of this fidelity at the polynucleotide level may be obtained by looking at the specificity of hybridization of *in vitro* and *in vivo* synthesized RNA, which are found to be essentially comparable (*110–112, 136, 149, 264*). The ability to form RNA–DNA hybrids which have stabilities comparable to those of DNA and RNA duplexes implies that the synthesized RNA chain is antiparallel to the chain of complementary sequence in its DNA template. However, the definitive test of this notion is provided by the X-ray analysis of a DNA–RNA hybrid duplex made by synthesizing RNA on a template of single-strand linear phage f1 DNA (*252*). It therefore appears certain that the recognition element of propagation is complementary base-pairing of the Watson–Crick type.

The fidelity of hybridization also implies that chain propagation is regularly coupled to progression along the DNA template in normal, four nucleoside triphosphate-requiring RNA transcription. If, on the one hand, the enzyme-native DNA–RNA complex containing the nascent polynucleotide chain persists throughout synthesis (*40*), then variations in the nature of the synthesized product can be studied as manifestations of processes controlling the initiation of transcription. On the other hand, failures of this regular propagation, which lead to slippage, can be produced (*57, 205, 244*). They are of great practical importance in the establishment of codon sequences (Chapter IV).

The propagation rate of *E. coli* RNA polymerase on various DNA templates has been measured. Unfortunately only measurements at low ionic strength have been reported in detail, but others have been referred to briefly or are yet to be published. Initial rates of chain elongation have been claimed to be as great as 100 nucleotides/second; rates of 15–35 nucleotides/second have also been reported at 30° and 37°C (*54, 74, 174, 300–302*). At low ionic strength, chain propagation slows down markedly as nascent RNA chains become longer, and it is thought that the inhibition of RNA polymerase is due to the nascent chains themselves (*40, 103, 209, 300*). The main conclusion to be drawn from the currently available data is that *in vivo* and *in vitro* propagation rates can be made quantitatively comparable, i.e., differing far less than an order magnitude. The contrast with replication is striking (Section III,B).

E. TERMINATION OF SYNTHESIS AND RELEASE OF RNA CHAINS

The known facts about the control and specificity of RNA synthesis in functioning cells require the existence of mechanisms for the termination of RNA chains. However, there is little evidence about the existence of such mechanisms operating *in vitro*. It is true that RNA synthesis *in vitro* with highly purified RNA polymerase slows down relatively rapidly and ultimately stops. Part of the inhibition may be due to the interaction of the synthesized RNA with enzyme and template (*103*, *209*) and is probably not due to specific arrest of RNA transcription occurring at specific termination sites on the DNA template. With endonuclease-free RNA polymerase, the polynucleotide chains synthesized *in vitro* are much larger than their *in vivo* counterparts and much longer than the average spacing of initiation sequences of transcription (*10*, *40*, *300*). Moreover in these circumstances, enzyme molecules and product remain attached to the template. Although the argument is by no means complete, nevertheless it seems more than likely that termination signals are not recognized in such *in vitro* RNA syntheses on native DNA templates.

Release of enzyme molecules from single-strand DNA templates does appear to occur during *in vitro* transcription: the synthesized RNA consists of large numbers of small chains (*57*), and the number of chains synthesized per RNA polymerase molecule is much greater with single-strand than with double-strand DNA templates (*54*, *232*). Thus, termination release appears to be secondary structure-dependent. It is as though the anticoding strand of the DNA template acted as a guide rail which holds, or helps to hold, RNA polymerase onto the nascent chain–template complex. One may conclude that the secondary structure of ordered helical DNA restricts both the termination–release and binding–initiation steps of RNA polymerization.

The most concrete illustration of this notion has been presented by Chamberlin who has studied RNA synthesis on templates of synthetic polydeoxyribonucleotides in great detail (*55*). Among the homopolymer helical complexes the thermodynamic stability relationships are such that RNA synthesis on the poly(dA:dT) template yields separable poly ribo A chains but synthesis on the poly dG:dC templates displaces dG chains from the template and yields poly rG:dC hybrids. In this latter synthesis which involves permanent dissociation of the template, there is clear evidence that enzyme is released from the template and that continuous initiation of new RNA chains occurs: (1) The number of chain termini as detected by incorporation of γ ^{32}P-labeled nucleoside triophosphates is very large and (2) competition between templates for enzyme occurs during reaction (54) (see Section V,C).

F. Interaction of RNA Polymerase with the Secondary Structure of the Template

It is important to be aware of the possibilities inherent in the interaction of RNA polymerase with the secondary structure of its template, even though the available information is sparse. In fact, information on the interaction of proteins with the secondary structure of the nucleic acids is only now beginning to accumulate. It is known that different proteins can stabilize or destabilize the secondary structure of the nucleic acids. For example, pancreatic RNase whose endonuclease activity is specific for single-strand polyribonucleotides, binds preferentially to single-strand DNA. By so doing, it lowers the transition temperature of double-strand DNA (98). Polylysine, histones, and other basic polypeptides stabilize the secondary structure of DNA to varying extents (2, 222, 276, 358). What is definitely established regarding RNA polymerase is that, if any disordering of DNA secondary structure occurs, it must be local and cannot involve the entire DNA template (110, 300). As was stated above, it is also known that the total number of polymerase binding sites increases upon DNA denaturation and that the initiation of RNA synthesis on these new sites involves predominantly purine nucleotides just as does initiation on helical DNA. For these new binding sites, denaturation clearly favors binding. Since enzyme binding in the absence of nucleoside triphosphates is reversible, it is likely that the reverse is also true: enzyme binding favors local helix destabilization at the binding site. Thus far it has not proved possible to demonstrate this directly (300).

Facts about the relation of the DNA template to the RNA product are also pertinent. RNA chains can be completely dissociated from T7 and T4 DNA–enzyme–RNA complexes by SDS (sodium dodecyl sulfate) (40, 299). Yet when ϕX174 I (doubly continuous, circular) DNA is used for RNA synthesis with *E. coli* polymerase, a detectable fraction of the RNA remains attached to the template in a form that is isolated after phenol and SDS purification as RNase resistant (147). Pulse-chase experiments show that it is the growing end of the RNA chain which is complexed to DNA and is RNase resistant. Throughout synthesis, a part of the product is found in this state, but at late times, when RNA synthesis slows down, the absolute amount of nuclease-resistant RNA decreases. It is not clear how to reconcile the divergent behavior of T4,

T7, and ϕX174 DNA templates except by assuming that the properties of the latter are specific to the closed circular DNA template. The supercoiled circular DNA is known to have unique secondary structure properties—the secondary structure stability of 10 nucleotide pairs is reduced for each turn of the super helix (*360*). It may be that the detection of the RNA–DNA complex reflects the kinetics of unfurling of the nascent RNA chain, a process for which the DNA strand complementary to the template is perhaps required (*147*). One must also suppose that (1) in the closed circular DNA this process is much slower than with linear DNA templates and that (2) the unfurling requires the presence of enzyme to keep the two complementary strands arranged in a specific way. These two suppositions would account for the facts that as propagation slows down, the amount of RNA in the complex decreases; and the isolated RNA–DNA complex does not spontaneously dissociate, whereas in the enzyme complex, RNA and DNA dissociate behind the growing point.

G. Modifications of Initiation and Propagation of RNA Transcription

Studies of the inhibition of *in vitro* transcription by RNA polymerase with a variety of agents are extensive (see review by Reich and Goldberg, *288*). Recently these studies have begun to be recast in the more detailed terms of effects on initiation, propagation, and differential transcription of RNA (*43, 300*). Experiments on actinomycin D, proflavine, rifamycin, and tRNA are particularly interesting. At sufficiently low concentrations, actinomycin D inhibition of RNA synthesis is incomplete. Richardson (*300*) has shown that neither binding of enzyme to template nor initiation of RNA chains is substantially decreased under these conditions. However, the average chain length of RNA molecules synthesized in a given time, and therefore the propagation rate, does decrease. It has also been reported that actinomycin D at low concentrations changes average composition and nearest neighbor frequencies of transcribed RNA, so that changes in propagation rate must be nonuniform among different transcription sites (*190*). Presumably this arises because of the preferential binding of actinomycin D at G (*113, 120*). By contrast, proflavine restricts the attachment of enzyme to the template. Once the enzyme is attached and has initiated RNA synthesis, proflavine is not an effective inhibitor of RNA synthesis. Uncharged and charged tRNA also act in this manner by competitively inhibiting attachment to the DNA template (*43, 99*).

Rifamycins inhibit RNA synthesis in bacteria. In contrast with actino-mycin D and proflavine, the rifamycins interact with RNA polymerase rather than DNA. Rifamycin resistant mutants of *E. coli* and *B. subtilis* with altered RNA polymerase that is not inhibited by rifamycin can be isolated. While the identity of the *E. coli* RNA polymerase subunit(s) that interact(s) with rifamycin is not known, subunit σ does not con-stitute an essential element for rifamycin binding. Rifamycin prevents initiation but not propagation of RNA synthesis both *in vitro* and *in vivo*. In this sense one may regard it as a generalized repressor, and its utility in studying bacterial physiology and virus development has been quickly appreciated (*15a, 81, 95, 110a, 140, 255, 255a, 328, 356a, 385*).

Histones and many other polycations inhibit *in vitro* RNA synthesis. It is also well known that nucleohistones are much poorer transcription templates than the DNA extracted from them (*3*). The desire to cor-relate these facts about *in vitro* RNA synthesis with the properties of chromatin in intact cells has prompted much speculation about the mechanism and specificity of histone action. The key property of such a regulation should be its differential effect on transcription: Either histones should preferentially inhibit transcription on certain portions of the template or else specific modifiers of histone–DNA interaction should play a role [for reviews and numerous references, see Bonner and Ts'o (*38*) and Frenster (*101*)]. The existence of differential effects on tran-scription was only recently established by Paul and Gilmour (*281*) who analyzed hybridization of synthesized RNA to the DNA template and showed that relative to the derived native DNA, calf thymus chromatin serves as restricted template for *M. lysodeikticus* RNA polymerase. While the histones of this chromatin affect its template activity quan-titatively, the restrictive quality which leads to partial transcription is evidently due to other proteins and perhaps also to RNA (*21, 118, 162*). However, histones and basic polypeptides differentially affect the tran-scription of different portions of bacterial DNA templates. This is shown by the changes of nearest neighbor frequencies accompanying inhibition of RNA synthesis by these polycations (*331*). It is known that polylysine and histones interact preferentially with A–T containing DNA (*2, 68, 222, 227, 276*), while polyarginine shows a preference for long dG:dC sequences. The polypeptide–DNA interaction is cooperative (*2, 222*). Perhaps this is why changes of transcription observed through nearest neighbor frequency analysis show a considerable degree of individuality for different DNA–polypeptide pairs (*331*). Information about whether the basic polypeptides and histones block RNA synthesis *in vitro* at the propagation or initiation level, or both, is not available.

The evidence for the simultaneity and coupling of transcription and translation is diverse and abundant; a little of the evidence has been discussed in the previous section and more is presented in Chapter IV. Attempts to look for the evidence, mechanism, and consequences of this coupling in systems synthesizing RNA *in vitro* have been guided by theoretical considerations (*339*) which involve a prominent role for ribosomes. The first observations along these lines were that ribosomes could associate with enzyme–DNA–nascent RNA complexes (*31, 50*). Subsequently, Shin and Moldave (*323, 324*) and Zillig and co-workers (*389*) demonstrated a stimulation of RNA synthesis with purified RNA polymerase as a result of ribosome addition. Recently Revel and Gros (*290*) have isolated and partly purified a factor (Chapter IV) that promotes the attachment of purified ribosomes to *in vitro* synthesized T4 RNA chains *in situ* and stimulates RNA synthesis under certain conditions. Perhaps these observations represent a beginning in working out relationships of transcription and translation *in vitro*. However, there is a basic uncertainty about what *in vivo* effects should be reproduced. *In vitro* studies thus far have relied primarily on the criterion of amount of RNA synthesized, without regard to pinpointing effects on initiation, release, or propagation. Yet it now appears that the most rapid *in vitro* propagation with purified polymerase and *in vivo* propagation of RNA chains may be within less than an order of magnitude of each other. The most distinctive effects of ribosomes and ribosome-bound proteins on *in vitro* transcribing systems may be not only on the rate of elongation of RNA chains but on other features of the synthesis such as repeated transcription, enzyme and RNA chain release, or quantitative modulation of initiation (*75, 183*).

H. THE *lac* AND λ REPRESSORS

Elucidating the mode of action of repressors must certainly be regarded as one of the keys to a detailed understanding of transcription. The effects of the action of the *trp* repressor have been described in Section IV of this chapter. Recent isolation of the *lac* and λ repressors (*115, 116, 286, 287*) has provided important facts and greatly advanced the possibilities for further experiments. Before describing the current status of work on these repressors, the general nature of their action as inferred from genetic and physiological experiments should be summarized (*94*).

The *E. coli lac* repressor is the product of a regulator gene (*i*) which is adjacent to the site of its action, the *lac* operator. The repressor is converted reversibly from an active to an inactive form by inducers like lactose or isopropyl β-thiogalactoside (IPTG). In its active form, the repressor blocks the initiation of transcription of the *lac* operon (the evidence was summarized in Section IV,C). Jacob and Monod (*177, 178*) postulated that the repressor acts directly on the DNA, but alternative schemes could be imagined and were then, and subsequently, proposed. Mutants defective in regulation of the *lac* operon are of two general kinds: (1) *i* gene mutants which make defective repressor or none at all; this repressor may be defective either in repression or in derepression; (2) operator mutants whose defect is at the site of action of the *lac* repressor.

The λ repressor is the product of a gene of phage λ (denoted as c_I). Its site of action is the immunity region of the phage chromosome and it restricts the transcription of viral genes at the beginning of the temporal sequence that culminates in maturation of progeny phage and lysis. A number of related phages (the lambdoid phages) are known to possess immunity regions that are nonidentical and to interact only with their homologous repressors. Hybrid phages can be constructed in which the immunity region of another phage (for example, phage 434) replaces the immunity region of phage λ (to constitute the hybrid λ imm 434). The development of such a hybrid phage is not susceptible to the action of the λ repressor.

Gilbert and Müller Hill (*115, 116*) have partially purified a protein from *E. coli* that binds the *lac* inducer, IPTG, and is made by *lac* i^+, but not by i^- cells. The IPTG-binding proteins isolated from two strains with different repression characteristics, have different affinities for the inducer. Thus the IPTG-binding protein has the characteristics of the *lac* repressor. It also binds to native DNA carrying the *lac* operator.* Moreover, the tightness of binding to such DNA is correlated with the genetic and physiological characteristics of the operator locus. In the presence of IPTG, binding to the *lac* DNA is abolished.

Ptashne (*286, 287*) has isolated the λ repressor and identified it as the product of the c_I gene in two ways: it is not made in c_I chain-terminating (*sus*) mutants and the defective (missense) product of a c_I mutant gene can be distinguished chromatographically. The λ repressor binds to λ DNA but not to λ imm 434 DNA. Unlike RNA polymerase, λ repressor binds to native and not to denatured DNA. Its binding to DNA is not inhibited by RNA.

Thus, both the *lac* and the λ repressors are seen to interact directly

---

* The DNA is that of a defective bacteriophage, $\phi80$, carrying the *lac* genes. The phage chromosome is only 1/70th the size of the *E. coli* chromosome, and this has the effect of correspondingly concentrating that DNA segment with which a specific interaction is sought.

and specifically with DNA. This is the mode of repressor action which was originally postulated by Jacob and Monod; one can write the equation

$$\begin{array}{c} \text{Active repressor} + \text{Operator (DNA)} \rightleftarrows \text{Repressor–operator complex} \\ + \\ \text{Inducer} \\ \Updownarrow \\ \text{Inactive repressor} \end{array}$$

It only remains to show how repressors selectively restrict RNA polymerization on the transcription units that they control: (1) A repressor may block RNA polymerase binding to DNA at the site appropriate for its conjugate transcription unit. (2) Binding may occur, but initiation of transcription may be blocked. (3) Initiation may occur, but the progress of RNA polymerization may be blocked very close to the initiation site. Direct biochemical evidence on this issue is, of course, not now available. However, detailed genetic mapping of the *lac* operon (*173*) evokes possibilities (2) and (3) as plausible alternatives to (1): a site *p* (promotor), which controls the maximum induced level of expression of the *lac* operon, can be distinguished from the operator (*o*) which, as the site of action of the *lac* repressor, controls the minimum rate of expression; *o* is between *p* and the proixmal structural gene of the lac operon. Mutations in *p* and *o* are both cis dominant. The above-stated possibilities revolve around whether *p* is a site for RNA polymerase binding or for the initiation of transcription and whether attachment of the repressor at *o* affects interactions at *p*.

I. Specificity of RNA Synthesis *in Vitro:* Control of Relative Initiation

In vivo RNA synthesis involves the selection both of the coding DNA strand and the correct initiation sequences, presumably at the same step. Different transcription units function at different rates, and, under varying circumstances, differential rates of synthesis can be changed drastically. Consequently, one should ask the following questions about *in vitro* transcription: Are one or two strands of the DNA template transcribed? Are certain portions of the genome excluded from transcription? Is synthesis initiated at all or some of the same loci as *in vivo*? Are the relative rates of transcription at different loci the same? Can a temporal sequence of *in vitro* transcription be detected? The answers to these questions are, at least in part, experimentally accessible now and are being worked out quite rapidly. The reasons for interest in this aspect of transcription are clear: (1) Selective *in vitro* transcription possesses at least some of the specific properties of *in vivo* RNA synthesis. (2) Such systems are useful for classifying positive and negative control elements of transcription. (3) They are required for the study of the mechanism of action of repressors and other control elements of transcription. At this moment only partial answers are available. In emphasizing those things that are beginning to be understood now, we are placing an emphasis that will, we hope, seem quite lopsided several years from now.

The experimental methods of analyzing selective transcription involve fractionation of the DNA template, either by use of suitably selected episomes (*168, 169, 186, 386*), by mechanical fractionation of the DNA template as is possible with λ DNA (*154, 362*), or by separation of complementary DNA strands (*14, 15, 67, 211*). Alternatively, RNA synthesized *in vitro* may be compared with RNA made in functioning cells by hybridization competition or RNA–RNA hybridization. Other usually less sensitive methods include determination of composition and nearest neighbor frequency. Only rarely, as in the case of ϕX174 and λ DNA transcription are these sufficiently distinctive to be useful (*149*); more often they have led to misconceptions about the selectivity of transcription.

Most of the pertinent results have been obtained with the use of viral DNA templates. Transcription on a variety of these templates with *E. coli* RNA polymerase is either partly or completely asymmetric. When tested, the messenger coding strand *in vivo* is also found to be the preferred transcribing strand *in vitro* (*62, 108, 111, 112, 148, 329, 349, 350, 357*). This property is not confined to *E. coli* polymerase transcribing homologous viral DNA templates but is possessed also by other, less purified or almost crude, RNA polymerases (*64*). The asymmetric transcription requires native DNA templates, though making only modest demand on their integrity (*112*).

Further information comes from a comparison of *in vivo* and *in vitro* transcription in programmed systems such as those involving the phage T4 and λ DNA templates during viral development. The pertinent information on RNA synthesis in cells infected by T-even phages has been presented in Section IV,D. The comparison of the *in vivo* and *in vitro* transcription on the T2 and T4 templates has thus far proved particularly instructive. It can be shown that native DNA templates serve for the synthesis of early T2 and T4 messenger selectively and that almost all early messenger species can be synthesized *in vitro*. Either RNA synthesis cannot be initiated at all at the late phage cistrons, or propagation at these loci is so ineffective that no appreciable quantities of late RNA are synthesized. In a striking way, then, the enzyme makes its selection on the template *in vitro* similar to that which occurs *in vivo* (*107, 111, 198, 199*).

The T4 DNA template has a molecular weight of 130×10^6 and the number of early T4 genes is not known with certainty. Approximately 40% of the 70 identified complementation groups are devoted to early functions as identified by a variety of criteria—phenotype or kinetics of 5 FU phenotypic rescue—and these early genes are spread over approximately 60% of the recombination map (*87, 88, 92, 93, 259*). On the basis of the molecular weight of the DNA, the total number of genes could be as much as three times greater than the 70 now known. Early and late functions are disposed in relatively large blocks on the chromosome. There may, in fact, be no more than four such blocks, two early regions, and two late ones. Certain sections of the template are evidently inaccessible to effective RNA synthesis, *in vitro,* just as they are in the early stages of viral development. Native DNA is required for selective asymmetric transcription. This is consistent with what we have said about initiation and enzyme binding and with the opening up of many more sites to RNA polymerase binding when DNA is denatured.

Similar results have been reported thus far for λ DNA templates. Table III shows that the correspondence between the nearest neighbor frequency of λ DNA and its RNA transcription product is less strikingly good than that of the other two bacterial templates. This has more recently been recognized as resulting from nonuniform transcription. Approximately three-quarters of the synthesized RNA is homologous to the AT-rich half of the DNA molecule which contains most of the early λ genes and only one-quarter is homologous to the GC-rich half which contains the late λ genes and only a small number of early genes (*62, 232, 264*). A number of auxiliary experimental tools—strand separation, DNA fractionation, and defective viruses—provide the opportunity for detailed analysis of *in vitro* synthesis on the λ DNA template and much important information will undoubtedly be provided by this system.

These experiments distinguish portions of the T4 and λ DNA templates that are open to transcription by polymerase and establish a correlation between open segments of the template and the early viral genes. They also suggest a considerable fidelity for the *in vitro* system. With both DNA templates, one can formulate stringent tests for the correspondence between *in vitro* and *in vivo* transcription because much is now known about the transcription of the early genes of both phages (*206, 249, 264, 312, 330, 350*). The normal state of the minimum system—T4 DNA segments for late genes and bacterial polymerase—is apparently closed, and transcription of the late viral genes requires one or more positive control elements. *In vivo* and *in vitro* experiments should ultimately permit one to distinguish whether such positive control elements of late viral gene transcription include a new RNA polymerase, a modification of the RNA polymerase of the host, a modification of the DNA template competent for late messenger transcription, or some other modifier of transcription-translation (*48, 133, 221, 304, 334*).

The heterologous *M. lysodeikticus* and *B. subtilis* RNA polymerases are restricted on the T4 DNA template in somewhat the same manner, although with this enzyme there is a certain amount of late message transcription. The specificity of transcription-selection is not, therefore, exclusively a property of homologous polymerase–template pairs (*46, 108*).

Finally, it should be possible to compare *in vivo* and *in vitro* transcription in another way. The functioning cell maintains and controls its wide range of abundances of different RNA species in a variety of ways. It is probable that a large part of the range of abundances is due to variations of transcription rates that are not determined by repressor–inducer interactions and by concomitant protein synthesis, but by intrinsically different rates of initiation of RNA synthesis on different DNA template sequences. This is the promotor hypothesis, which has not yet been tested critically *in vitro*. The available data on the composition of RNA synthesized on bacterial DNA templates *in vitro* (e.g., Table III), clearly specify that ribosomal RNA cannot be an overwhelmingly abundant product of *in vitro* transcription, but are not adequate to exclude the possibility that an appreciable fraction of the template is transcribed very infrequently or not at all. A test (*108*) suggests that transcription of the T4 template by *E. coli* polymerase *in vitro* is, under some conditions, capable of yielding early T4 messenger in relative abundances that correspond roughly to the relative abundances of different T4 early RNA species present in infected cells during the early period. If this should prove to be correct, it will provide a means of finding out how the basic (open) level of activity of a gene is set.

ACKNOWLEDGMENTS

I should like to express my thanks to numerous colleagues who communicated unpublished results, or patiently read the original manuscript: D. Glaser, M. Chamberlin, M. Goulian, R. Kaempfer, V. Aposhian, R. Zimmerman, W. Salser, N. Sueoka, K. G. Lark, S. Tomizawa, F. Imamoto, P. Hanawalt, R. Haselkorn, E. N. Brody, W. Epstein, M. Stodolsky, M. Schaechter, B. K. Zimmerman, B. Strauss, H. Manor, J. Richardson, and R. Milette.

118   E. P. GEIDUSCHEK

I should also like to acknowledge a Research Career Development Award of the U.S. Public Health Service.

REFERENCES

1. M. Abe and J. Tomizawa, Replication of the *Escherichia coli* K12 chromosome. *Proc. Natl. Acad. Sci. U. S.* **58**, 1911 (1967).
2. E. O. Akinrimisi, J. Bonner, and P. O. P. Ts'o, Binding of basic proteins to DNA. *J. Mol. Biol.* **11**, 128 (1965).
3. V. G. Allfrey and A. E. Mirsky, Evidence for the complete DNA-dependence of RNA synthesis in isolated Thymus nuclei. *Proc. Natl. Acad. Sci. U. S.* **48**, 1590 (1962).
4. V. G. Allfrey and A. E. Mirsky, Role of histone in nuclear function. *In* "The Nucleohistones" (J. Bonner and P. O. P. Ts'o, eds.), p. 267. Holden-Day, San Francisco, California, 1964.
5. D. H. Alpers and G. M. Tomkins, The order of introduction and deinduction of the enzymes of the lactose operon in *E. coli*. *Proc. Natl. Acad. Sci. U. S.* **53**, 797 (1965).
6. D. H. Alpers and G. M. Tomkins, Sequential transcription of the genes of the lactose operon and its regulation by protein synthesis. *J. Biol. Chem.* **241**, 4434 (1966).
7. B. N. Ames and P. E. Hartman, The histidine operon. *Cold Spring Harbor Symp. Quant. Biol.* **28**, 349 (1963).
8. D. D. Anthony, E. Zeszotek, and D. A. Goldthwait, Initiation by the DNA-dependent RNA polymerase. *Proc. Natl. Acad. Sci. U. S.* **56**, 1026 (1966).
8a. H. V. Aposhian, Private communication (1967).
9. H. V. Aposhian and A. Kornberg, Enzymatic synthesis of DNA. IX. The polymerase formed after T2 bacteriophage infection of *Escherichia coli*. A new enzyme. *J. Biol. Chem.* **237**, 519 (1962).
10. K. Asano, Size heterogeneity of T2 messenger RNA. *J. Mol. Biol.* **14**, 71 (1965).
11. G. Attardi, S. Naono, I. Rouvière, F. Jacob, and F. Gros, Production of messenger RNA and regulation of protein synthesis. *Cold Spring Harbor Symp. Quant. Biol.* **28**, 363 (1963).
12. G. Attardi, P. C. Huang, and S. Kabat, Recognition of ribosomal RNA sites in DNA. I. Analysis of the *E. coli* system. *Proc. Natl. Acad. Sci. U. S.* **53**, 1490 (1965).
13. G. Attardi, H. Parnas, M. H. Hwang, and B. Attardi, Giant size rapidly labeled nuclear RNA and cytoplasmic messenger RNA in immature duck erythrocytes. *J. Mol. Biol.* **20**, 145 (1966).
14. S. Aurisicchio, A. Coppo, P. Donini, C. Frontali, F. Graziosi, and G. Toschi, No. ISS, 61/33. Res. Rept. Phys. Lab., Istituto Superiore di Sanita, Rome, 1961.
15. S. Aurisicchio, E. Dore, C. Frontali, F. Gaeta, and G. Toschi, Chromatographic purification of one of the complementary strands of DNA of phage α. *Biochim. Biophys. Acta* **80**, 514 (1964).
15a. C. Babinet and H. Condamine, Mutants résistants à la Rifampicine, modifiés dans leur DNA-RNA polymerase. *Compt. Rend.* **D267**, 231 (1968).
16. U. Bachrach and S. Persky, Inhibition of messenger RNA synthesis by oxidized spermine. *Biochem. Biophys. Res. Commun.* **24**, 135 (1966).
16a. R. F. Baker and C. Yanofsky, The periodicity of RNA polymerase initiations— A new regulatory feature of transcription. *Proc. Natl. Acad. Sci. U. S.* **60**, 313 (1968).

17. M. I. Baldi and R. Haselkorn, Private communication, unpublished (1965).

18. R. L. Baldwin, Appendix: Molecular weight and sedimentation coefficient of DNA polymerase. *J. Biol. Chem.* **239**, 231 (1964).

19. Y. Becker and W. K. Joklik, Messenger RNA in cells infected with vaccinia virus. *Proc. Natl. Acad. Sci. U. S.* **51**, 577 (1964).

20. J. Beckwith and E. R. Signer, Transposition of the *Lac* region of *E. coli.* I. Inversion of the Lac operon and transduction of *Lac* by Φ80. *J. Mol. Biol.* **19**, 254 (1966).

21. I. Bekhor, G. M. Kung, and J. Bonner, Sequence-specific interaction of DNA and chromosomal protein. *J. Mol. Biol.* **39**, 351 (1969).

22. A. Bendich and H. S. Rosenkranz, Some thoughts on the double-stranded model of DNA. *Progr. Nucleic Acid Res.* **1**, 219 (1963).

23. F. Ben-Hamida and D. Schlessinger, Stability of β-galactosidase messenger ribonucleic acid in *E. coli. J. Bacteriol.* **90**, 1611 (1965).

24. C. Berg and L. C. Caro, Chromosome replication in *E. coli.* I. Lack of influence of the integrated F factor. *J. Mol. Biol.* **29**, 419 (1967).

25. D. Berg, K. Barnett, D. Hinckle, J. McGrath, and M. J. Chamberlin, A subunit of RNA polymerase involved in chain initiation. *Federation Proc.* **28**, 659 (1969).

26. P. Berg, R. D. Kornberg, H. Fancher, and M. Dieckmann, Competition between RNA polymerase and DNA polymerase for the DNA template. *Biochem. Biophys. Res. Commun.* **18**, 932 (1965).

27. M. J. Bessman, Private communication (1967).

28. R. Beukers and W. Berends, Isolation and identification of the irradiation product of thymine. *Biochim. Biophys. Acta* **41**, 550–551 (1960).

29. R. Beukers, J. Ijlstra, and W. Berends, The effect of ultraviolet light on some components of the nucleic acids. VI. The origin of the U.V. sensitivity of DNA. *Rec. Trav. Chim.* **79**, 101 (1960).

30. J. P. Bilezikian, R. O. R. Kaempfer, and B. Magasanik, Mechanism of tryptophanase induction in *Escherichia coli. J. Mol. Biol.* **27**, 495 (1967).

31. H. A. Bladen, R. Byrne, L. G. Levin, and M. W. Nirenberg, An electron microscopic study of a DNA-ribosome complex formed *in vitro. J. Mol. Biol.* **11**, 78 (1965).

32. A. Bolle, R. H. Epstein, W. A. Salser, and E. P. Geiduschek, Transcription during bacteriophage T4 development: Requirements for late messenger synthesis. *J. Mol. Biol.* **33**, 338 (1968).

33. F. J. Bollum, Studies on the nature of calf thymus DNA-polymerase products. *Cold Spring Harbor Symp. Quant. Biol.* **28**, 21 (1963).

34. F. J. Bollum, "Primer" in DNA polymerase reactions. *Progr. Nucleic Acid Res.* **1**, 1 (1963).

35. E. T. Bolton and B. J. McCarthy, A general method for the isolation of RNA complementary to DNA. *Proc. Natl. Acad. Sci. U. S.* **48**, 1390 (1962).

36. E. T. Bolton and B. J. McCarthy, Fractionation of complementary RNA. *J. Mol. Biol.* **8**, 201 (1964).

37. F. Bonhoeffer and A. Gierer, On the growth mechanism of the bacterial chromosome. *J. Mol. Biol.* **7**, 534 (1963).

38. J. Bonner and P. O. P. Ts'o, eds., "The Nucleohistones." Holden-Day, San Francisco, California, 1964.

39. J. Bonner and R. C. Huang, Role of histone in chromosomal RNA synthesis.

In "The Nucleohistones" (J. Bonner and P. O. P. Ts'o eds.), p. 251. Holden-Day, San Francisco, California, 1964.

40. H. Bremer and M. W. Konrad, A complex of enzymatically synthesized RNA and template DNA. *Proc. Natl. Acad. Sci. U. S.* **51,** 807 (1964).

41. H. Bremer, M. W. Konrad, and R. Bruner, Capacity of T4 DNA to serve as template for purified *E. coli* RNA polymerase. *J. Mol. Biol.* **16,** 104 (1966).

42. H. Bremer, M. W. Konrad, K. Gaines, and G. S. Stent, Direction of chain growth in enzymic RNA synthesis. *J. Mol. Biol.* **13,** 540 (1965).

43. H. Bremer and G. S. Stent, Binding of *E. coli* transfer RNA to *E. coli* RNA polymerase. *Acta Biochim. Polon.* **13,** 367 (1966).

44. H. Bremer and D. Yuan, Chain growth rate of messenger RNA in *Escherichia coli* infected with bacteriophage T4. *J. Mol. Biol.* **34,** 527 (1968).

45. S. Brenner, F. Jacob, and M. Meselson, An unstable intermediate carrying information from genes to ribosomes for protein synthesis. *Nature* **190,** 576 (1961).

46. E. N. Brody, Private communication (1966).

46a. G. G. Brownlee, F. Sanger, and B. G. Barrell, The sequence of 5 S ribosomal RNA. *J. Mol. Biol.* **34,** 379 (1968).

47. R. R. Burgess, The subunit structure of *E. coli* DNA-dependent RNA polymerase. *Federation Proc.* **27,** 295 (1968).

48. R. R. Burgess, A. A. Travers, J. J. Dunn, and E. K. F. Bautz, Factor stimulating transcription by RNA polymerase. *Nature* **221,** 43 (1969).

49. K. Burton, M. R. Lunt, G. B. Peterson, and J. C. Siebke, Studies of nucleotide sequences in DNA. *Cold Spring Harbor Symp. Quant. Biol.* **28,** 27 (1963).

50. R. Byrne, J. G. Levin, H. A. Bladen, and M. W. Nirenberg, The *in vitro* formation of a DNA-ribosome complex. *Proc. Natl. Acad. Sci. U. S.* **52,** 140 (1964).

51. J. Cairns, The bacterial chromosome and its manner of replication as seen by autoradiography. *J. Mol. Biol.* **6,** 208 (1963).

52. J. Cairns, The chromosome of *E. coli. Cold Spring Harbor Symp. Quant. Biol.* **28,** 43 (1963).

53. J. Cairns, Replicons in mammalian cells. *J. Mol. Biol.* **15,** 372 (1966).

54. M. J. Chamberlin, Private communication (1967).

55. M. J. Chamberlin, Comparative properties of DNA, RNA, and hybrid homopolymer pairs. *Federation Proc.* **24,** 1446 (1965).

56. M. J. Chamberlin and P. Berg, DNA directed synthesis of RNA by an enzyme from *E. coli. Proc. Natl. Acad. Sci. U. S.* **48,** 81 (1962).

57. M. J. Chamberlin and P. Berg, Mechanism of RNA polymerase action: Formation of DNA-RNA hybrids with single-stranded templates. *J. Mol. Biol.* **8,** 297 (1964).

58. S. Champe, Bacteriophage reproduction. *Ann. Rev. Microbiol.* **17,** 87 (1963).

59. K. S. Chiang, Meiotic DNA replication mechanism in *Chlamydomonas Reinhardtii.* Ph.D. thesis, Princeton University (University Microfilms, Inc., Ann Arbor, Michigan) (1965).

60. K. S. Chiang and N. Sueoka, Replication of chloroplast DNA in *Chlamydomonas reinhardtii* during vegetative cell cycle: Its mode and regulation. *Proc. Natl. Acad. Sci. U. S.* **57,** 1506 (1967).

61. E. H. L. Chun and J. W. Littlefield, The separation of the light and heavy strands of Bromouracil-substituted mammalian DNA. *J. Mol. Biol.* **3,** 668 (1961).

62. S. N. Cohen and J. Hurwitz, Transcription of Complementary Strands of phage λ DNA *in vivo* and *in vitro. Proc. Natl. Acad. Sci. U. S.* **57,** 1759 (1967).

63. S. S. Cohen, The biochemistry of viruses. *Ann. Rev. Biochem.* **32,** 83 (1963).

64. A. J. E. Colvill, L. Kanner, G. P. Tocchini-Valentini, M. T. Sarnat, and E. P. Geiduschek, Asymmetric RNA synthesis *in vitro:* Heterologous DNA-enzyme systems; *E. Coli* RNA polymerase. *Proc. Natl. Acad. Sci. U. S.* **53,** 1140 (1965).

65. G. Contesse, S. Naono, and F. Gros, Effet des mutations polaires sur la transcription de l'operon lactose chez. *E. coli. Compt. Rend.* **D263,** 1007 (1966).

66. S. Cooper and C. E. Helmstetter, Chromosome replication and the division cycle of *E. coli* B/r. *J. Mol. Biol.* **31,** 519 (1968).

67. S. Cordes, H. T. Epstein, and J. Marmur. Some properties of the DNA of phage α. *Nature* **191,** 1097 (1961).

68. C. F. Crampton, R. Lipschitz, and E. Chargaff, Studies on nucleoproteins. II. Fractionation of DNA through fractional dissociation of their complexes with basic proteins, *J. Biol. Chem.* **211,** 125 (1954).

69. L. V. Crawford and P. H. Black, The nucleic acid of simian virus 40. *Virology* **24,** 388 (1964).

70. L. V. Crawford, E. M. Crawford, J. P. Richardson, and H. S. Slayter, The binding of RNA polymerase to polyoma and papilloma DNA. *J. Mol. Biol.* **14,** 593 and 597 (1965).

71. F. H. C. Crick, Codon anticodon pairing: The wobble hypothesis. *J. Mol. Biol.* **19,** 548 (1966).

72. F. Cuzin and F. Jacob, Existence of a genetic unit of transmission composed of different replicons in *E. coli* 12. *Ann. Inst. Pasteur* **112,** 529 (1967).

73. V. Daniel, S. Sarid, and U. Z. Littauer, Amino acid acceptor activity of bacteriophage T4 transfer RNA. *FEBS Letters* **2,** 39 (1968).

74. J.-L. Darlix, A. Sentenac, and P. Fromageot, Etude du RNA synthétisé *in vitro. Biochim. Biophys. Acta* **166,** 438 (1968).

75. J. Davidson. L. M. Pilarski, and H. Echols, A factor that stimulates RNA synthesis by purified RNA polymerase. *Proc. Natl. Acad. Sci. U. S.* **63,** 168 (1969).

76. P. F. Davison, D. Freifelder, and B. W. Holloway, Interruptions in the polynucleotide strands of bacteriophage DNA. *J. Mol. Biol.* **8,** 1 (1964).

77. M. Delbruck and G. S. Stent, On the mechanism of DNA replication. *In* "The Chemical Basis of Heredity" (W. D. McElroy and B. Glass, eds.), p. 699. Johns Hopkins Press, Baltimore, Maryland, 1957.

78. H. Denis, Gene expression in amphibian development. II. Release of the genetic information in growing embryos. *J. Mol. Biol.* **22,** 285–304 (1966).

79. E. S. Dennis and R. G. Wake, Autoradiography of the *Bacillus subtilis* chromosomes. *J. Mol. Biol.* **15,** 435 (1966).

80. A. de Waard, A. V. Paul, and I. R. Lehman, The structural gene for DNA polymerase in bacteriophages T4 and T5. *Proc. Natl. Acad. Sci. U. S.* **54,** 1241 (1965).

81. E. di Mauro, L. Snyder, P. Marino, A. Lamberti, A. Coppo, and G. P. Tocchini-Valentini, Rifampicin sensitivity of the components of DNA-dependent RNA polymerase. *Nature* **222,** 533 (1969).

82. B. Djordjevic and W. Szybalski, Genetics of human cell lines. III. Incorporation of 5-Bromo- and 5-Iododeoxyuridine into DNA of human cells and its effect on radiation sensitivity. *J. Exptl. Med.* **112,** 509 (1960).

83. W. F. Dove, Action of the λ chromosome. I. Control of functions late in bacteriophage development. *J. Mol. Biol.* **19**, 187 (1966).
84. P. Driskell-Zamenhof, Bacterial episomes. *In* "The Bacteria" (I. C. Gunsalus and R. Y. Stanier, eds.), Vol. 5, p. 155. Academic Press, New York, 1964.
85. C. F. Earhart, G. Y. Tremblay, M. J. Daniels, and M. Schaechter, DNA replication studied by a new method for the isolation of cell membrane-DNA complexes. *Cold Spring Harbor Symp. Quant. Biol.* **33**, 707 (1968).
86. H. Eberle and K. G. Lark, Chromosome replication in *Bacillus subtilis* cultures growing at different rates. *Proc Natl. Acad. Sci. U. S.* **57**, 95 (1967).
87. R. S. Edgar, and W. B. Wood, Morphogenesis of bacteriophage T4 in extracts of mutant-infected cells. *Proc. Natl. Acad. Sci. U. S.* **55**, 498 (1966).
88. G. Edlin, Gene regulation during bacteriophage T4 development. I. Phenotypic reversion of T4 amber mutants by 5-Fluorouracil. *J. Mol. Biol.* **12**, 363 (1965).
89. H. A. Eisen, C. R. Fuest, L. Siminovitch, R. Thomas, L. Lambert, L. Pereira DaSilva, and F. Jacob, Genetics and physiology of defective lysogeny in K12 (lambda); studies of early mutants. *Virology* **30**, 224–241 (1966).
90. D. Elson, Metabolism of nucleic acids (macromolecular DNA and RNA). *Ann. Rev. Biochem.* **34**, 449 (1965).
91. P. T. Englund, M. P. Deutscher, T. M. Jovin, R. B. Kelley, N. R. Cozzarelli, and A. Kornberg, Structural and functional properties of *Escherichia coli* DNA polymerase. *Cold Spring Harbor Symp. Quant. Biol.* **33**, 1 (1968).
92. R. H. Epstein and A. Bolle, To be published (1969).
93. R. H. Epstein, A. Bolle, C. M. Steinberg, E. Kellenberger, E. Boy de la Tour, R. Chevalley, R. S. Edgar, M. Susman, G. H. Denhardt, and A. Lielausis, Physiological studies of conditional lethal mutants of bacteriophage T4D. *Cold Spring Harbor Symp. Quant. Biol.* **28**, 375 (1963).
94. W. Epstein and J. Beckwith, Regulation of gene expression. *Ann. Rev. Biochem.* **37**, 411 (1968).
95. D. H. Ezekiel and J. E. Hutchins, Mutations affecting RNA polymerase associated with Rifampicin resistance in *Escherichia coli. Nature* **220**, 276 (1968).
96. D. P. Fan, A. Higa, and C. Levinthal, Messenger RNA decay and protection. *J. Mol. Biol.* **8**, 210 (1964).
97. G. C. Fareed and C. C. Richardson, Enzymatic breakage and joining of deoxyribonucleic acid. II. The structural gene for polynucleotide ligase in bacteriophage T4. *Proc. Natl. Acad. Sci. U. S.* **58**, 665 (1967).
98. G. Felsenfeld, G. Sandeen, and P. H. von Hippel, The destabilizing effect on ribonuclease on the helical DNA structure. *Proc. Natl. Acad. Sci. U. S.* **50**, 644 (1963).
99. C. F. Fox, R. I. Gumport, and S. B. Weiss, The enzymatic synthesis of RNA. V. The interaction of RNA polymerase with nucleic acid. *J. Biol. Chem.* **240**, 2101 (1965).
100. E. J. Freeman and O. W. Jones, Binding of RNA polymerase to T7 DNA: Evidence for minimal number of polymerase molecules required to cause retention of polymerase-T7 DNA complex on membrane filters. *Biochem. Biophys. Res. Commun.* **29**, 45 (1967).
101. J. H. Frenster, Control of DNA strand separations during selective transcription and asynchronous replication. *Proc. Intern. Conf. Cell Nucleus: Metab. Radiosensitivity, Rijswijk, Netherlands, 1966.* Taylor & Francis, London, 1966.

102. J. D. Friesen, Control of messenger RNA synthesis and decay in *E. coli*. *J. Mol. Biol.* **20**, 559 (1966).
103. E. Fuchs, R. L. Millette, W. Zillig, and G. Walter, Influence of salts on RNA synthesis by DNA dependent RNA polymerase from *E. coli*. *European J. Biochem.* **3**, 183 (1967).
104. J. J. Furth, J. Hurwitz, and M. Anders, The role of DNA in ribonucleic acid synthesis. I. The purification and properties of RNA polymerase. *J. Biol. Chem.* **237**, 2611 (1962).
105. L. P. Gage and E. P. Geiduschek, Repression of early viral messenger transcription in the development of a bacteriophage. *J. Mol. Biol.* **30**, 435 (1967).
106. A. T. Ganesan and J. Lederberg, A cell-membrane bound fraction of bacterial DNA. *Biochem. Biophys. Res. Commun.* **18**, 824 (1965).
107. E. P. Geiduschek, Some aspects of RNA synthesis on a DNA template. *Bull. Soc. Chim. Biol.* **47**, 1571 (1965).
108. E. P. Geiduschek, E. N. Brody, and D. L. Wilson, Some aspects of RNA transcription. *In* "Molecular Associations in Biology" (B. Pullman, ed.), Academic Press, New York, 1968. p. 163.
109. E. P. Geiduschek and R. Haselkorn, Messenger RNA. *Ann. Rev. Biochem.* **38**, 647 (1969).
110. E. P. Geiduschek, T. Nakamoto, and S. B. Weiss, The enzymatic synthesis of RNA: Complementary interaction with DNA. *Proc. Natl. Acad. Sci. U. S.* **47**, 1405 (1961).
110a. E. P. Geiduschek and J. Sklar, Continual requirement for a host RNA polymerase component in a bacteriophage development. *Nature* **221**, 833 (1969).
111. E. P. Geiduschek, L. Snyder, A. J. E. Colvill, and M. T. Sarnat, Selective synthesis of T-even bacteriophage early messenger *in vitro*. *J. Mol. Biol.* **19**, 541 (1966).
112. E. P. Geiduschek, G. P. Tocchini-Valentini, and M. T. Sarnat, Asymmetric synthesis of RNA *in vitro*: Dependence on DNA continuity and conformation. *Proc. Natl. Acad. Sci. U. S.* **52**, 486 (1964).
113. M. Gellert, C. E. Smith, D. Neville, and G. Felsenfeld, Actinomycin binding to DNA: Mechanism and specificity. *J. Mol. Biol.* **11**, 445 (1965).
114. D. Giacomoni and S. Spiegelman, Origin and biologic individuality of the genetic dictionary. *Science* **138**, 1328 (1962).
115. W. Gilbert and B. Müller-Hill, Isolation of the *Lac* Repressor. *Proc. Natl. Acad. Sci. U. S.* **56**, 1891 (1966).
116. W. Gilbert and B. Müller-Hill, The *lac* operator in DNA. *Proc. Natl. Acad. Sci. U. S.* **58**, 2415 (1967).
117. D. Gillespie and S. Spiegelman, A quantitative assay for DNA-RNA hybrids with DNA immobilized on a membrane. *J. Mol. Biol.* **12**, 829 (1965).
118. R. S. Gilmour and J. Paul, The nature of the specific restriction of template activity in the chromatin of animal cells. *Biochem. J.* **104**, 27 (1967).
119. V. R. Glisin, M. V. Glisin, and P. Doty, The nature of messenger RNA in the early stages of sea urchin development. *Proc. Natl. Acad. Sci. U. S.* **56**, 285 (1966).
120. I. H. Goldberg, M. Rabinowitz, and E. Reich, Basis of antinomycin action. I. DNA binding and inhibition of RNA-polymerase synthetic reactions by actinomycin. *Proc. Natl. Acad. Sci. U. S.* **48**, 2094 (1962).

121. A. Goldstein, J. Kirschbaum, and A. Roman, Direction of synthesis of messenger RNA in cells of *E. coli. Proc. Natl. Acad. Sci. U. S.* **54**, 1669 (1965).

122. H. M. Goodman and A. Rich, Formation of DNA-soluble RNA hybrid and its relation to the origin, evolution and degeneracy of s RNA. *Proc. Natl. Acad. Sci. U. S.* **48**, 2101 (1962).

123. M. Goulian, Initiation of the replication of single-stranded DNA by *Escherichia coli* DNA polymerase. *Cold Spring Harbor Symp. Quant. Biol.* **33**, 11 (1968).

124. M. Goulian and A. Kornberg, Enzymatic synthesis of DNA. XXIII. Synthesis of the replicative form of bacteriophage ϕX174 DNA. *Proc. Natl. Acad. Sci. U. S.* **58**, 1723 (1967).

125. M. Goulian, A. Kornberg, and R. L. Sinsheimer, Enzymatic synthesis of DNA. XXIV. Synthesis of infectious bacteriophage ϕX174 DNA. *Proc. Natl. Acad. Sci. U. S.* **58**, 2321 (1967).

126. M. Goulian, Z. J. Lucas, and A. Kornberg, Enzymatic synthesis of DNA. XXV. Purification and properties of DNA polymerase induced by infection with phage T4+. *J. Biol. Chem.* (1969) (in press).

127. M. H. Green, Strand selective transcription of T4 DNA *in vitro. Proc. Natl. Acad. Sci. U. S.* **52**, 1388 (1964).

128. H. Greenberg and S. Penman, Methylation and processing of ribosomal RNA in HeLa cells. *J. Mol. Biol.* **21**, 527 (1966).

129. R. G. Greenberg, New dUTPase and dUDPase activities after infection of E. coli by T2 bacteriophage. *Proc. Natl. Acad. Sci. U. S.* **56**, 1226 (1966).

130. F. Gros, W. Gilbert, H. H. Hiatt, G. Attardi, P. F. Spahr, and J. D. Watson, Molecular and biological characterization of messenger RNA. *Cold Spring Harbor Symp. Quant. Biol.* **26**, 111 (1961).

131. F. Gros, H. H. Hiatt, W. Gilbert, C. G. Kurland, R. W. Risebrough, and J. D. Watson, Unstable ribonucleic acid revealed by pulse labelling of *E. coli. Nature* **190**, 581 (1961).

132. W. R. Guild and M. Robison, Evidence for message reading from a unique strand of pneumoccal DNA. *Proc. Natl. Acad. Sci. U. S.* **50**, 106 (1963).

133. G. Hager, B. D. Hall, K. Fields, and F. Beguin, Altered transcription specificity of highly purified RNA polymerase. *6th Meeting European Biochem. Soc., Madrid, 1969* Abstr. No. 703.

134. B. D. Hall, M. Green, A. P. Nygaard, and J. Boezi, The copying of DNA in T2 infected *E. coli. Cold Spring Harbor Symp. Quant. Biol.* **28**, 201 (1963).

135. B. D. Hall, A. P. Nygaard, and M. Green, Control of T2-specific RNA synthesis. *J. Mol. Biol.* **9**, 143 (1964).

136. B. D. Hall and S. Spiegelman, Sequence complementarity of T2-DNA and T2-specific RNA. *Proc. Natl. Acad. Sci. U. S.* **47**, 137 (1961).

137. J. B. Hall and R. L. Sinsheimer, The structure of the DNA of bacteriophage ϕX174. IV. Pyrimidine sequences. *J. Mol. Biol.* **6**, 115 (1963).

138. P. C. Hanawalt, O. Maaløe, D. J. Cummings, and M. Schaechter, The normal DNA replication cycle. II. *J. Mol. Biol.* **3**, 156 (1961).

139. P. C. Hanawalt and D. S. Ray, Isolation of the growing point in the bacterial chromosome. *Proc. Natl. Acad. Sci. U. S.* **52**, 125 (1964).

140. G. Hartmann, K. O. Honikel, F. Knüsel, and J. Nuesch, The specific inhibition of the DNA directed RNA synthesis by Rifamycin. *Biochim. Biophys. Acta* **145**, 843 (1967).

141. L. H. Hartwell and B. Magasanik, The molecular basis of histidase induction in *B. subtilis. J. Mol. Biol.* **7**, 401 (1963).

142. L. H. Hartwell and B. Magasanik, The mechanism of histidase induction and formation in *B. subtilis. J. Mol. Biol.* **10,** 105 (1964).

143. R. Haselkorn, Physical and chemical properties of plant viruses. *Ann. Rev. Plant Physiol.* **17,** 137 (1966).

144. R. Haselkorn, M. I. Baldi, and J. Doskočil, Synthesis of messenger RNA in T4-infected *E. coli* B. *In* "The Biochemistry of Virus Replication," Pp. 79–92, Universitetsforlaget, Oslo, 1968.

145. S. Hattman and P. H. Hofschneider, Interference of bacteriophage T4 in the reproduction of RNA phage M12. *J. Mol. Biol.* **29,** 173 (1967).

146. W. F. Haut and J. H. Taylor, Studies of Bromouracil deoxyriboside substitution in DNA of bean roots (Vicia faba). *J. Mol. Biol.* **26,** 389 (1967).

147. M. Hayashi, A DNA-RNA complex as an intermediate of *in vitro* genetic transcription. *Proc. Natl. Acad. Sci. U. S.* **54,** 1736 (1965).

148. M. Hayashi, M. N. Hayashi, and S. Spiegelman, Restriction of *in vivo* genetic transcription to one of the complementary strands of DNA. *Proc. Natl. Acad. Sci. U. S.* **50,** 664 (1963).

149. M. Hayashi, M. N. Hayashi, and S. Spiegelman, DNA circularity and the mechanism of strand selection in the generation of genetic messages. *Proc. Natl. Acad. Sci. U. S.* **51,** 351 (1964).

150. M. Hayashi, S. Spiegelman, N. Franklin, and S. E. Luria, Separation of the RNA message transcribed in response to a specific inducer. *Proc. Natl. Acad. Sci. U. S.* **49,** 729 (1963).

151. D. H. Hayes, F. Hayes, and M. F. Guerin, Association of rapidly labelled bacterial RNA with ribosomal RNA in solutions of high ionic strength. Part II. *J. Mol. Biol.* **18,** 499 (1966).

152. W. Hayes, "The Genetics of Bacteria and Their Viruses." Wiley, New York, 1964.

153. C. E. Helmstetter, S. Cooper, O. Pierucci, and E. Revelas, On the bacterial life sequence. *Cold Spring Harbor Symp. Quant. Biol.* **33,** 809 (1968).

154. A. D. Hershey and E. Burgi, Complementary structure of interacting sites at the ends of λ DNA molecules. *Proc. Natl. Acad. Sci. U. S.* **53,** 325 (1965).

155. B. Hirt, Evidence for semiconservative replication of circular polyoma DNA. *Proc. Natl. Acad. Sci. U. S.* **55,** 997 (1966).

156. D. S. Hogness, W. Doerfler, J. Egan, and L. Black, The position and orientation of genes in λ and λ dg DNA. *Cold Spring Harbor Symp. Quant. Biol.* **31,** 129 (1966).

157. D. S. Hogness and J. R. Simmons, Breakage of λ DNA: Chemical and genetic characterization of each isolated half-molecule. *J. Mol. Biol.* **9,** 411 (1964).

158. N. H. Horowitz and R. L. Metzenberg, Biochemical aspects of genetics. *Ann. Rev. Biochem.* **34,** 527 (1965).

159. J. Hosoda, A mutant of bacteriophage T4 defective in α-glucosyl transferase. *Biochem. Biophys. Res. Commun.* **27,** 294 (1967).

160. P. Howard-Flanders and R. P. Boyce, DNA repair and genetic recombination. *Radiation Res.* Suppl. 6, 156 (1966).

161. W. T. Hsu and S. B. Weiss, Selective translation of T4 template RNA by ribosomes from T4 infected *Escherichia coli. Proc. Natl. Acad. Sci. U. S.* (1969) (in press).

162. R. C. Huang and P. C. Huang, Effect of protein-bound RNA associated with chick embryo chromatin on template specificity of the chromatin. *J. Mol. Biol.* **39,** 365 (1969).

163. J. A. Huberman, Visualization of replicating mammalian and T4 bacteriophage DNA. *Cold Spring Harbor Symp. Quant. Biol.* **33**, 509 (1968).

163a. J. A. Huberman and A. D. Riggs, On the mechanism of DNA replication in mammalian chromosomes. *J. Mol. Biol.* **32**, 327 (1968).

164. J. Hurwitz and J. T. August, The role of DNA in RNA synthesis. *Progr. Nucleic Acid. Res.* **1**, 59 (1963).

165. J. Hurwitz, A. Bresler, and R. Diringer, The enzymic incorporation of ribonucleotides and the effect of DNA. *Biochem. Biophys. Res. Commun.* **3**, 15 (1960).

166. J. Hurwitz, J. J. Furth, M. Anders, and A. H. Evans, The role of DNA in RNA synthesis. II. The influence of DNA on the reaction. *J. Biol. Chem.* **237**, 3752 (1962).

167. F. Imamoto, On the initiation of transcription of the tryptophan operon in *Escherichia coli. Proc. Natl. Acad. Sci. U. S.* **60**, 305 (1968).

167a. F. Imamoto and N. Morikawa, Private communication (1968).

168. F. Imamoto, N. Morikawa, K. Sato, S. Mishima, T. Nishimura, and A. Matsushiro, On the transcription of the tryptophan operon in E. coli. II. Production on the specific messenger RNA. *J. Mol. Biol.* **13**, 157 (1965).

169. F. Imamoto, N. Morikawa, and K. Sato, On the transcription of the tryptophan operon in *E. coli.* III. Multicistronic messenger RNA and polarity for transcription. *J. Mol. Biol.* **13**, 169 (1965).

170. F. Imamoto and C. Yanofsky, On the transcription of the tryptophan operon in polarity mutants of *E. coli.* I. Characterization of the *tryp-* mRNA of polar mutants. *J. Mol. Biol.* **28**, 1 (1967).

171. F. Imamoto and C. Yanofsky, On the transcription of the tryptophan operon in polarity mutants of *E. coli.* II. Evidence for normal production of *tryp-*mRNA molecules and for premature termination of transcription. *J. Mol. Biol.* **28**, 25 (1967).

172. R. B. Inman, C. L. Schildkraut, and A. Kornberg, Enzymic synthesis of DNA. XX. Electron microscopy of products primed by native templates. *J. Mol. Biol.* **11**, 285 (1965).

173. K. Ippen, J. H. Miller, J. Scaife, and J. Beckwith, A new initiation site in the *Lac* operon of *E. coli. Nature* **217**, 825 (1968).

174. A. Ishihama and T. Kameyama, The molecular mechanism of the enzymic reaction in RNA synthesis. *Biochim. Biophys. Acta* **138**, 480 (1967).

175. J. Ito and F. Imamoto, Sequential derepression and repression of the tryptophan operon in *E. coli. Nature* **220**, 441 (1968).

176. F. Jacob, S. Brenner, and F. Cuzin, On the regulation of DNA replication in bacteria. *Cold Spring Harbor Symp. Quant. Biol.* **23**, 329 (1963).

177. F. Jacob and J. J. Monod, Genetic regulatory mechanisms in the synthesis of protein. *J. Mol. Biol.* **3**, 318 (1961).

178. F. Jacob and J. J. Monod, On the regulation of gene activity. *Cold Spring Harbor Symp. Quant. Biol.* **26**, 193 (1961).

179. F. Jacob and E. L. Wollman, "Sexuality and the Genetics of Bacteria." Academic Press, New York, 1961.

180. H. S. Jansz and P. H. Pouwels, Structure of the replicating form of bacteriophage ϕX174. *Biochem. Biophys. Res. Commun.* **18**, 589 (1965).

181. W. K. Joklik, The poxviruses. *Bacteriol. Rev.* **30**, 33 (1966).

182. O. W. Jones and P. Berg, Studies on the binding of RNA polymerase to polynucleotides. *J. Mol. Biol.* **22**, 199 (1966).

183. O. W. Jones, M. Dieckmann, and P. Berg, Ribosome-induced dissociation of RNA from an RNA polymerase–DNA–RNA complex. *J. Mol. Biol.* **31,** 177 (1968).
184. E. Jordan and M. Meselson, A discrepancy between genetic and physical lengths on the chromosome of bacteriophage lambda. *Genetics* **51,** 77 (1965).
185. J. Josse, A. D. Kaiser, and A. Kornberg, Enzymatic synthesis of DNA. VIII. Frequencies of nearest neighbor base sequences in DNA. *J. Biol. Chem.* **236,** 864 (1961).
186. A. Joyner, L. N. Isaacs, H. Echols, and S. W. Sly, DNA replication and messenger RNA production after induction of wild type lambda. *J. Mol. Biol.* **19,** 174 (1966).
187. M. Kadoya, H. Mitsui, Y. Takagi, E. Otaka, H. Suzuki, and S. Osawa, A deoxyribonucleic acid-protein complex having DNA-polymerase and RNA-polymerase activities in cell-free extracts of *E. coli. Biochim. Biophys. Acta* **91,** 36 (1964).
188. R. O. R. Kaempfer and B. Magasanik, Effect of infection with T-even phage on the inducible synthesis of β-galactosidase in *E. coli. J. Mol. Biol.* **27,** 453 (1967).
189. R. O. R. Kaempfer and B. Magasanik, Mechanism of β-galactosidase induction in *Escherichia coli. J. Mol. Biol.* **27,** 475 (1967).
190. E. Kahan, F. M. Kahan, and J. Hurwitz, The role of deoxyribonucleic acid in ribonucleic acid synthesis. VI. Specificity of action of actinomycin D. *J. Biol. Chem.* **238,** 2491 (1963).
191. A. S. Kaplan and T. Ben-Porat, Mode of replication of pseudorabies virus DNA. *Virology* **23,** 90 (1964).
192. D. Kennell, Titration of the gene sites on DNA by DNA–RNA hybridization. II. The *Escherichia coli* chromosome. *J. Mol. Biol.* **34,** 85 (1968).
193. A. Kepes, Kinetics of induced enzyme synthesis. Determination of the mean life of galactosidase-specific messenger RNA. *Biochim. Biophys. Acta* **70,** 293 (1963).
194. A Kepes, Kinetic analysis of the early events in induced enzyme synthesis. *Cold Spring Harbor Symp. Quant. Biol.* **28,** 325 (1963).
195. A. Kepes, Sequential transcription and translation in the lactose operon of *E. coli. Biochim. Biophys. Acta* **138,** 107 (1967).
196. A. Kepes and S. Beguin, Hydroxylamine, an inhibitor of peptide chain initiation. *Biochem. Biophys. Res. Commun.* **18,** 377 (1965).
197. R. B. Khesin and M. F. Shemyakin, Some properties of information RNA's and their complexes with DNA. *Biokhimiya* **27,** 761 (1962).
198. R. B. Khesin, M. F. Shemyakin, Zh. M. Gorlenko, S. L. Bogdanova, and T. P. Afanaseva, RNA polymerase in *E. coli* cells infected with T2 phage. *Biokhimiya* **27,** 1092 (1962).
199. R. B. Khesin, Zh. M. Gorlenko, M. F. Shemyakin, I. A. Bass, and A. A. Prozorov, Connection between protein synthesis and regulation of messenger RNA formation in *E. coli* B cells upon development of T2 phage. *Biokhimiya* **28,** 1070 (1963).
200. C. Kidson, Deoxyribonucleic acid secondary structure in the region of the replication point. *J. Mol. Biol.* **17,** 1 (1966).
201. N. O. Kjeldgaard, Regulation of nucleic acid and protein formation in bacteria. *In* "Advances in Microbial Physiology" (A. H. Rose and J. F. Wilkinson, eds.), Vol. 1, p. 39. Academic Press, New York, 1967.

202. A. K. Kleinschmidt, A. Burton, and R. L. Sinsheimer, Electron microscopy of the replicative form of the DNA of the bacteriophage ϕX174. *Science* **142,** 961 (1963).

203. A. K. Kleinschmidt, S. J. Kass, R. C. Williams, and C. A. Knight, Cyclic DNA of Shope papilloma virus. *J. Mol. Biol.* **13,** 749 (1965).

204. A. K. Kleinschmidt, D. Lang, D. Jacherts, and R. K. Zahn, Preparation and length measurements of the total DNA content of T2 bacteriophages. *Biochim. Biophys. Acta* **61,** 857 (1962).

205. A. Kornberg, L. L. Bertsch, J. F. Jackson, and H. G. Khorana, Enzymatic synthesis of DNA. XVI. Oligonucleotides as templates and the mechanism of their replication. *Proc. Natl. Acad. Sci. U. S.* **51,** 315 (1964).

206. P. Kourilsky, L. Marcaud, P. Sheldrick, D. Luzzati, and F. Gros, Studies on the messenger RNA of bacteriophage λ. I. Various species synthesized early after induction of the prophage. *Proc. Natl. Acad. Sci. U. S.* **61,** 1013 (1968).

207. A. W. Kozinski, Molecular recombination in the ligase negative T4 amber mutant. *Cold Spring Harbor Symp. Quant. Biol.* **33,** 375 (1968).

208. A. W. Kozinski and P. B. Kozinski, Fragmentary transfer of ^{32}P-labeled parental DNA to progeny phage. II. The average size of the transferred parental fragment. Two-cycle transfer. Repair of the polynucleotide chain after fragmentation. *Virology* **20,** 213 (1963).

209. J. S. Krakow, Azotobacter vinelandii ribonucleic acid polymerase. II. Effect of ribonuclease on polymerase activity. *J. Biol. Chem.* **241,** 1830 (1966).

210. J. S. Krakow, K. Daley, M. Karstadt, *Azotobacter Vinelandii* RNA polymerase. VIII. Enzyme transitions during unprimed r[I-C] synthesis. *Proc. Natl. Acad. Sci. U. S.* **62,** 432 (1969).

211. H. Kubinski, Z. Opara-Kubinska, and W. Szybalski, Patterns of interaction between polyribonucleotides and individual DNA strands derived from several vertebrates, bacteria and bacteriophages. *J. Mol. Biol.* **20,** 313 (1966).

212. H. E. Kubitschek, H. E. Bendigkeit, and M. R. Loken, Onset of DNA synthesis during the cell cycle in chemostat cultures. *Proc. Natl. Acad. Sci. U. S.* **57,** 1611 (1967).

213. C. G. Kurland, Molecular characterization of ribonucleic acid from *E. coli* ribosomes. I. Isolation and molecular weights. *J. Mol. Biol.* **2,** 83 (1960).

214. C. Lark, Regulation of deoxyribonucleic acid synthesis in *E. coli:* Dependence on growth rates. *Biochim. Biophys. Acta* **119,** 517 (1966).

215. K. G. Lark, Regulation of chromosome replication and segregation in bacteria. *Bacteriol. Rev.* **30,** 3 (1966).

216. K. G. Lark, Initiation and control of DNA synthesis. *Ann. Rev. Biochem.* **38,** 569 (1969).

217. K. G. Lark, T. Repko, and E. J. Hoffman, The effect of amino acid deprivation on subsequent deoxyribonucleic acid replication. *Biochim. Biophys. Acta* **76,** 9 (1963).

218. R. Lavallée, Private communication (1966).

219. L. Leive, Some effects of inducer on synthesis and utilization of β-galactosidase messenger RNA in actinomycin-sensitive *E. coli. Biochem. Biophys. Res. Commun.* **20,** 321 (1965).

220. L. Leive, RNA degradation and the assembly of ribosomes in actinomycin-treated *E. coli. J. Mol. Biol.* **13,** 862 (1965).

221. K. J. Lembach, A. Kuninaka, and J. M. Buchanan, The relationship of DNA replication to the control of protein synthesis in protoplasts of T4-infected *Escherichia coli* B. *Proc. Natl. Acad. Sci. U. S.* **62**, 446 (1969).

222. M. Leng and G. Felsenfeld, The preferential interactions of polylysine and polyarginine with specific base sequences in DNA. *Proc. Natl. Acad. Sci. U. S.* **56**, 1325 (1966).

223. C. Levinthal, D. P. Fan, A. Higa, and R. A. Zimmerman, The decay and protection of messenger RNA in bacteria. *Cold Spring Harbor Symp. Quant. Biol.* **28**, 183 (1963).

224. C. Levinthal, J. Hosoda, and D. Shub, The control of protein synthesis after phage infection. *In* "The Molecular Biology of Viruses" (S. J. Colter and W. Paranchych, eds.), p. 71. Academic Press, New York, 1967.

225. C. Levinthal, A. Keynan, and A. Higa, Messenger RNA turnover and protein synthesis in *B. subtilis* inhibited by actinomycin D. *Proc. Natl. Acad. Sci. U. S.* **48**, 1631 (1962).

226. T. Lindahl, D. D. Henley, and J. R. Fresco, Molecular weight and molecular weight distribution of unfractionated yeast transfer RNA. *J. Am. Chem. Soc.* **87**, 4961 (1965).

227. J. A. Lucy and J. A. V. Butler, Fractionation of deoxyribonucleoprotein. *Biochim. Biophys. Acta* **16**, 431 (1955).

228. S. E. Luria, Asymmetric transcription of T4 phage DNA by purified RNA polymerase. *Biochem. Biophys. Res. Commun.* **18**, 735 (1965).

229. O. Maaløe, The control of normal DNA replication in bacteria. *Cold Spring Harbor Symp. Quant. Biol.* **26**, 45 (1961).

230. O. Maaløe and P. C. Hanawalt, Thymine deficiency and the normal DNA replication cycle. I. *J. Mol. Biol.* **3**, 144 (1961).

231. O. Maaløe and N. O. Kjeldgaard, "Control of Macromolecular Synthesis." Benjamin, New York, 1966.

232. U. Maitra, S. N. Cohen, and J. Hurwitz, Specificity of initiation and synthesis of RNA from DNA templates. *Cold Spring Harbor Symp. Quant. Biol.* **31**, 113 (1966).

233. U. Maitra and J. Hurwitz, The role of DNA in RNA synthesis. IX. Nucleoside triphosphate termini in RNA polymerase products. *Proc. Natl. Acad. Sci. U. S.* **54**, 815 (1965).

234. U. Maitra and J. Hurwitz, The role of DNA in RNA synthesis. XIII. Modified purification and additional properties of RNA polymerase from *E. coli. J. Biol. Chem.* **242**, 4897 (1967).

235. G. Mangiarotti and D. Schlessinger, Polyribosome metabolism in *E. coli.* II. Formation and lifetime of messenger RNA molecules, ribosomal subunit couples, and polyribosomes. *J. Mol. Biol.* **29**, 395 (1967).

236. G. Mangiarotti, D. Apirion, D. Schlessinger, and L. Silengo, Biosynthetic precursors of 30 S and 50 S ribosomal particles in *Escherichia coli. Biochemistry* **7**, 456 (1968).

237. H. Manor, D. Goodman, and G. S. Stent, RNA chain growth rates in *Escherichia coli. J. Mol. Biol.* **39**, 1 (1969).

238. H. Manor and R. Haselkorn, Properties of ribonucleic acid components in ribonucleoprotein particle preparations obtained from *E. coli* RC relaxed strain. *J. Mol. Biol.* **24**, 269 (1967).

239. J. Marmur and P. Doty, Thermal renaturation of deoxyribonucleic acids. *J. Mol. Biol.* **3**, 585 (1961).

240. R. G. Martin, The one-operon-one messenger theory of transcription. *Cold Spring Harbor Symp. Quant. Biol.* **28,** 357 (1963).

241. H. R. Massie and B. H. Zimm, Molecular weight of the DNA in the chromosome of *E. coli* and *B. subtilis. Proc. Natl. Acad. Sci. U. S.* **54,** 1636 (1965).

242. B. R. McAuslan and J. R. Kates, Regulation of virus-induced deoxyribonucleases. *Proc. Natl. Acad. Sci. U. S.* **55,** 1581 (1966).

243. B. J. McCarthy and E. T. Bolton, Interaction of complementary RNA and DNA. *J. Mol. Biol.* **8,** 184 (1964).

244. B. D. Mehrotra and H. G. Khorana, Studies on polynucleotides. XL. Synthetic deoxyribopolynucleotides as templates for ribonucleic acid polymerase: The influence of temperature on template function. *J. Biol. Chem.* **240,** 1750 (1965).

245. M. Meselson and F. W. Stahl, The replication of DNA in *Escherichia coli. Proc. Natl. Acad. Sci. U. S.* **44,** 671 (1958).

246. M. Meselson, F. W. Stahl, and J. Vinograd, Equilibrium sedimentation of macromolecules in density gradients. *Proc. Natl. Acad. Sci. U. S.* **43,** 581 (1957).

247. M. Meselson and J. J. Weigle, Chromosome breakage accompanying genetic recombination in bacteriophage. *Proc. Natl. Acad. Sci. U. S.* **47,** 857 (1961).

248. A. M. Michelson, "The Chemistry of Nucleosides and Nucleotides." Academic Press, New York, 1963.

249. G. Milanesi, E. N. Brody, and E. P. Geiduschek, Sequence of the *in vitro* transcription of T4 DNA. *Nature* **220,** 1014 (1969).

250. R. L. Milette, Private communication (1966).

251. R. L. Milette, E. Fuchs, and W. Zillig, Private communication (1966).

252. G. Milman, R. Langridge, and M. J. Chamberlin, The structure of a DNA-RNA hybrid. *Proc. Natl. Acad. Sci. U. S.* **57,** 1804 (1967).

253. S. Mitra and A. Kornberg, Enzymatic mechanisms of DNA replication. *J. Gen. Physiol.* **49,** 59 (1966).

254. S. Mitra, P. Reichard, R. B. Inman, L. L. Bertsch, and A. Kornberg, Enzymic synthesis of deoxyribonucleic acid. XXII. Replication of a circular single-stranded DNA template by DNA polymerase of *Escherichia coli. J. Mol. Biol.* **24,** 429 (1967).

255. S. Mizuno, H. Yamazaki, K. Nitta, and H. Umezawa, Inhibition of DNA-dependent RNA polymerase reaction of *E. coli* by an antimicrobial antibiotic, streptovaricin. *Biochim. Biophys. Acta* **157,** 322 (1967).

255a. S. Mizuno, H. Yamazaki, K. Nitta, and H. Umezawa, Inhibition of initiation of DNA-dependent RNA synthesis by an antibiotic B44P. *Biochem. Biophys. Res. Commun.* **30,** 379 (1968).

256. A. Monroy, R. Maggio, and A. M. Rinaldi, Experimentally induced activation of the ribosomes of the unfertilized sea urchin egg. *Proc. Natl. Acad. Sci. U. S.* **54,** 107 (1965).

257. P. B. Moore and K. Asano, Divalent cation-dependent binding of messenger to ribosomal RNA. *J. Mol. Biol.* **18,** 21 (1966).

257a. D. E. Morse, R. F. Baker, and C. Yanofsky, Translation of the tryptophan messenger RNA of *E. coli. Proc. Natl. Acad. Sci. U. S.* **60,** 1428 (1968).

258. D. E. Morse and C. Yanofsky, Private communication (1968).

259. G. Mosig, Distances separating genetic markers in T4 DNA. *Proc. Natl. Acad. Sci. U. S.* **56,** 1177 (1966).

260. D. Nakada and D. P. Fan, Protection of β-galactosidase messenger RNA from decay during anaerobiosis in *E. coli. J. Mol. Biol.* **8**, 223 (1964).
261. D. Nakada and B. Magasanik, The roles of inducer and catabolic repressor in the synthesis of β-galactosidase by *Escherichia coli. J. Mol. Biol.* **8**, 105 (1964).
262. T. Nakamoto, C. F. Fox, and S. B. Weiss, Enzymatic synthesis of ribonucleic acid. I. Preparation of ribonucleic acid polymerase from extracts of *Micrococcus lysodeikticus. J. Biol. Chem.* **239**, 167 (1964).
263. T. Nakamoto and S. B. Weiss, The biosynthesis of RNA: Priming by polyribonucleotides. *Proc. Natl. Acad. Sci. U. S.* **48**, 880 (1962).
264. S. Naono and F. Gros, Control and selectivity of λ DNA transcription in lysogenic bacteria. *Cold Spring Harbor Symp. Quant. Biol.* **31**, 363 (1966).
265. S. Naono, J. Rouvière, and F. Gros, Messenger RNA forming capacity in ribosome depleted bacteria. *Biochim. Biophys. Acta.* **129**, 271 (1966).
266. M. Nass, The circularity of mitochondrial DNA. *Proc. Natl. Acad. Sci. U. S.* **56**, 1215 (1966).
267. A. P. Nygaard and B. D. Hall, A method for the detection of RNA-DNA complexes. *Biochem. Biophys. Res. Commun.* **12**, 98 (1963).
268. A. P. Nygaard and B. D. Hall, Formation and properties of RNA-DNA complexes. *J. Mol. Biol.* **9**, 125 (1964).
269. K. I. Oda and W. K. Joklik, Hybridization and sedimentation studies on "Early" and "Late" vaccinia messenger RNA. *J. Mol. Biol.* **27**, 395 (1967).
270. M. Oishi, Studies of DNA replication *in vivo.* I. Isolation of the first intermediate of DNA replication in bacteria as single-stranded DNA. *Proc. Natl. Acad. Sci. U. S.* **60**, 329 (1968).
271. M. Oishi, H. Yoshikawa, and N. Sueoka, Synchronous and dichotomous replication of the *B. subtilis* chromosome during spore germination. *Nature* **204**, 1069 (1964).
272. M. Oishi and N. Sueoka, Location of genetic loci of ribosomal RNA on the *B. subtilis* chromosome. *Proc. Natl. Acad. Sci. U. S.* **54**, 483 (1965).
273. R. Okazaki, T. Okazaki, K. Sakabe, and K. Sugimoto, Mechanism of DNA replication: Possible discontinuity of DNA chain growth. *Japan. J. Med. Sci. Biol.* **20**, 255 (1967).
274. R. Okazaki, T. Okazaki, K. Sakabe, K. Sugimoto, and A. Sugino, Mechanism of DNA chain growth. I. Possible discontinuity and unusual secondary structure of newly synthesized chains. *Proc. Natl. Acad. Sci. U. S.* **59**, 598 (1968).
275. T. Okazaki and A. Kornberg, Enzymatic synthesis of deoxyribonucleic acid. XV. Purification and properties of a polymerase from *B. subtilis. J. Biol. Chem.* **239**, 259 (1964).
276. D. E. Olins, A. L. Olins, and P. H. von Hippel, Model nucleo-protein complexes: Studies on the interaction of cationic homopolypeptides with DNA. *J. Mol. Biol.* **24**, 157 (1967).
277. Z. Opara-Kubinska, H. Kubinski, and W. Szybalski, Interaction between denatured DNA, polyribonucleotides and ribosomal RNA: Attempts at preparative separation of the complementary DNA strands. *Proc. Natl. Acad. Sci. U. S.* **52**, 923 (1964).
278. C. W. M. Orr, S. T. Herriott, and M. J. Bessman, The enzymology of virus-infected bacteria. VII. A new deoxyribonucleic acid polymerase induced by bacteriophage T5. *J. Biol. Chem.* **240**, 4652 (1965).
279. A. O'Sullivan and N. Sueoka, Sequential replication of the *Bacillus subtilis*

chromosome. IV. Genetic mapping by density transfer experiment. *J. Mol. Biol.* **27**, 349 (1967).

280. A. B. Pardee and L. S. Prestidge, The initial kinetics of enzyme induction. *Biochim. Biophys. Acta* **49**, 77 (1961).

281. J. Paul and R. S. Gilmour, Template activity of DNA is restricted in chromatin. *J. Mol. Biol.* **16**, 242 (1966).

282. D. E. Pettijohn and P. C. Hanawalt, Evidence for repair-replication of UV damaged DNA in bacteria. *J. Mol. Biol.* **9**, 395 (1964).

283. D. E. Pettijohn and T. Kamiya, The interaction of RNA polymerase with polyoma DNA. *J. Mol. Biol.* **29**, 275 (1967).

284. W. Plaut, D. Nash, and T. Fanning, Ordered replication of DNA in polytene chromosomes of *Drosophila melanogaster. J. Mol. Biol.* **16**, 85 (1966).

285. M. Polsinelli, G. Milanesi, and A. T. Ganesan, *In vivo* replication of the DNA of *Bacillus subtilis* phage SPP-1. *Proc. Natl. Acad. Sci. U. S.* 1969 (in press).

286. M. Ptashne, Isolation of the λ phage repressor. *Proc. Natl. Acad. Sci. U. S.* **57**, 306 (1967).

287. M. Ptashne, Specific binding of the λ phage repressor to λ DNA. *Nature* **214**, 232 (1967).

288. E. Reich and I. H. Goldberg, Actinomycin and nucleic acid function. *Progr. Nucleic Acid Res. Mol. Biol.* **3**, 183 (1964).

289. H. Reiter and B. S. Strauss, Repair of damage induced by a monofunctional alkylating agent in a transformable ultraviolet-sensitive strain of *Bacillus subtilis. J. Mol. Biol.* **14**, 179 (1965).

290. M. Revel, M. Herzberg, A. Becarevic, and F. Gros, Role of a protein factor in the functional binding of ribosomes to natural messenger RNA. *J. Mol. Biol.* **33**, 231 (1968).

291. C. C. Richardson, The 5'-terminal nucleotides of T7 bacteriophage deoxyribonucleic acid. *J. Mol. Biol.* **15**, 49 (1966).

292. C. C. Richardson, R. B. Inman, and A. Kornberg, Enzymic synthesis of deoxyribonucleic acid. XVIII. The repair of partially single-stranded DNA templates by DNA polymerase. *J. Mol. Biol.* **9**, 46 (1964).

293. C. C. Richardson and A. Kornberg, A deoxyribonucleic acid phosphatase-exonuclease from *Escherichia coli.* I. Purification of the enzyme and characterization of the phosphatase activity. *J. Biol. Chem.* **239**, 242 (1964).

294. C. C. Richardson, I. R. Lehman, and A. Kornberg, A deoxyribonucleic acid phosphatase-exonuclease from *Escherichia coli.* II. Characterization of the exonuclease activity. *J. Biol. Chem.* **239**, 251 (1964).

295. C. C. Richardson, Y. Masamune, T. R. Live, A. Jacquemin-Sablon, B. Weiss, and G. Fareed, Studies on the joining of DNA by polynucleotide ligase of phage T4. *Cold Spring Harbor Symp. Quant. Biol.* **33**, 151 (1968).

296. C. C. Richardson, C. L. Schildkraut, and A. Kornberg, Studies on the replication of DNA by DNA polymerase. *Cold Spring Harbor Symp. Quant Biol.* **28**, 9 (1963).

297. C. C. Richardson, C. L. Schildkraut, H. V. Aposhian, and A. Kornberg, Enzymatic synthesis of deoxyribonucleic acid XIV. Further purification and properties of deoxyribonucleic acid polymerase of *Escherichia coli. J. Biol. Chem.* **239**, 222 (1964).

298. J. P. Richardson, Some physical properties of RNA polymerase. *Proc. Natl. Acad. Sci. U. S.* **55**, 1616 (1966).

299. J. P. Richardson, The binding of RNA polymerase to DNA. *J. Mol. Biol.* **21**, 83 (1966).

300. J. P. Richardson, Enzymatic synthesis of RNA from T7 DNA. *J. Mol. Biol.* **21**, 115 (1966).

301. J. P. Richardson, Private communication (1967–1969).

302. J. P. Richardson, RNA polymerase and the control of RNA synthesis. *Prog. Nucleic Acid Mol. Biol.* **9**, 75 (1969).

303. S. Riva and A. Cascino, To be published.

304. S. Riva and E. P. Geiduschek, Replication-coupled transcription in T4 development. *Federation Proc.* **28**, 660 (1969).

305. R. B. Roberts, Studies of macromolecular biosynthesis. *Carnegie Inst. Wash. Publ.* **624**, Sect. VI (1964).

306. R. Rosset, J. Julien, and R. Monier, Ribonucleic acid composition of bacteria as a function of growth rate. *J. Mol. Biol.* **18**, 308 (1966).

307. R. Rosset, and R. Monier, A propos de la présence d'acide ribonucleique de faible poids moleculaire dans les ribosomes d'*Escherichia coli*. *Biochim. Biophys. Acta* **68**, 653 (1963).

308. R. Rosset, R. Moiner, and J. Julien, Les ribosomes d'*Escherichia coli* I. Mise en évidence d'un RNA ribosome de faible poids moleculaire. *Bull. Soc. Chim. Biol.* **46**, 87 (1964).

309. T. F. Roth and M. Hayashi, Allomorphic forms of bacteriophage øX174 replicative DNA. *Science* **154**, 658 (1966).

310. A. Ryter and F. Jacob, Etude au microscope électronique de la liaison entre noyau et mesosome chez *B. subtilis*. *Ann. Inst. Pasteur* **107**, 384 (1964).

311. B. P. Sagik, M. H. Green, M. Hayashi, and S. Spiegelman, Size distribution of "Informational" RNA. *Biophys. J.* **2**, 409 (1962).

312. W. A. Salser, R. H. Epstein, and A. Bolle, Transcription during bacteriophage T4 development: A demonstration that distinct subclasses of the "early" RNA appear at different times and that some are "turned off" at late times. *J. Mol. Biol.*, in press (1970).

313. W. A. Salser, J. Janin, and C. Levinthal, Measurement of the unstable RNA in exponentially growing cultures of *B. subtilis* and *E. coli*. *J. Mol. Biol.* **31**, 237 (1968).

314. M. Schaechter, M. W. Bentzon, and O. Maaløe, Synthesis of deoxyribonucleic acid during the division cycle of bacteria. *Nature* **183**, 1207 (1959).

315. M. Schaechter, E. P. Previc, and M. E. Gillespie, Messenger RNA and polyribosomes in *Bacillus megaterium*. *J. Mol. Biol.* **12**, 119 (1965).

316. K. Scherrer, L. Marcaud, F. Zajdela, I. M. London, and F. Gros, Patterns of RNA metabolism in a differentiated cell: A rapidly labeled, unstable 60 S RNA with messenger properties in duck erythroblasts. *Proc. Natl. Acad. Sci. U. S.* **56**, 1377 (1966).

317. C. L. Schildkraut, C. C. Richardson, and A. Kornberg, Enzymic synthesis of deoxyribonucleic acid. XVII. Some unusual physical properties of the product primed by native DNA templates. *J. Mol. Biol.* **9**, 24 (1964).

318. M. Sekiguchi and S. S. Cohen, The synthesis of messenger RNA without protein synthesis. II Synthesis of phage-induced RNA and sequential enzyme production. *J. Mol. Biol.* **8**, 638 (1964).

319. R. B. Setlow, Physical changes and mutagenesis. *J. Cellular Comp. Physiol.* **64**, Suppl. 1, 51 (1964).

320. R. B. Setlow and W. L. Carrier, The disappearance of thymine dimers from

DNA: An error-correcting mechanism. *Proc. Natl. Acad. Sci. U. S.* **51**, 226 (1964).

321. R. B. Setlow and W. L. Carrier, The excision of pyrimidine dimers *in vivo* and *in vitro. Proc. Intern. Conf. Repl. Recomb. Genet. Mater., Canberra, Australia, 1967.*

322. H. T. Shigeura and G. E. Boxer, Incorporation of 3'-deoxyadenosine-5' triphosphate into RNA by RNA polymerase from *Micrococcus Lysodeikticus. Biochem. Biophys. Res. Commun.* **17**, 758 (1964).

323. D. H. Shin and K. Moldave, The stimulation of RNA synthesis by ribosomes *in vitro. Biochem. Biophys. Res. Commun.* **22**, 232 (1966).

324. D. H. Shin and K. Moldave, Effect of ribosomes on the biosynthesis of ribonucleic acid *in vitro. J. Mol. Biol.* **21**, 231 (1966).

325. E. H. Simon, Transfer of DNA from parent to progeny in a tissue culture line of human carcinoma of the cervix (strain HeLa). *J. Mol. Biol.* **3**, 101 (1961).

326. J. H. Sinclair and B. J. Stevens, Circular DNA filaments from mouse mitochondria. *Proc. Natl. Acad. Sci. U. S.* **56**, 508 (1966).

327. M. F. Singer and P. Leder, Messenger RNA: An evaluation. *Ann. Rev. Biochem.* **35**, 195 (1966).

328. A. Sippel and G. Hartmann, Mode of action of rifamycin on the RNA polymerase reaction. *Biochim. Biophys. Acta* **157**, 218 (1968).

329. A. Skalka, Regional and temporal control of genetic transcription in phage lambda. *Proc. Natl. Acad. Sci. U. S.* **55**, 1190 (1966).

330. A. Skalka, B. Butler, and H. Echols, Genetic control of transcription during development of phage λ. *Proc. Natl. Acad. Sci. U. S.* **58**, 576 (1967).

331. A. Skalka, H. Fowler, and J. Hurwitz, The effect of histones on the enzymatic synthesis of ribonucleic acid. *J. Biol. Chem.* **241**, 588 (1966).

332. D. A. Smith, A. M. Martinez, R. L. Ratliff, D. L. William, and F. N. Hayes, Template-induced dissociation of RNA polymerase. *Biochemistry* **6**, 3057 (1967).

333. I. Smith, D. Dubnau, P. Morell, and J. Marmur, Chromosomal location of DNA base sequences complementary to transfer, 5 S, 16 S and 23 S ribosomal RNA in *Bacillus subtilis. J. Mol. Biol.* **33**, 123 (1967).

334. S. S. Snyder and E. P. Geiduschek, *In vitro* synthesis of T4 late messenger RNA. *Proc. Natl. Acad. Sci. U. S.* **59**, 459 (1968).

335. J. H. Spencer and E. Chargaff, Studies on the nucleotide arrangement in deoxyribonucleic acids. VI. Pyrimidine nucleotide clusters: Frequency and distribution in several species of the AT-type. *Biochim. Biophys. Acta* **68**, 18 (1963).

336. J. F. Speyer, Mutagenic DNA polymerase. *Biochem. Biophys. Res. Commun.* **21**, 6 (1965).

337. N. W. Stead and O. W. Jones, Jr., The binding of RNA polymerase to DNA: Stabilization by nucleoside triphosphates. *Biochim. Biophys. Acta* **145**, 679 (1967).

338. G. S. Stent, "Molecular Biology of the Bacterial Viruses." Freeman, San Francisco, California, 1963.

339. G. S. Stent, The operon: On its third anniversary. Modulation of transfer RNA species can provide a workable model of an operator-less operon. *Science* **144**, 816 (1964).

340. N. Sternberger and A. Stevens, Studies of complexes of RNA polymerase and λ DNA. *Biochem. Biophys. Res. Commun.* **24**, 937 (1966).

341. A. Stevens, Incorporation of the adenine ribonucleotide into RNA by cell fractions from *E. coli B. Biochem. Biophys. Res. Commun.* **3**, 92 (1960).

342. A. Stevens, A. J. Emery, Jr., and N. Sternberger, Sedimentation properties of *E. coli* RNA polymerase and its complexes with polyuridylic acid. *Biochem. Biophys. Res. Commun.* **24**, 929 (1966).

343. A. Stevens and A. J. Henry, Jr. Studies of the ribonucleic acid polymerase from *Escherichia coli.* I. Purification of the enzyme and studies of ribonucleic acid formation. *J. Biol. Chem.* **239**, 196 (1964).

344. B. S. Strauss, DNA repair mechanisms and their relation to mutation and recombination. *Current Topics Microbiol. Immunol.* **44**, 1 (1968).

345. F. W. Studier, Sedimentation studies of the size and shape of DNA. *J. Mol. Biol.* **11**, 373 (1965).

346. N. Sueoka, Mitotic replication of DNA in *Chlamydomonas Reinhardi. Proc. Natl. Acad. Sci. U. S.* **46**, 83 (1960).

347. N. Sueoka, Mechanisms of replication and repair of nucleic acid. *In* "Molecular Genetics" (J. H. Taylor, ed.), Part 2, p. 1. Academic Press, New York, 1967.

348. N. Sueoka, T. Kano-Sueoka, and W. J. Gartland, Modification of sRNA and regulation of protein synthesis. *Cold Spring Harbor Symp. Quant. Biol.* **31**, 571 (1966).

348a. K. Sugimoto, T. Okazaki, Y. Imae, and R. Okazaki, Mechanism of DNA chain growth. III. Equal annealing of T4 nascent short DNA chains with the separated complementary strands of the phage DNA. *Proc. Natl. Acad. Sci. U. S.* (1969) (in press).

349. W. Summers and W. Szybalski, Totally asymmetric transcription of coliphage T7 *in vivo:* Correlation with poly G binding sites. *Virology* **34**, 9 (1968).

350. W. Szybalski, Initiation and patterns of transcription during phage development. *Proc. 8th Can. Cancer Conf., 1968* p. 183. Pergamon Press, Oxford, 1969.

351. W. Szybalski, H. Kubinski, and P. Sheldrick, Pyrimidine clusters on the transcribing strand of DNA and their possible role in the initiation of RNA synthesis. *Cold Spring Harbor Symp. Quant. Biol.* **31**, 123 (1966).

352. J. H. Taylor, Replicating units in mammalian DNA. *In* "Molecular Genetics" (J. H. Taylor, ed.), Part 1, p. 65. Academic Press, New York, 1963.

353. J. H. Taylor, The duplication of chromosomes. *In* "Probleme der Biologischen Reduplikation" (von P. Sitte, ed.), Vol. 3, p. 9. Springer, Berlin, 1966.

354. J. H. Taylor, P. S. Woods, and W. L. Hughes, The organization and duplication of chromosomes as revealed by autoradiographic studies using tritium-labeled thymidine. *Proc. Natl. Acad. Sci. U. S.* **43**, 122 (1957).

355. K. Taylor, Z. Hradecna, and W. Szybalski, Asymmetric distribution of the transcribing regions on the complementary strands of coliphage λ DNA. *Proc. Natl. Acad. Sci. U. S.* **57**, 1618 (1967).

356. A. Tissières, S. Bourgeois, and F. Gros, Inhibition of RNA polymerase by RNA. *J. Mol. Biol.* **7**, 100 (1963).

356a. G. P. Tocchini-Valentini, P. Marino, and A. J. E. Colvill, Mutant of *E. coli* containing an altered DNA-dependent RNA polymerase. *Nature* **220**, 275 (1968).

357. G. P. Tocchini-Valentini, M. Stodolsky, A. Aurisicchio, M. T. Sarnat, F. Graziosi, S. B. Weiss, and E. P. Geiduschek, On the asymmetry of RNA synthesis *in vivo. Proc. Natl. Acad. Sci. U. S.* **50**, 935 (1963).

357a. A. A. Travers and R. R. Burgess, Cyclic reuse of the RNA polymerase sigma factor. *Nature* **222,** 537 (1969).

358. M. Tsuboi, K. Matsuo, and P. O. P. Ts'o, Interaction of poly-L-lysine and nucleic acids. *J. Mol. Biol.* **15,** 256 (1966).

359. E. F. J. VanBruggen, P. Borst, G. J. C. M. Ruttenberg, M. Gruber, and A. M. Kroon, Circular mitochondrial DNA. *Biochim. Biophys. Acta* **119,** 437 (1966).

360. J. Vinograd and J. Lebowitz, Physical and topological properties of circular DNA. *J. Gen. Physiol.* **49,** 103 (1966).

361. A. Wacker, H. Dellweg, and D. Weinblum, Strahlen-chemische Veränderungen der Bakterien-DNS *in vivo. Naturwissenschaften* **47,** 477 (1960).

362. J. C. Wang, U. S. Nandi, D. S. Hogness, and N. Davidson, Isolation of λ *dg* deoxyribonucleic acid halves by Hg (II) binding and CS_2SO_4 density-gradient centrifugation. *Biochemistry* **4,** 1697 (1965).

363. H. R. Warner and J. E. Barnes, Evidence for a dual role for the bacteriophage T4-induced deoxycytidine triphosphate nucleotidehydrolase. *Proc. Natl. Acad. Sci. U. S.* **56,** 1233 (1966).

364. J. D. Watson and F. H. C. Crick, The structure of DNA. *Cold Spring Harbor Symp. Quant. Biol.* **18,** 123 (1953).

365. B. Weiss and C. C. Richardson, Enzymatic breakage and joining of deoxyribonucleic acid. I. Repair of single-strand breaks in DNA by an enzyme system from *Escherichia coli* infected with T4 bacteriophage. *Proc. Natl. Acad. Sci. U. S.* **57,** 1021 (1967).

365a. S. B. Weiss, W. T. Hsu, J. W. Foft, and N. H. Scherberg, Transfer RNA coded by the T4 bacteriophage genome. *Proc. Natl. Acad. Sci. U. S.* **61,** 114 (1968).

366. S. B. Weiss and T. Nakamoto, On the participation of DNA in RNA biosynthesis. *Proc. Natl. Acad. Sci. U. S.* **47,** 694 (1961).

367. S. B. Weiss, Private communication (1967).

368. S. B. Weiss and T. Nakamoto, The enzymatic synthesis of RNA: Nearest-neighbor base frequencies. *Proc. Natl. Acad. Sci. U. S.* **47,** 1400 (1961).

369. S. B. Weiss and T. Nakamoto, Net synthesis of ribonucleic acid with a microbial enzyme requiring deoxyribonucleic acid and four ribonucleoside triphosphates. *J. Biol. Chem.* **236,** 18 (1961).

370. R. Werner, Distribution of growing points in DNA of bacteriophage T4. *J. Mol. Biol.* **33,** 679 (1968).

371. R. Werner, Initiation and propagation of growing points in the DNA of phage T4. *Cold Spring Harbor Symp. Quant. Biol.* **33,** 501 (1968).

372. A. H. Whiteley, B. J. McCarthy, and H. R. Whiteley, Changing populations of messenger RNA during sea urchin development. *Proc. Natl. Acad. Sci. U. S.* **55,** 519 (1966).

373. J. S. Wiberg, Mutants of bacteriophage T4 unable to cause breakdown of host DNA. *Proc. Natl. Acad. Sci. U. S.* **55,** 614 (1966).

374. R. Winslow and R. A. Lazzarini, Amino acid regulation of the rates of synthesis and chain elongation of ribonucleic acid in *E. coli. J. Biol. Chem.* **244,** 3387 (1969).

375. H. G. Wittmann and C. Scholtissek, Biochemistry of viruses, *Ann. Rev. Biochem.* **35,** 299 (1966).

376. B. Wolf, A. Newman, and D. A. Glaser, The origin and direction of replication of the *E. coli* K12 chromosome. *J. Mol. Biol.* **32,** 611 (1967).

377. W. B. Wood and P. Berg, Influence of DNA secondary structure on DNA-dependent polypeptide synthesis. *J. Mol. Biol.* **9,** 452 (1964).

378. W. B. Wood and R. S. Edgar, Building a bacterial virus. *Sci. Am.* **217,** 60 (1967).
379. S. A. Yankofsky and S. Spiegelman, The identification of the ribosomal RNA cistron by sequence complementarity. I. Specificity of complex formation. *Proc. Natl. Acad. Sci. U. S.* **48,** 1069 (1962).
380. S. A. Yankofsky and S. Spiegelman, The identification of the ribosomal RNA cistron by sequence complementarity. II. Saturation of and competitive interaction at the RNA cistron. *Proc. Natl. Acad. Sci. U. S.* **48,** 1466 (1962).
381. S. A. Yankofsky and S. Spiegelman, Distinct cistrons for the two ribosomal RNA components. *Proc. Natl. Acad. Sci. U. S.* **49,** 538 (1963).
382. C. Yanofsky and S. Ito, Nonsense codons and polarity in the tryptophan operon. *J. Mol. Biol.* **21,** 313 (1966).
383. M. Yoneda and F. J. Bollum, Deoxynucleotide-polymerising enzymes of calf thymus gland. I. Large scale purification of terminal and replicative deoxynucleotidyl transferases. *J. Biol. Chem.* **240,** 3385 (1965).
384. H. Yoshikawa and N. Sueoka, Sequential replication of *Bacillus subtilis* chromosome. *Proc. Natl. Acad. Sci. U. S.* **49,** 559 (1963).
385. T. Yura and K. Igarashi, RNA polymerase mutants of *Escherichia coli.* I. Mutants resistant to streptovaricin. *Proc. Natl. Acad. Sci U. S.* **61,** 1313 (1968).
386. T. Yura and M. Imai, Genetic transcription of the tryptophan operon in *Escherichia coli. Japan. J. Med. Sci. Biol.* **20,** 254 (1967).
387. T. Zehavi-Willner and D. G. Comb, Studies on the relationship between transfer and 5 S RNA. *J. Mol. Biol.* **16,** 250 (1966).
388. W. Zillig, E. Fuchs, and R. Millette, *Procedures Nucleic Acid Res.* **1,** 323 (1966).
389. W. Zillig, R. L. Millette, E. Fuchs, G. Walter, P. Palm, and H. Priess, Observations on the structure and function of DNA-dependent RNA-polymerase. *Proc. 7th Intern. Congr. Biochem., Tokyo, 1967* Abstracts I, Symp. II, p. 97, Sci. Council, Japan, Tokyo, 1968.
390. B. K. Zimmerman, Purification and properties of deoxyribonucleic acid polymerase from micrococcus lysodeikticus. *J. Biol. Chem.* **241,** 2035 (1966).
391. R. A. Zimmermann and C. Levinthal, Messenger RNA and RNA transcription time. *J. Mol. Biol.* **30,** 349 (1967).

Abbreviated List of General References

In addition to a conventional (though far from exhaustive) bibliography, it might be helpful to provide a more restricted list of auxiliary readings. The choice presented is by no means unique. It relies, wherever possible, on general texts, reviews, and symposia since these are the easiest routes to the literature.

General

(a) Textbooks and monographs
Scientific American
W. Hayes, "The Genetics of Bacteria and their Viruses." Blackwell, Oxford, 1964.
O. Maaløe and N. O. Kjeldgaard, "Control of Macromolecular Synthesis." Benjamin, New York, 1966.
A. M. Michelson, "The Chemistry of Nucleosides and Nucleotides." Academic Press, New York, 1965.
G. Stent, "Molecular Biology of Bacterial Viruses." Freeman, San Francisco, California, 1963.

J. D. Watson, "The Molecular Biology of the Gene." Benjamin, New York, 1965.

(b) Symposia and review series

Annual Reviews of Biochemistry—recent volumes—each have one or more pertinent articles.

Cold Spring Harbor Symposia Quantitative Biology, Volumes **18, 26, 28, 31, 33** (1953, 1961, 1963, 1966, 1968).

S. J. Colter and W. Paranchych, eds., "The Molecular Biology of Viruses." Academic Press, New York, 1963.

J. N. Davidson and W. E. Cohn, eds., "Progress in Nucleic Acids and Molecular Biology." Academic Press, New York, 1963 and foll.

(c) Other articles not included in the above lists.

M. J. Chamberlin, *Fed. Proc.* **24**, 1446 (1965).

P. Howard-Flanders and R. P. Boyce, *Radiation Res.* Suppl. 6, 156 (1966).

N. O. Kjeldgaard, *In* "Advances in Microbial Physiology" (A. H. Rose and J. F. Wilkinson, eds.), Vol. 1, p. 39. Academic Press, New York, 1967.

S. Mitra and A. Kornberg, *J. Gen. Physiol.* **49**, 59 (1966).

B. S. Strauss, *Current Topics Microbiol. Immunol.* **44**, 1 (1968).

N. Sueoka, *In* "Molecular Genetics" (J. H. Taylor, ed.), Part 2, p. 1. Academic Press, New York, 1967.

III

Transfer RNA and Amino Acid Activation

Gunter von Ehrenstein

I. INTRODUCTION

A. THE NATURE OF THE PROBLEM

In addition to the chemical problem of peptide bond formation, the biosynthesis of the primary structure of proteins involves the translation of a sequence of nucleic acid bases in the template RNA into the amino acid sequence of a particular protein. To effect the translation elaborate biochemical machinery is required, and it is not surprising that protein synthesis is much more complicated than the synthesis of nucleic acid which is a relatively simple process not involving the translation from one chemical language into another.

B. THE ADAPTOR HYPOTHESIS

The side groups of many amino acids seem to be unable to interact specifically and efficiently enough with the purine and pyrimidine bases in the templates to meet the requirements for accuracy and speed in the assembly of a polypeptide chain. This dilemma led F. H. C. Crick in 1958 to the formulation of the adaptor hypothesis: "This hypothesis states that the amino acid does not find the correct place on the template by specifically adsorbing to it, but is carried there by a special molecule, named an adaptor, that itself can fit onto the right places on the template and to which the amino acid is joined by a special enzyme" (*42, 43*). Since the discovery of transfer RNA (tRNA) and its amino acid specificity, the hypothesis was reformulated in biochemical terms by M. Hoagland (80): The amino acids are activated by a set of amino acid-specific activating enzymes, the aminoacyl–tRNA synthetases, and covalently attached through their carboxyl groups to the ends of specific transfer RNA molecules (the adaptors) which carry them to the template RNA. The correct location on the template is accomplished by specific base pairing between a base sequence on the adaptor and a complementary base sequence on the template RNA.

The role of tRNA as the adaptor has been demonstrated experimentally in the following way (36, 165). The amino acid cysteine was attached to its normal transfer RNA by the cysteinyl–tRNA synthetase. It was then possible to convert the cysteine by desulfuration with Raney nickel into alanine, another natural amino acid, without breaking the linkage between the amino acid and the tRNA, producing the hybrid alanyl–tRNA^{Cys} (Fig. 1). The coding properties of the hybrid were tested in a cellfree protein synthesizing system using a synthetic polyribonucleotide that contained only U and G as a template, poly (U$_5$,G). This polymer normally promotes the incorporation into polypeptide of cysteine but not

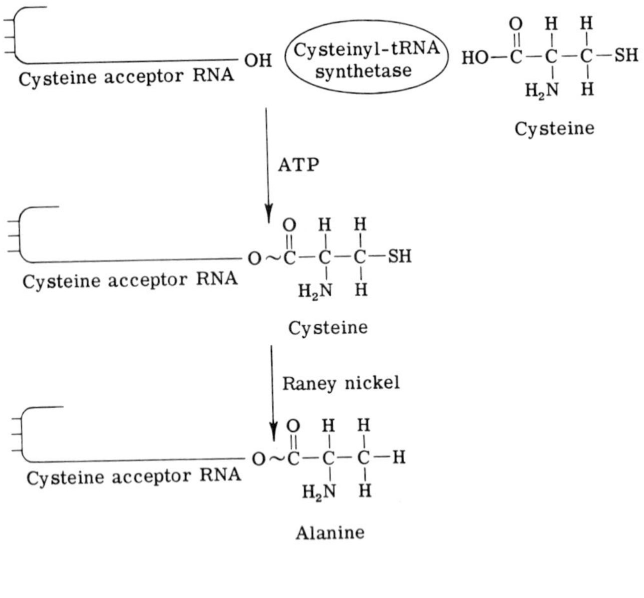

FIG. 1. Proof of the adaptor hypothesis. Cysteine is attached to its tRNA by the cysteinyl–tRNA synthetase. Raney nickel converts the cysteine into alanine producing the hybrid alanyl–tRNA^{Cys}. Alanine attached to tRNA^{Cys} is incorporated into protein as if it were cysteine (after Chapeville et al., 36).

alanine (Part A, Chapter IV). The alanine attached to the cysteine tRNA was incorporated as if it were cysteine (*36*).

This result with the artificial template was extended to the synthesis of a natural protein, i.e., hemoglobin. In this experiment the hybrid ^{14}C-alanyl–tRNA^{Cys} from *E. coli* was allowed to label tryptic peptides of hemoglobin. The only peptide containing radioactivity was one normally containing cysteine and no alanine. Conversely, no radioactivity was found in any of the other peptides that normally contain alanine but no cysteine. The control experiments showed that ^{14}C-cysteinyl–tRNA^{Cys} incorporated radioactivity also only into the cysteine-containing peptide. In contrast, ^{14}C-alanyl-tRNA^{Ala} did not label the cysteine-containing peptide but did label most of the alanine-containing peptides (*165*).

In summary, once an amino acid is attached to the tRNA it no longer participates in information transfer. The amino acid goes where the tRNA dictates and has no further control over its own destination. Thus, it follows that the code (i.e., the correspondence between nucleotide sequence in template RNA and the amino acid sequence in protein) is embodied in the precise structures and interrelationships of the set of transfer RNA adaptors and aminoacyl–tRNA synthetases. It also follows that the overall specificity of protein synthesis, i.e., the discrimination between amino acids, cannot be greater than the specificity with which the aminoacyl–tRNA complexes are formed.

In the aminoacyl–tRNA complex the carboxyl group of the amino acid is activated. The group potential of this complex is sufficiently high to form a peptide bond. Thus matter, energy, and information are fed jointly into the pathway leading to the ordered polymerization of amino acids.

Aminoacyl–tRNA complexes, containing the activated and adapted amino acids, function as direct amino acid precursors in protein synthesis. In addition, tRNA and aminoacyl–tRNA complexes also have regulatory functions in this process. For example, they provide the start signal and may also govern the rate of polypeptide chain assembly (modulation). Furthermore, aminoacyl–tRNA's are more directly involved in repression of biosynthetic pathways than are the free amino acids. They probably also serve as a link in the close coupling between the overall rate of RNA synthesis and the intracellular level of amino acids. This chapter describes the structure and function of transfer RNA.

II. ISOLATION AND PURIFICATION OF TRANSFER RNA

In view of its function, tRNA may be expected to be present in all living organisms, and it has been isolated from a wide variety of species.

It is the smallest of the natural nucleic acids having a molecular weight between 25,000–30,000 (indicating a chain length of the order of 70–80 nucleotides). Transfer RNA is also called soluble RNA (sRNA) because it is mainly present in the soluble portion of the cell in contrast to the particle-bound high molecular weight ribosomal and messenger ribonucleic acids (r- and mRNA, respectively). Alternative names describing its function such as (amino acid-) acceptor RNA, or adaptor RNA, may be encountered in the literature. Transfer RNA accounts for between 10 and 15% of the total RNA of *E. coli* cells. Particularly good sources of tRNA are yeast, *E. coli*, and mammalian liver.

The most generally applicable method for the preparation of tRNA from unbroken cells, tissue homogenates, or cell fractions is extraction with aqueous phenol. This technique can be used for large-scale preparation of tRNA from yeast or *E. coli* (*115*). Proteins, especially nucleases, are rapidly denatured by shaking with phenol thus preventing the degradation of the RNA by the ubiquitous nucleases. Upon sedimentation 2 phases are obtained, a phenol phase containing denatured proteins and lipids and an aqueous phase containing most of the nucleic acids of the cell contaminated with polysaccharides and small molecules such as nucleoside polyphosphates, amino acids, etc. The nucleic acids can be precipitated from this aqueous phase by the addition of alcohol. This procedure will remove most of the contaminating small molecules which are soluble in alcohol. Transfer RNA is then separated from the other nucleic acid fractions (mainly rRNA) by extraction with 1 M NaCl, in which the tRNA is soluble. Further purification can be achieved by chromatography on DEAE cellulose (an anion exchanger prepared from a cellulose matrix) or by fractional precipitation with isopropanol.

The single most important rule to observe when preparing and handling nucleic acids is to prevent contamination with nucleases. Traces of nucleases can be removed by adsorption onto Bentonite (a form of clay). Alternatively, nuclease inhibitors can be added. These are mostly polyanions such as polyvinylsulfate or polyglucosesulfate which probably compete with nucleic acids for the active sites of the enzymes (nucleic acids are polyanions by virtue of their polyribosephosphate backbone).

One of the advantages in working with tRNA, in contrast to other nucleic acids, is the availability of a convenient assay for one of its biochemical functions. Thus, the purity and quality of a particular preparation can be easily assessed by measuring its ability to accept amino acids. Table I shows the acceptor activity for the 20 amino acids of a typical sample of *E. coli* tRNA prepared by the procedure described in this section. Details of this reaction will be discussed in a later section. At this stage it is only important to note that a preparation of this type repre-

TABLE I

AMINO ACID ACCEPTOR ACTIVITY OF A REPRESENTATIVE
PREPARATION OF *E. coli* tRNA

Amino acid	cpm[a]/ mg tRNA	Millimicromoles of amino acid/ mg tRNA	Mole of amino acid/ mole tRNA[b]
Alanine	9,790	1.10	0.034
Arginine	17,620	1.98	0.061
Aspartic acid	8,190	0.92	0.029
Cysteine	2,840	0.32	0.010
Glutamic acid	6,680	0.75	0.023
Glycine	9,540	1.07	0.033
Histidine	2,580	0.29	0.009
Isoleucine	8,450	0.95	0.029
Leucine	23,600	2.65	0.082
Lysine	15,410	1.70	0.053
Methionine	8,200	0.92	0.029
Phenylalanine	9,800	1.10	0.034
Proline	2,320	0.26	0.008
Serine	10,780	1.21	0.038
Threonine	7,200	0.81	0.025
Tryptophan	12,000	1.35	0.042
Tyrosine	11,850	1.33	0.041
Valine	21,000	2.36	0.073
		21.06	0.653

[a] The specific activity of the ^{14}C-amino acids was 10 μC per micromole; 1 mμC gave 890 cpm.

[b] The molecular weight of tRNA was 31,000.

sents a mixture containing at least 1 tRNA for each of the 20 amino acids. This assay can also be used for monitoring the purification of a particular species of tRNA. Consequently rapid progress has been made in the large scale purification and fractionation of specific tRNA molecules.

III. FRACTIONATION OF TRANSFER RNA

Numerous techniques for the fractionation of crude tRNA into individual species have been used with variable success. They may be divided into 2 groups; the physical and the chemical methods. The useful physical methods involve column chromatography, countercurrent distribution, and recently electrophoresis on polyacrylamide gels (*78*).

A. PHYSICAL TECHNIQUES

The physical methods for separation rely mainly on 3 fundamental
structural parameters; chain length, base composition, and base sequence.
Differences in these three may lead to further differences between individual chains in such secondary features as secondary and tertiary structure and charge under a variety of pH conditions. Differences in folding
may be accentuated by the variation of temperature and salt concentration and by the use of hydrogen-bond-breaking solvents such $7 M$ urea
or formamide. It must be pointed out, however, that even quite extensive
dissimilarities in primary structure may not necessarily lead to differences
in those parameters which are being exploited in a particular fractionation procedure. Conversely, it has been shown that 2 tRNA molecules
which are apparently identical in primary structure may be separable
by a procedure that differentiates between three-dimensional configurations. A difference in three-dimensional structure could be generated by
such incidentals as, for example, the binding of metal ions and small molecules (1, 57, 63, 118). The method of isolation or other events in the history of a particular preparation could become important variables. The
trivial possibility that heterogeneity can be obtained by partial nuclease
degradation of the tRNA should also be kept in mind. Thus the demonstration of heterogeneity in a particular preparation of tRNA with any
given fractionation procedure is only of limited significance unless it can
be demonstrated that the heterogeneity is due to a difference in primary
structure or that it is, at least, accompanied by a difference in function.

It follows from this discussion that the fractionation of tRNA is still
more or less an empirical art. This is reflected in the great number of
methods that have been described.

1. Column Chromatography

The most commonly used column techniques are ion exchange chromatography, chromatography on methylated serum albumin–Kieselguhr
(MAK) columns, and partition chromatography.

The anion exchangers DEAE–cellulose and particularly DEAE–Sephadex have been most frequently used for ion-exchange chromatography
(37, 65, 124). These materials have a high capacity for nucleic acids
which makes them potentially useful for preparative purposes. The net
charge on a tRNA molecule is a function of the fraction of bases proton-
atable under acidic conditions and of the chain length. The interaction
of the RNA with the polysaccharide matrix of the column material also
seems to be an important factor in fractionation. The interaction of aromatic groups on the tRNA with aromatic substituents on the cellulose,
such as the benzoyl group in the case of the benzoylated DEAE–cellulose

(BD–cellulose), has recently been exploited with great success for the fractionation of tRNA (65).

The MAK column has been a very useful analytical tool for demonstrating heterogeneity among tRNA's (109, 158, 159). However, the use of such columns for preparative purposes is limited due to their low capacity. For the preparation of a MAK column, the carboxyl groups of bovine serum albumin are methylated, leaving the cationic amino groups unreacted. The modified protein is then adsorbed onto a Kieselguhr matrix. This type of column can therefore be thought of mainly as an anion exchanger with a very low charge density.

The fractionation principle in partition chromatography is essentially the same as that operating in countercurrent distribution, i.e., the unequal partition of different molecular species between 2 phases of a suitable solvent system. In partition chromatography 1 phase is immobilized on a column by adsorption onto a suitable matrix such as Sephadex or diatomaceous earth. The sample is dissolved in the mobile phase and allowed to equilibrate with the stationary phase while percolating through the column. Partition chromatography combines the high resolution obtainable in countercurrent distribution with the technical simplicity of a column procedure. It has been used for preparative purposes with a variety of solvent systems (92, 93, 119, 128).

2. Countercurrent Distribution

Countercurrent distribution (CCD) until recently has been the most successful, reproducible, and generally applicable technique for the fractionation of tRNA (6, 44, 66, 162). Its high resolution and capacity make it very useful for both analytical and preparative purposes. Its main disadvantages are the bulkiness and high cost of a CCD machine and the tediousness of its operation. Up to several grams of starting material can be handled. The most widely used solvent system consists of isopropanol, formamide, and phosphate buffer. Alanine–, serine–, tyrosine–, valine–, and phenylalanine–tRNA from yeast have been prepared by this method in sufficient quantity and purity for sequence studies. Recently a combination of chromatography on BD-cellulose (65) followed by chromatography on DEAE–Sephadex in the presence of high salt (124) has replaced CCD as the standard method for the preparation of pure tRNA for sequence analysis.

B. Chemical Techniques

The specific aminoacylation of a particular tRNA species with its amino acid by the corresponding aminoacyl–tRNA synthetase has been used for the purification of amino acid-specific tRNA molecules.

F<small>IG</small>. 2. The free cis-diol grouping of the ribose moiety of the 3′-terminal adenosine of tRNA chains is oxidized to a dialdehyde by sodium metaperiodate. Right: The attached amino acid protects the ribose from oxidation.

In one method the cis-diol grouping (Fig. 2) of the unfilled attachment sites of the other tRNA's is oxidized with sodium metaperiodate (Section V,B,3). The resultant dialdehyde can be converted into a hydrazone derivative or into a Schiff base. The altered tRNA chains can then be selectively removed from the protected aminoacylated tRNA chain. In another method, the aminoacylated tRNA chains are altered by combining the bound amino acid with a polypeptide. The resultant complex is then separated by physical methods. In a third method, a phenoxyacetyl derivative is formed via the free amino group of the bound amino acid. This aromatic group is specifically retarded on BD-cellulose. A high degree of purification can be achieved by this relatively simple procedure.

These chemical methods suffer from such technical difficulties as the quantitative removal from the tRNA of all unwanted bound amino acids prior to derivatization, quantitative oxidation of the cis-diol groups, and the presence of contaminating activity in the aminoacyl–tRNA synthetase used for charging. Moreover, in the many cases where more than 1 species of tRNA exist that can accept the same amino acid, one of the physical methods must still be used for final separation.

IV. MULTIPLE SPECIES OF TRANSFER RNA

When a preparation of *E. coli* tRNA is subjected to 400 transfers in the countercurrent distribution apparatus, assay of the amino acid acceptor activity in each fraction shows that the tRNA's for different amino acids can be separated from one another (Fig. 3). This is in agreement with the postulate of the adaptor hypothesis that there is at least 1 specific tRNA for each amino acid. For many amino acids, more than 1 tRNA acceptor can be demonstrated by this method. As many as 5 varieties of leucine tRNA have been detected. With an increase in the number of transfers, or with different separation techniques, separable acceptor activities can be demonstrated for even more amino acids (*20, 66*). Thus the presence of more than one tRNA for a given amino acid seems to be the rule rather than the exception. This conclusion is confirmed by methods other than physical separation, such as DNA–tRNA hybridization, and by chemical and genetic evidence (see below).

In some cases the multiple tRNA's for a particular amino acid are present in approximately equivalent amounts whereas in other cases minor

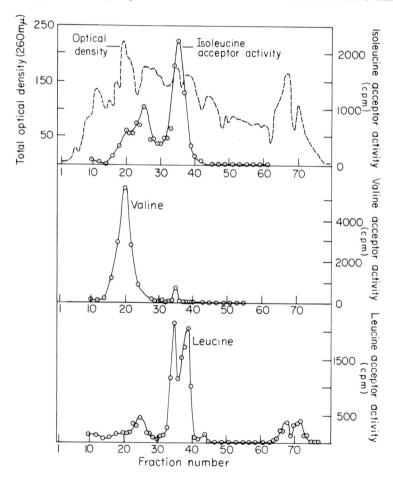

FIG. 3. Countercurrent distribution (CCD) of *E. coli* RNA; 400 transfers were performed. Transfer RNA's for different amino acids can be separated from each other.

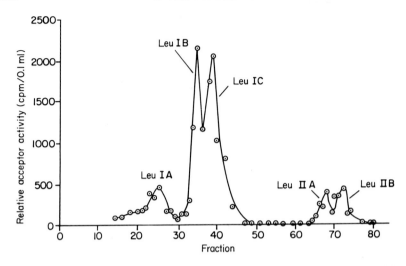

FIG. 4. Assay of the leucine acceptor activity of the CCD run shown in Fig. 3. Five leucine tRNA's can be separated from each other in this way (after Weisblum *et al.*, *172*).

components may be present. For example, 1 or 2 major leucine tRNA's are accompanied by 3 minor ones (Fig. 4).

Multiple tRNA's have been found not only in *E. coli* but also in other sources such as yeast and rat liver. Only a few systematic studies comparing the fractionation patterns of tRNA's from different organisms under identical conditions have been reported. For example, the partition coefficients of tRNA^{Tyr} from rat liver and from yeast are different. Organisms may also differ in regard to the relative abundances of tRNA species. Thus a major tRNA in one organism may constitute a minor one in a second organism, and vice versa. Further evidence for species variations and for differences in the composition of the tRNA complement between individuals of the same species will be discussed later in Sections XI and XII.

A. NOMENCLATURE IN CURRENT USE

The amino acid acceptor specificity is shown as a superscript (for example, tRNA^{Leu} for the leucine tRNA). When the amino acid is bound to the acceptor site of the transfer RNA the resulting compound is designated as leucyl–tRNA^{Leu}. In the case of multiple species, the peak number obtained in a particular fractionation procedure is added as a subscript (e.g., tRNA$^{Leu}_{2b}$). If the codon to which this tRNA responds is known it can be employed rather than the peak number (e.g., tRNA$^{Leu}_{UUG}$, tRNA$_{CUG}^{Leu}$). Other functions in which tRNA's for a particular amino acid differ can be indicated as subscripts (e.g., tRNA$_f^{Met}$ for the formylatable species and tRNA$_m^{Met}$ for the nonformylatable species of methionine tRNA). Indications of its source may be given in parentheses [e.g., (*E. coli*) leucyl-tRNA$_{UUG}^{Leu}$ or leucyl-tRNA$_{UUG}^{Leu}$ (*E. coli*)].

B. Chemical Differences between Specific tRNA Molecules

As illustrated by the following 3 examples, a chemical difference between specific tRNA molecules is often not detectable by a functional or physical difference. The sequences of the 2 (yeast) serine tRNA's studied by Zachau and his collaborators differ in only 3 nucleotides (*183*) (Fig. 10). Consequently extensive countercurrent distribution was necessary to obtain clear-cut evidence for the 2 varieties. Indeed, they were not completely separated from one another even after 1500 transfers (*184*). Since yeast is a diploid organism, it is uncertain whether the 2 structural genes for serine tRNA's are on the same chromosome (assuming that the tRNA was isolated from a pure strain of brewer's yeast). In the case of tyrosine tRNA (*E. coli*) there is evidence for at least 3 structural genes on 1 chromosome specifying 2 principal tyrosine tRNA sequences which can be separated by column chromatography on DEAE Sephadex. They are present in proportions of 60 and 40%. They differ only in a sequence of 2 nucleotides, UCACAG versus UCAUCG (Fig. 10) (*67*). Two *N*-formyl methionine tRNAs have been found in *E. coli* differing by only 1 nucleotide. The major tRNA$_f^{Met}$ has a 7-methylguanylic acid (7MeG) which, in the minor tRNA $_f^{Met}$, is replaced by adenylic acid (A) (Fig. 10). The proportions of these 2 tRNA's are estimated as 75% with 7MeG and 25% with A. The 2 tRNA's were not separated from one another by a combination of chromatography on DEAE–Sephadex and BD–cellulose (*45*).

Several recent examples illustrate that a difference in the common CpCpA end group may produce separable multiple tRNA's. Two components resulting from the countercurrent distribution of tRNA^{Phe} from yeast have been found to have the sequences . . . CpApCpC and . . . CpApCpCpA, respectively, at the amino acid-acceptor ends of the chains. After enzymatic addition of a terminal adenosine to the tRNA^{Phe} component lacking this group, the 2 forms could no longer be separated by countercurrent distribution. Thus, it would appear that a single tRNA^{Phe} species can be separated into 2 components due to the presence or absence of the adenosine end group (*135*). The CpCpA terminus has also been shown to differ in 2 components of (yeast) tRNA^{Tyr} (*106a*) and of (yeast) tRNA^{Ser} (*99*).

Other structural variations involving the minor bases may produce separable tRNA's having otherwise identical sequences of the 4 major bases (*102*).

C. DIFFERENCES IN ADAPTOR OR ACCEPTOR FUNCTION

The second criterion by which the presence of multiple species of tRNA can be established is difference in function. The functional difference may involve coding properties (i.e., the adaptor function) or responses to a heterologous aminoacyl–tRNA synthetase (i.e., the acceptor function). Difference in the acceptor function between tRNA molecules will be described in Section XI,B,2. The coding properties of multiple tRNA species will be discussed in Section XII,B,1.

A functional difference is expected to be accompanied in most cases by a difference in nucleotide sequence, particularly where differences in coding exist between otherwise indistinguishable tRNA molecules. The best examples are the (*E. coli*) su_{III}^{+} and su_{III}^{-} tyrosine tRNA's. The 2 tyrosine tRNA's differ by a single base in their anticodon, CUA versus GUA. This change enables $tRNA_{su_{III}^{+}}^{Tyr}$ to recognize the amber codon UAG whereas $tRNA_{su_{III}^{-}}^{Tyr}$ recognizes the tyrosine codons UAU and UAC (*67*). Differences in recognition by heterologous aminoacyl–tRNA synthetases have not been correlated to specific sequence differences. Comparison of the sequences of the yeast and *E. coli* tyrosine tRNA's show very striking differences (Fig. 10) (*67, 106*). Neither yeast nor *E. coli* tyrosine tRNA can be acylated by the heterologous synthetase. This is consistent with the assumption that differences in acceptor activity also are caused by differences in nucleotide sequence.

V. CHEMISTRY AND SEQUENCE DETERMINATION OF TRANSFER RNA

A. BASE COMPOSITION

The base compositions of tRNA preparations can be determined after alkaline hydrolysis (*113*). The mixture of mononucleotides can be resolved on ion exchange columns or by two-dimensional electrophoresis and chromatography on paper or thin-layer plates. Some of the minor bases undergo destruction and transformation, complicating the determination of base composition. Thus, under alkaline conditions, 1-methyladenosine isomerizes to 6-methyladenosine, the imidazol ring in 7-methylguanosine is cleaved, and dihydropyrimidine derivatives are broken down. To avoid these complications snake venom phosphodiesterase and *E. coli* alkaline phosphatase have been used to hydrolyze tRNA completely into nucleosides.

The base compositions of tRNA's from a wide variety of sources are very similar but different from those of other types of RNA. Their most distinguishing feature is the high content of guanylic and cytidylic acid. The ratios A/U and G/C are near unity (*26*).

1. *Minor Constituents*

In addition to the 4 major nucleosides, tRNA's from all species contain a relatively large proportion of minor nucleosides. The list of unusual nucleosides contains to date some 20 different compounds isolated from tRNA (Fig. 5). Since the original discovery of 5-ribosyluracil (also named pseudouridine and abbreviated ψ) and many other minor bases, the characterization of unusual constituents in transfer RNA has proceeded steadily and it is unlikely that the list is yet complete (*21, 34, 35, 46, 70–72, 72a, 100, 101, 107*). Many of the unusual nucleosides are methylated derivatives of the normal nucleosides. Some of these undoubtedly cannot take part in base pairing because the methyl groups take the place of the hydrogen atoms necessary for hydrogen bond formation (Fig. 6). Here 5,6-dihydrouracil seems to occur almost as frequently as 5-ribosyluracil. The absence of the 5,6 double bond has 2 important consequences; the ring is no longer planar, nor does it absorb ultraviolet light near 260 mμ. This latter property explains why the base escaped discovery until Robert Holley found it in yeast alanine transfer RNA (*107*). Although the minor nucleosides create problems of characterization, their presence has proved to be of great value in the determination of base sequences in tRNA's since they act as identifying markers in oligonucleotides facilitating the determination of overlaps between fragments. The biosynthesis and function of the minor constituents of tRNA will be discussed in Section IX,B.

B. END GROUPS AND CHAIN LENGTH

1. *Alkaline Hydrolysis*

In all tRNA preparations from different sources examined so far, the 3'-acceptor end of each active chain is terminated by the sequence . . . pCpCpA. When such preparations are hydrolyzed, under alkaline conditions, to yield a mixture of mononucleotides, the 3'-terminal adenosyl

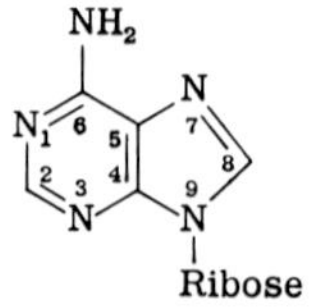

Adenosine

1-Methyladenosine
2-Methyladenosine
N^6-Methyladenosine
N^6-Dimethyladenosine

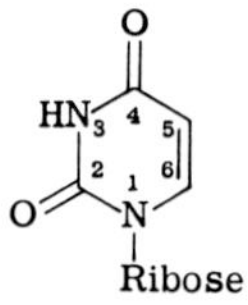

Uridine

3-Methyluridine
5-Methyluridine
(ribothymidine)

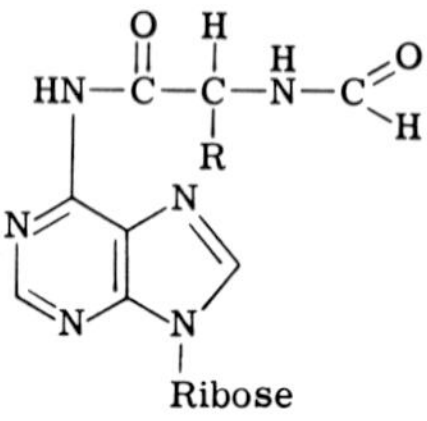

N^6-(N-Formyl-α-aminoacyl) adenosine

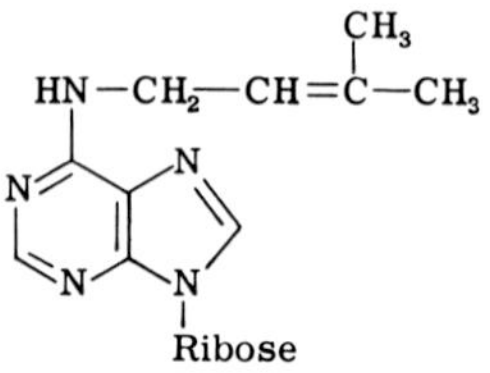

N^6-γ,γ-Dimethyl-allyladenosine (iso-pentenyladenosine)

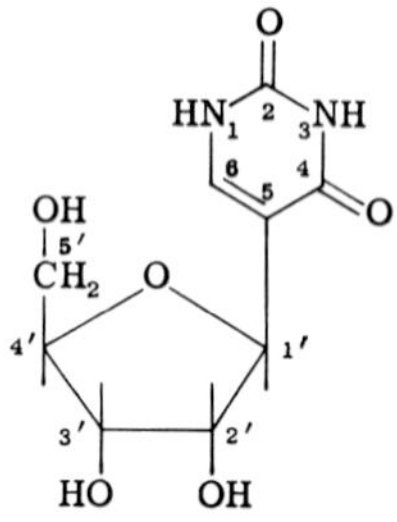

N^4-Acetylcytidine

5-β-D-Ribofuranosyl-uracil (pseudouridine)

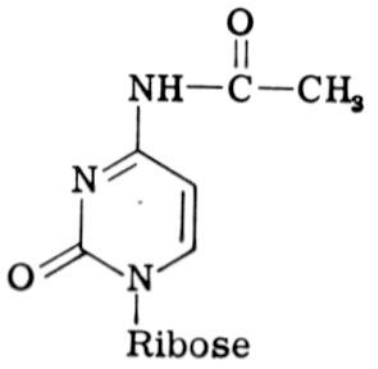

residue is released as adenosine. The 5'-terminal ends of the tRNA chains carry a 5'-phosphate group and yield (2' or 3')-5'-nucleoside diphosphates (*112*). About 78% of all tRNA chains from yeast have a pGp end group at the 5'-terminus, 10% terminate in pUp, 7% in pAp, and 5% in pCp (*13*). The internal nucleotides are released as 2' or 3'-nucleoside monophosphates. The average chain length can be estimated

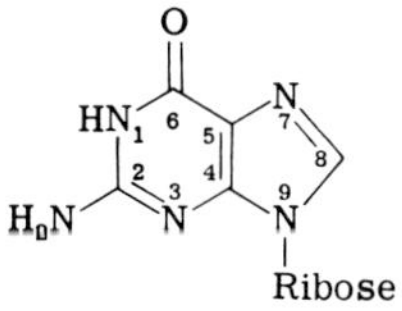

Guanosine

1-Methylguanosine
7-Methylguanosine
N^2-Methylguanosine
N^2-Dimethylguanosine

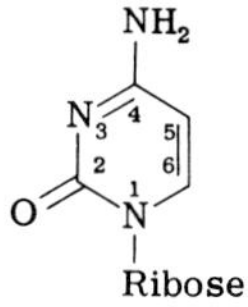

Cytidine

3-Methylcytidine
5-Methylcytidine

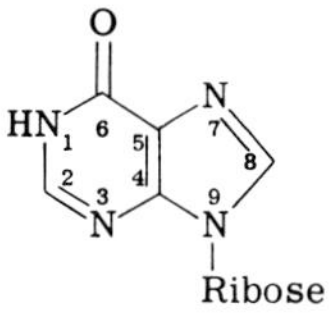

Inosine and
1-methylinosine

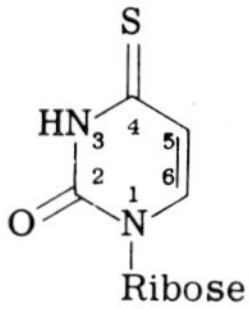

4-Thiouridine and
2-thiouridine

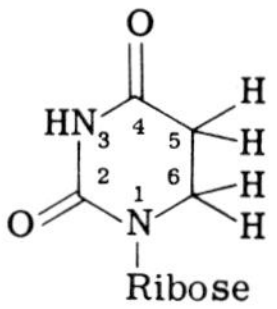

5,6-Dihydrouridine

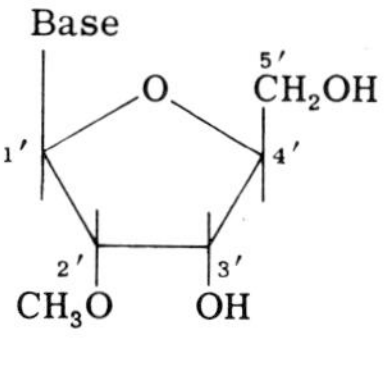

Base

2'-O-Methylriboside

Adenine

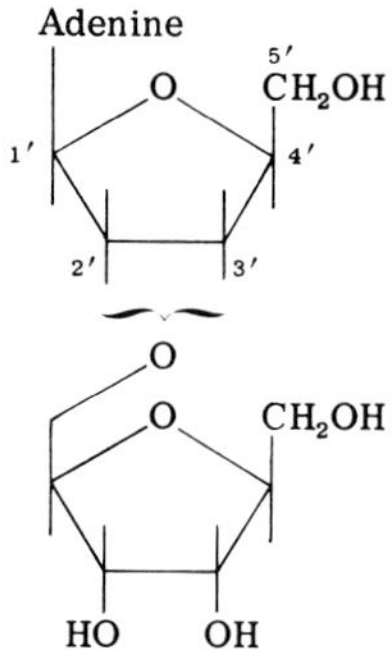

9[2'-(3')-O-Ribosyl-β-D-
ribofuranosyl]adenine

FIG. 5. The minor nucleosides of tRNA.

from the ratio of nucleoside and nucleoside diphosphate to internal nucleoside monophosphates. The chemical chain length thus determined varies between 62 and 98 internal nucleotides per end group. If one assumes the average molecular weight of a mononucleotide to be about

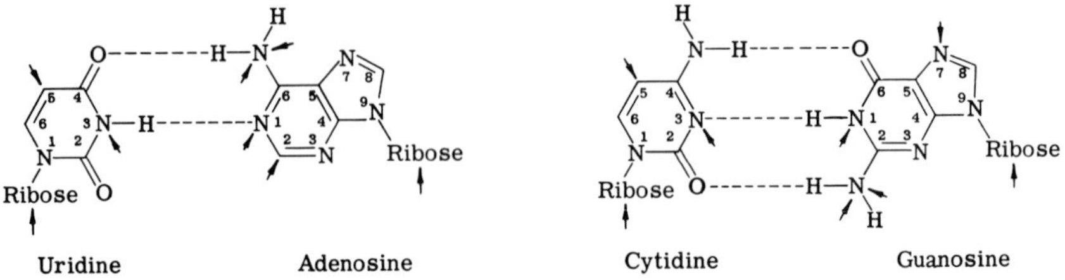

FIG. 6. The impairment of base pairing by methyl groups (after Zachau *et al.,* *185*). Another effect of methylation of the ring is to alter the keto–enol equilibrium.

340, the chemical molecular weight can then be calculated. Molecular weights range from **21,100** to **33,300** in reasonable agreement with the physically determined molecular weights, **23,300** to **35,000** for different preparations from various sources (*26*).

2. Other End Group Techniques

Various other methods have been used for end group analysis of tRNA's and of the oligonucleotides produced by hydrolysis with specific nucleases.

a. Enzymic Hydrolysis. Exonucleases degrade tRNA and oligonucleotides into the constituent mononucleotides. Snake venom phosphodiesterase attacks from the 3′-hydroxyl end with the stepwise release of nucleoside-5′-monophosphates. The 5′-terminus of an oligonucleotide, if not phosphorylated, will be released as a nucleoside. Spleen phosphodiesterase starts from the 5′-hydroxyl end with the stepwise release of nucleoside-3′-monophosphates. In this degradation the unphosphorylated 3′-terminal base is released as a nucleoside. The snake venom enzyme is inhibited if the 3′-hydroxyl terminus carries a phosphate. Conversely, the spleen enzyme is inhibited if the 5′-hydroxyl is phosphorylated. Since the terminal base (5′-OH in the case of venom diesterase and 3′-OH in the case of the spleen enzyme) must be free of phosphate in order to be recognizable as the end group, nucleotide chains whose end groups are to be determined by exonucleases must be dephosphorylated by pretreatment with *E. coli* alkaline phosphomonoesterase. The complete hydrolysis of tRNA by snake venom phosphodiesterase for the determination of labile minor nucleosides has been mentioned above.

The alkali-resistant internucleotide linkages involving those nucleotides with a 2′-*O*-methylribose group are cleaved by this enzyme, permitting the identification of the corresponding nucleosides. The application of partial exonuclease degradation in the sequence determination of oligonucleotides will be discussed in Section V,C,3.

b. Chemical End Group Labeling Methods. The 2 techniques for end group determination discussed above require relatively large amounts of starting material. Moreover small amounts of end group have to be determined in the presence of large amounts of nucleotides from internal positions. These difficulties are avoided by radioactive labeling of the end groups.

Two techniques involve labeling of the 5'-phosphate end of transfer RNA. In one, the end group is labeled with ^{14}C-aniline by conversion to the phosphoroanilidate (136). In the second, ^{14}C-methyl phosphoro-morpholidate is used to form a ^{14}C-methylpyrophosphate derivative of the 5'-phosphate end group (137). The enzymic ^{32}P phosphorylation of the 5'-phosphate end groups of polynucleotides with polynucleotide kinase should be mentioned (139, 174). This method has not been applied extensively to tRNA.

Techniques for labeling the 3'-terminus in tRNA rely on the presence of a free 2',3'-diol. The dialdehyde formed after oxidation with sodium metaperiodate can be coupled to various radioactive reagents. One of these techniques has been applied successfully to yeast tRNA. It involves the reduction of the 2',3'-dialdehyde with ^{3}H-sodium boro-hydride (135). The reaction is quantitative, and the labeled derivative formed is stable under alkaline conditions.

3. Stepwise Degradation

The chemical stepwise degradation of RNA or oligonucleotides starts with a chain carrying free hydroxyl groups on the 2'- and 3'-carbon atoms of the ribose of the terminal nucleoside. The degradation involves 3 separate steps (Fig. 7). The cis-diol group is first oxidized with periodate to give a dialdehyde. In the second step the terminal phosphodiester bond is cleaved, probably by a β-elimination which is catalyzed by primary amines. A Schiff's base addition product between the dialdehyde and the amine has been isolated. In the third step the terminal base is liberated together with its degraded ribose moiety by an unknown reaction mechanism (35). The released base can then be identified. The RNA chain is now 1 unit shorter and carries a 3'-phosphate end group. This phosphate end group must be removed with a phosphomonesterase to permit the initiation of a second cycle according to the following scheme:

$$\text{tRNA} \ldots \ldots \text{pCpCpA} \xrightarrow[\text{elimination}]{\text{Periodate}} \text{tRNA} \ldots \ldots \text{pCpCp} + \text{adenine} \qquad (1)$$

$$\text{tRNA} \ldots \ldots \text{pCpCp} \xrightarrow{\text{Phosphatase}} \text{tRNA} \ldots \ldots \text{pCpC} + \text{iP} \qquad (2)$$

$$\text{tRNA} \ldots \ldots \text{pCpC} \xrightarrow[\text{cycle}]{\text{Second}} \text{tRNA} \ldots \ldots \text{pCp} + \text{cytosine, etc.} \qquad (3)$$

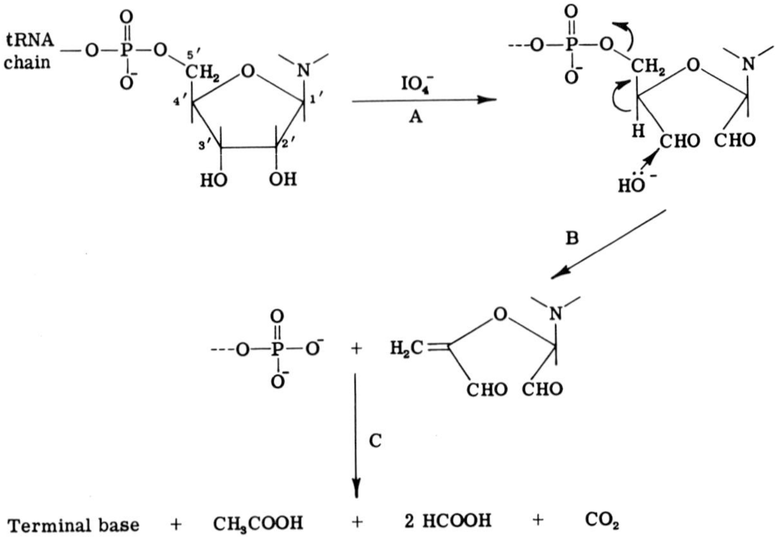

FIG. 7. Stepwise degradation of polynucleotides by periodate oxidation: (A) Cleavage of the cis-diol gives a dialdehyde. (B) Cleavage of the terminal phosphodiester bond by a β-elimination. (C) Release of the terminal base requires a large excess of periodate (35).

This method has been used for terminal sequence analysis of tRNA and TMV-RNA. In principle, the stepwise procedure can be used repeatedly leading to the sequential liberation of all the bases of the original chain. In practice, however, it has not been carried beyond 3 or 4 cycles.

4. Terminal Sequence Analysis of Bulk tRNA

The common -CpCpA end group of tRNA can be removed by the enzyme tRNA pyrophosphorylase in the presence of pyrophosphate [Reaction (4) below]. Only the terminal 3 nucleotides are removed from all chains. The 3'-termini can then be labeled with a $C^{32}P$ group by treatment with the same enzyme in the presence of CTP which is labeled with ^{32}P in the α- or ester-phosphate group of the ribose. Two equivalents of $C^{32}PPP$ are incorporated per tRNA chain [Reaction (5)].

$$\text{tRNA} \ldots \ldots \text{pXpCpCpA} + 3\text{PP} \rightleftharpoons \text{tRNA} \ldots \ldots \text{pX} + 2\text{CTP} + \text{ATP} \quad (4)$$
$$\text{tRNA} \ldots \ldots \text{pX} + 2\text{C}^{32}\text{PPP} \quad \rightarrow \text{tRNA} \ldots \ldots \text{pX}^{32}\text{PC}^{32}\text{PC} + 2\text{PP} \quad (5)$$

This procedure labels the acceptor end of tRNA molecules, which can then be analyzed in various ways: (1) Digestion with alkali transfers the ^{32}P-label to the fourth nucleotide. (2) Hydrolysis with pancreatic

RNase gives labeled oligonucleotides having sequences extending back along the chain to the next pyrimidine residue. (3) Digestion with T_1-RNase gives labeled 3'-terminal sequences extending back to the first guanylic acid residue.

This procedure has been refined for studying the terminal sequence of specific aminoacyl-tRNA's without prior fractionation. Bulk tRNA is charged with a particular amino acid using the corresponding amino-acyl–tRNA synthetase. In this way chains specific for the amino acid are protected against oxidation with periodate. The remaining tRNA chains, not containing a protective aminoacyl group, are then oxidized with periodate. Oxidized tRNA's with 2',3'-dialdehyde groups in the ribose moieties of the adenosyl termini undergo β-elimination, leaving the unprotected chains with a 3'-phosphomonoester group. These chains do not serve as substrates for the tRNA pyrophosphorylase. After removal of the amino acid the specific chains are pyrophosphorylyzed and labeled with $C^{32}PPP$ as before (16, 96).

An alternative to this specific labeling procedure utilizes the radio-active amino acid directly as the marker. Digestion with T_1-RNase and isolation of the terminal aminoacyl–oligonucleotides must be carried out under slightly acidic conditions to avoid cleavage of the alkali-labile aminoacyl linkage (77).

In bulk E. coli tRNA 68%, 24%, 8%, and less than 1% of the chains have adenylate, guanylate, uridylate, and cytidylate, respectively, in the fourth position. The fifth base is predominantly a pyrimidine. In T_1-RNase digests, half the chains end either in -GpCpCpA or -GpCpApCpCpA. In 10% of the chains the first G residue is in the ninth position. Similar heterogeneity of the sequences adjacent to the terminal -CpCpA has been found in bulk tRNA from yeast and from rat liver (96).

With the specific ^{32}P-labeling technique described above, followed by T_1-RNase digestion, the tRNA chains specific for isoleucine were found to terminate in the unique sequence -GpCp (UpC) pApCpCpA. Two terminal sequences for leucine specific chains were isolated, 75% of the tRNA^{Leu} chains terminating in -GpCpApCpCpA and 25% in -GpUpApCpCpA. This was the first demonstration that tRNA molecules specific for the same amino acid can differ in their nucleotide sequence (16).

After T_1-RNase degradation of various ^{14}C-aminoacyl–tRNA's from yeast, a single ^{14}C-aminoacyl–oligonucleotide was obtained from each of the serine-, tyrosine-, glycine-, threonine-, phenylalanine-, and alanine-acceptor tRNA's (149). Two different ^{14}C-valyl– and ^{14}C-leucyl–oligonucleotides were shown to be present in both yeast and rat liver tRNA. The 2 valyl–oligonucleotides from rat liver tRNA could be completely separated from the 2 valyl–oligonucleotides of yeast tRNA (*89, 116*). Thus tRNA's specific for the same amino acid can vary widely in their terminal nucleotide sequences [compare also the 2 tyrosine tRNA's from yeast and *E. coli* (Fig. 10) and the 2 valine tRNAs from 2 yeasts *Torulopsis utilis* and *Saccharomyces cerevisiae* (Fig. 10)].

C. General Sequence Methods

The complete nucleotide sequence of the alanine tRNA from yeast was worked out by Robert Holley and his collaborators in 1965 (*5, 82–84*). This achievement represents a major advance in nucleic acid chemistry and in the understanding of the mechanism of protein synthesis. Since then the nucleotide sequences of several other tRNA's from yeast, and *E. coli* have been reported (*11, 45, 67, 106, 134, 160a, 177a, 183*).

The general strategy of nucleotide sequence determination in RNA has been very similar to that used for amino acid sequence determination in proteins as first introduced by Sanger. The macromolecule is cleaved in several specific ways, either chemically or enzymatically, producing different sets of small fragments. The fragments in each set are then separated and the sequences of their monomers are determined. The order of the fragments within the total RNA chain is then deduced from a consideration of overlapping sequences.

The sequence analysis of proteins is facilitated by the presence of 20 different amino acids. This makes for greater selectivity in breaking the chain into fragments, and for easier separation and characterization of such fragments. Many different and unique fragments can be produced, facilitating the deduction of a unique sequence.

The situation in nucleic acids is very different. Because there are only 4 different common bases, degradation generally yields many short fragments with identical sequences from different parts of the chain, making it extremely difficult if not impossible to determine the order of the fragments. Transfer RNA represents a special case, since it contains a variety of minor bases. These have been used to determine overlaps of oligonucleotide fragments, particularly since most of the·minor bases occur only once in a tRNA chain.

The information thus obtained is still not enough to permit the reconstruction of the complete chain. Still larger fragments are required to provide further overlapping of sequences. Limited enzymic degradation of the original chain has been the most successful method for producing such large fragments.

1. Specificity of Nucleases

Pancreatic ribonuclease and T_1-ribonuclease are the 2 principal endo-nucleases used for hydrolyzing RNA chains reproducibly into sets of oligonucleotide fragments. They have a well-defined specificity, cleaving internucleotide bonds adjacent to certain bases. Several other nucleases have been used in special situations. Micrococcal nuclease, for example, has proven useful for the sequence determination of oligonucleotides (115, 133).

Pancreatic ribonuclease cleaves internucleotide linkages of the type pyrimidine-pN. It produces a mixture of pyrimidine-3'-phosphates and a series of oligonucleotides ending in a pyrimidine-3'-phosphate. Most phosphodiester bonds involving unusual pyrimidine bases such as pseudo-uridine, 5-methyluracil, and 5,6-dihydrouracil are also sensitive to pancreatic ribonuclease. However, linkages involving 3-methylpyrim-idines and 2'-O-methylpyrimidine nucleotides are resistant to the action of the enzyme.

The most specific ribonuclease currently available was isolated from Taka-Diastase, a commercial product of the mold *Aspergillus oryzae*. This enzyme, commonly known as T_1-ribonuclease, hydrolyzes inter-nucleotide bonds of the type GpN. It produces guanosine-3'-monophos-phate from polyguanylic acid sequences such as -GpGpN, together with a mixture of oligonucleotides ending in Gp.

During the hydrolysis of RNA with either of these enzymes, oligo-nucleotides bearing a 2',3'-cyclic phosphate end group are produced as intermediates. These cyclic phosphates are cleaved further by the same enzyme, at a slower rate, to 3'-phosphates exclusively. Bonds involv-ing the methylated guanosine derivatives, such as 1-methyl-GpN and N^2,N^2-dimethyl-GpN, are cleaved by T_1-ribonuclease only to the stage of the cyclic phosphate intermediate. IpN and 1-methyl-IpN are split to yield the corresponding cyclic phosphates, which are slowly hydro-lyzed to the 3'-phosphates.

2. Fractionation of Oligonucleotides

The general scheme for separating a mixture of oligonucleotides into its individual components involves: (1) Preliminary separation accord-ing to chain length, (2) separation of each group into compositional isomers, and (3) further separation into sequential isomers. The sep-aration problem in digests of tRNA is simplified because the frequency of long compositional or sequential isomers is much lower than in high molecular weight RNA.

The polyanionic nature of oligonucleotides, stemming from their phosphodiester backbone, has been utilized for their separation by anion-exchange chromatography. DEAE–cellulose in the form of columns and paper sheets and DEAE–Sephadex have been used most successfully as adsorbents (*142, 143, 163*).

a. Separation According to Chain Length. At neutral pH all oligo-nucleotides of a given chain length carry the same net negative charge and hence their order of elution from an anion exchanger should be a function of their chain length. However, secondary binding forces be-tween the nucleotide bases and the cellulose matrix of the adsorbent complicate the separation. Purine-rich (especially guanine-rich) oligo-nucleotides are generally held more tightly than pyrimidine-rich oligo-nucleotides. This results in considerable overlap of components of different chain length. The use of 7 *M* urea, which minimizes the charge-independent binding of oligonucleotides, has made possible separation into subgroups of identical chain length (*163*). In this solvent the main binding force is the electrostatic interaction of the phosphate anions with the quaternary ammonium groups on the cellulose. Purines, however, still retain some affinity for the cellulose. This phenomenon has been valuable in the subsequent isolation of individual components.

The separation of longer oligonucleotides is improved by removal of the terminal phosphate groups by treating the digest with (*E. coli*) alkaline phosphomonoesterase. This treatment results in the removal of 2 negative phosphate charges without shortening the chain appre-ciably. The formation of molecules with a more uniform charge distri-bution may be the reason for the improved chromatographic behavior.

b. Separation of Individual Oligonucleotides. DEAE–cellulose or DEAE–Sephadex, in the presence of 7 *M* urea, have been used to frac-tionate the isopliths further into individual components. Since the total negative charge of the phosphate groups is essentially constant, the pat-tern of elution is governed by the charged groups of the purine and pyrimidine rings: (1) Under acidic conditions, at pH 2.7, C and A are protonated and fractionation is achieved on the basis of C + A versus U content. Additional differentiation between C and A is obtained because the purine base A has a higher affinity to the DEAE–cellulose than does C. (2) Under alkaline conditions, between pH 8.5 to 9.5, the phenolic hydroxyl groups of U and G are partially ionized. Again, in combination with the greater retention of purines, a high degree of resolution can be achieved.

c. Mapping Techniques. Combinations of paper chromatography in one dimension followed by electrophoresis or chromatography in the second dimension have been used successfully for the fractionation of mixtures of oligonucleotides. In comparison to column chromatography, these techniques require much less starting material and are simple and fast. They may, however, not achieve the accuracy of column procedures for quantitative analysis. A particularly elegant procedure, using very small amounts of ^{32}P-labeled RNA digests, employs high voltage electrophoresis in $7\,M$ urea to minimize interaction between oligonucleotides at pH 3.5 on cellulose acetate strips in the first dimension and electrophoresis at pH 1.9 on DEAE–cellulose paper in the second dimension (143). The ^{32}P-labeled oligonucleotides are detected by radioautography. The base compositions and, in many cases, the base sequences of the components in a particular digest can be deduced directly by comparing their positions with reference maps (Fig. 8). The complete nucleotide sequences of 2 tyrosine tRNA's (67) and 2 N-formylmethionine tRNA's (45) of E. coli and of the low molecular weight (5 S) ribosomal RNA of E. coli (27) and human KB carcinoma cells (56a) have been worked out exclusively by this mapping system.

3. Sequence Analysis of Oligonucleotides

Base composition and end group determination of oligonucleotides by alkaline hydrolysis or by complete hydrolysis with either one or both of the exonucleases has been discussed above. These procedures are sufficient to establish the sequence of the di- and trinucleotides. The 3'-end can also be deduced from the specificity of the enzyme that had been used to produce the fragment.

Longer oligonucleotides can be cleaved into small fragments by one of the specific endonucleases described above. For example, pancreatic ribonuclease fragments can be split further with T_1-RNase. Conversely T_1-RNase fragments can be degraded into smaller fragments with pancreatic RNase. In favorable cases the sequence of a long fragment can be determined by the analysis of the shorter fragments.

For many of the larger fragments the methods discussed so far are still inadequate. Partial stepwise degradation with either snake venom phosphodiesterase or spleen phosphodiesterase has been applied to the sequence determination of long oligonucleotides (85, 143). The product is a mixture of oligonucleotides of successively smaller chain lengths. These are easily separated by column chromatography on DEAE-cellulose or by electrophoresis on DEAE-cellulose paper. In the paper technique the relative mobilities of the smaller oligonucleotides are characteristic of the base that was removed, and the sequence of the original fragment can be deduced directly. In the column technique only the 3'-terminal nucleoside of each fragment has to be identified after alkaline hydrolysis to establish the sequence of the original oligonucleotide. For example, the octanucleotide GpGpGpApGpApGpU, obtained from a pancreatic RNase digest of yeast tRNA^Ala, yielded the partial digestion products GpGpGpA, GpGpGpApG, GpGpGpApGpA, GpGpGpApGpApG, and a small amount of the original GpGpGpApGpApGpU (83).

4. Ordering of Oligonucleotide Fragments

The next step in the sequence analysis is to determine the order of the fragments in the original molecule. Because a 5′-phosphate group is

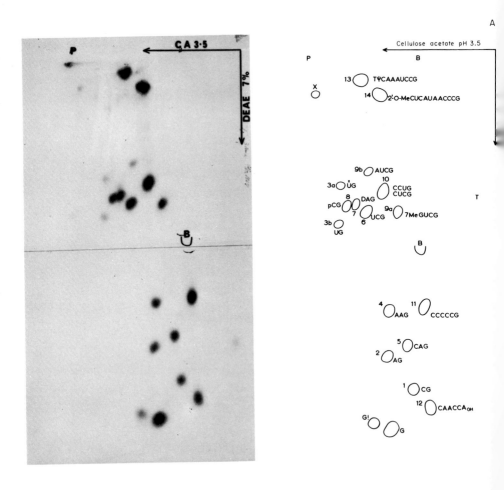

Fig. 8A

at one end of each tRNA molecule and a 3′-hydroxyl group at the other end, the 2 sequences that form the ends of a tRNA molecule can easily be selected from the pancreatic- or T_1-ribonuclease fragments. The presence of unusual nucleotides and unique short sequences allows the deduction of certain overlaps relating the 2 sets of fragments. For ex-

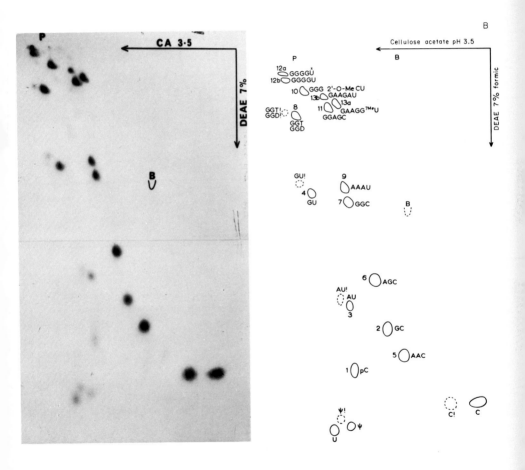

FIG. 8B

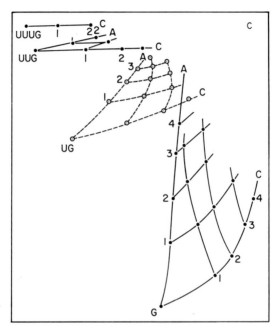

Fig. 8. Fingerprints of digests of N-formylmethionine tRNA from E. coli labeled with ^{32}P. (A) Radioautograph of a ribonuclease T₁ digest. The tracing indicates the composition of each oligonucleotide. (B) Radioautograph and its tracing of a pancreatic ribonuclease digest. Electrophoresis in the first dimension was on cellulose acetate at pH 3.5 in 7 M urea, electrophoresis in the second dimension was on DEAE paper in 7% formic acid. (C) Diagram showing how the composition of a nucleotide determines its position in the two-dimensional fractionation system. U residues have the greatest effect on mobility in the DEAE–paper dimension. Thus the lines joining the spots form three graticules corresponding to sections with two, one and no U residues, respectively. One axis on each graticule represents the number of C residues. [Parts A and B were kindly furnished by Dr. Shyam Dube, 8C after (27).]

ample, the single inosinic acid residue of yeast tRNA^{Ala} was present in the sequence IpGpCp from a pancreatic RNase digest and in the sequence CpUpCpCpCpUpUpIp from a T₁-RNase digest. These 2 sequences must therefore overlap to give CpUpCpCpCpUpUpIpGpCp. In this way a consolidated list of sequences can be constructed. In the case of yeast tRNA^{Ala} this list contained 16 sequences. Two of these constituted the ends of the molecule.

Determination of the relative positions of the intermediate 14 sequences was accomplished by the isolation of still larger fragments after controlled limited hydrolysis with T₁-RNase. Particularly useful was a very brief treatment of the tRNA^{Ala} at 0°C in the presence of magnesium ions which split the molecule at only 1 position indicated by the arrow

             Me

pG-G-G-C-G-U-G-U-G-G-

           Di   Me

           Me

C-G-C-U-C-C-C-U-U-Ip: C-I-ψ-G-G-G-A-G-A-G-   A-C-U-C-G-U-C-C-A-C-COH

                            Di   Me

                            Me

D-C-G-G-D-A-G-C-G-C-G-C-U-C-C-C-U-U-Ip: C-I-ψ-G-G-G-A-G-A-G-U*-C-U-C-C-G-G-T-ψ-C-G-A-U-U-C-C-G-G-A-C-U-C-G-U-C-C-A-C-COH

                                  Di   Me

                                  Me

U-A-G-D-C-G-G-D-A-G-C-G-C-G-C-U-C-C-C-U-U-Ip: C-I-ψ-G-G-G-A-G-A-G-U*-C-U-C-C-G-G-T-ψ-C-G-A-U-U-C-C-G-

                                          Di   Me

        Me   Me

pG-G-G-C-G-C-G-U-A-G-D-C-G-G-D-A-G-C-G-C-G-C-U-C-C-C-U-U-I-Gp: C-I-ψ-G-G-G-A-G-A-G-U*-C-U-C-C-G-G-T-ψ-C-G-A-U-U-C-C-G-G-A-C-U-C-G-U-C-C-A-C-COH

FIG. 9. A selected set of overlapping large fragments that establish the sequence of *E. coli* alanine tRNA. The fragments were isolated from partial T₁ ribonuclease digests (*84*).

in Fig. 10 (*129*). Analysis of the 2 fragments allowed the placement of the 14 sequences into either the right- or left-hand portion of the molecule. Somewhat more vigorous but still limited treatment of the tRNA with T_1-RNase produced an additional number of large fragments (*5*). Figure 9 shows a selected set of such large fragments. The information obtained from the analysis of these fragments, together with the data from the 2 larger fragments, was sufficient to establish the complete nucleotide sequence of alanine tRNA. The primary structures of several other tRNA's from yeast and *E. coli* have been worked out subsequently, using very similar procedures (*12, 47, 105*).

VI. NUCLEOTIDE SEQUENCE AND SECONDARY STRUCTURE OF TRANSFER RNA

The presently known nucleotide sequences of tRNA's are shown in Fig. 10. They are arranged in a cloverleaf structure. This cloverleaf structure is obtained when the tRNA molecules are folded so that the number of intramolecular Watson-Crick base pairs is a maximum. Despite the chemical individuality of each tRNA molecule the chains seem to be designed according to a general pattern. The existence of structural homologies also has led to considerable speculation about a common evolutionary origin of the tRNA's.

A generalized cloverleaf structure for RNA based on the above sequences is shown in Fig. 11. To facilitate orientation, the molecule has been subdivided into arms and loops; the amino acid arm, the dihydrouracil arm, the anticodon arm, the S-arm (so named because it was first recognized in serine tRNA), the G-T-ψ-C-arm, and the 3'-CCA terminus.

Since the completion of the manuscript the nucleotide sequences of five additional tRNA's were reported (*10a, 40a, 45a, 153a, 160b*). These sequences confirm the generalizations drawn from the nine tRNA's contained in Fig. 10.

Certain similarities and differences in the primary structures of these tRNA's are worth noting. The overall base compositions are similar except for tRNA[Ala], which has only 8 adenine residues as contrasted against 15 to 20 in the others. The low A content of tRNA[Ala] is most likely responsible for its hydrophilic character. In the Holley CCD system the A content seems to reflect the separation of tRNA's, tRNA[Ala] being the most hydrophilic (*6, 44*). Correlation between CCD and the content of the other 3 major bases in tRNA are very weak. Two of the minor components, ribothymidine and pseudouridine, and at least one of the various methylated guanine derivatives occur in all tRNA's. Dihydrouracil occurs in most of them. Alanine tRNA has no methylated C derivative, no $2'$-O-methylribose nucleoside, and no methylated A unless one considers 1-methylinosine to be derived from 1-methyladenosine by deamination. The 2 serine tRNA's (yeast) differ from each other in 3 residues; a C–U exchange in the S-region, and 2 A–G exchanges in the G-T-ψ-C-region. The 2 tyrosine tRNA's (*E. coli*) differ by 2 nucleotides and the 2 N-formylmethionine tRNA's (*E. coli*) by one. N^6-(Δ^2-Isopentenyl)-A has been found in tRNA^{Ser} and tRNA^{Tyr} from yeast. So far, N^6-acetyl-C has only been found in tRNA^{Ser}. As many as 6 dihydrouracils are present in tRNA^{Tyr} (yeast); tRNA^{Tyr} (yeast), and both tRNA$_f^{\text{Met}}$ (*E. coli*) have a pCp-group at the $5'$-terminus while the others have pGp-. The $3'$-terminus is the already familiar grouping -CpCpA common to all tRNA's. Base pairing near the $5'$- and $3'$-end of the tRNA molecules starts at the fifth base leaving 4 bases at the $3'$-end unpaired, except in tRNA$_f^{\text{Met}}$ where base pairing starts at base number 6 leaving 5 bases at the $3'$-end and the pCp at the $5'$-end unpaired. This unusual feature may be responsible for some of the special functions of this tRNA. The chain lengths vary from 75 nucleotides for tRNA$_1^{\text{Val}}$ (*Torulopsis utilis*) to 84 nucleotides for the 2 serine tRNA's. The additional nucleotides fall mainly· in the S-region. The amino acid arm contains 7 base pairs including, in most cases, 1 G–U pair. The dihydrouracil arm contains 3 to 4 base pairs and 8 to 12 nucleotides in the loop. The anticodon arm consists of 5 standard base pairs and 7 nucleotides in the loop. The S-arm is of variable length. A loop at the end of an RNA double helix need be no longer than 3 nucleotides. The G-T-ψ-C arm also has 5 base pairs in the stem, sometimes with 1 G–U pair, and 7 nucleotides in the loop.

In addition to the -CpCpA terminus, the sequence GpTpψpCpGp is common to most tRNA's except for the replacement, in tRNA$_2^{\text{Ser}}$ (yeast) and in both species of tRNA$_f^{\text{Met}}$ (*E. coli*), of the G on the $3'$-end of this portion of the sequence by A. These tRNA's have the same base sequence in the entire loop. The GpTpψpCp-sequence has been found in T$_1$-RNase digests of bulk tRNA from yeast, *E. coli,* and rat liver. The yield suggests that it might be present in all tRNA's of these organisms (*188*). In addition, the sequences ApGpH$_2$Up, 2-OMeGpGp, and GpCp in the dihydrouracil region are common to most tRNA's shown in Fig. 10. The CpUp sequence preceding the anticodon and the A^{modified} pAp sequence following it also are common features for most tRNA's. The presence of dimethyl-G at the $5'$-end of the anticodon region in four of the chains is

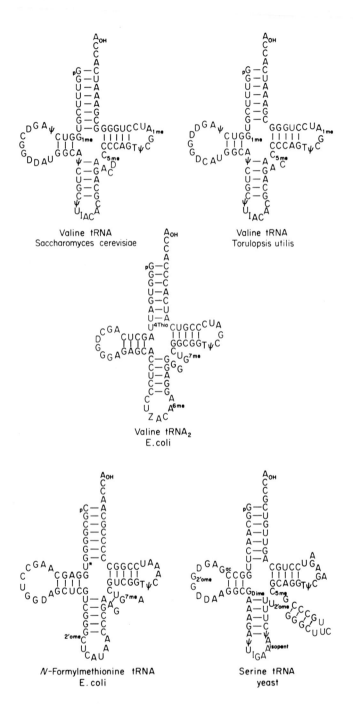

Valine tRNA
Saccharomyces cerevisiae

Valine tRNA
Torulopsis utilis

Valine tRNA₂
E.coli

N-Formylmethionine tRNA
E.coli

Serine tRNA
yeast

Fig. 10. Cloverleaf models of tRNA's. Explanation in text. Z in the *E. coli*

168

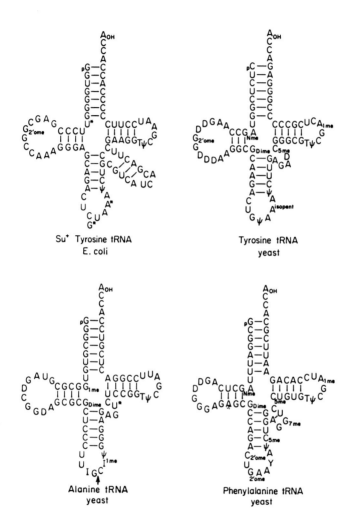

Su⁺ Tyrosine tRNA
E. coli

Tyrosine tRNA
yeast

Alanine tRNA
yeast

Phenylalanine tRNA
yeast

tRNA$^{Val}_1$ sequence is presumably a U derivative. A* in *E. coli* tyrosine tRNA has been shown to be 2-methylthio N^6-(Δ^2-isopentenyl)adenosine (*72a*).

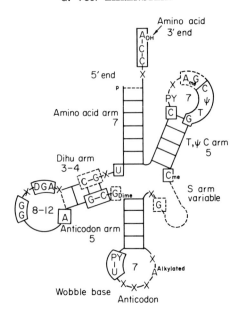

FIG. 11. Generalized cloverleaf structure of tRNA. A base pair is indicated by a line between 2 parts of a tRNA chain. A dashed line indicates a base pair occurring in some tRNA's. X indicates a nucleotide that varies with the tRNA species. The sequences enclosed in solid boxes are common to all tRNA's neglecting nucleotide modifications. The dotted boxes enclose sequences common to most tRNA's. The numbers in the center of the loops and the numbers next to the arms indicate the number of nucleotides in the loop and the number of base pairs in the arm, respectively (redrawn and modified from Fuller and Hodgson, 58).

also noteworthy. The modified nucleoside 2-methylthio N^6-(Δ^2-iso-pentenyl)adenosine has been shown to follow the anticodon in the tyrosine and phenylalanine tRNA's from *E. coli* (*10a, 72a*). It may also be present in *E. coli* cysteine tRNA and in one serine and one leucine tRNA (*E. coli*). Note that U is the first base of the codons of these amino acids.

A. Methods for Studying the Secondary Structure of tRNA

Several physical and chemical methods have been used to study the secondary structures of both bulk and specific tRNA in solution. The physical methods include the study of the size and shape of the molecule by measurement of the hydrodynamic properties (*90*), measurements of the hyperchromicity produced by thermal denaturation as a function of wavelength, optical rotatory dispersion (*54, 144*), tritium exchange, and X-ray diffraction studies (*58, 97*). The chemical methods have employed reagents which react preferentially with nucleotides in non-hydrogen bonded regions (for example, reaction with formaldehyde or oxidation of adenylic acid residues to form the 1-N oxide derivatives).

The susceptibility of certain regions of the tRNA molecule to degradation by various endo- and exonucleases under controlled conditions has proved to be another useful tool for obtaining information on secondary structure. The exonucleases snake venom diesterase, polynucleotide phosphorylase and tRNA pyrophosphorylase, and the endonucleases pancreatic RNase and T₁-RNase have been used for this purpose. The most notable example of this type of study has been discussed above as a crucial step in sequence determination. In these experiments, alanine tRNA was hydrolyzed into large fragments by partial digestion with T₁-RNase. The cleavage of the molecule into 2 halves at 0°C particularly suggested that the IpGpC sequence is at an exposed

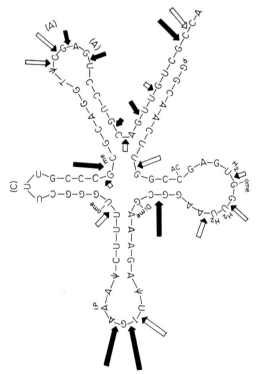

Fig. 12. The partial cleavage of yeast serine tRNA with pancreatic ribonuclease A or T₁ ribonuclease. The length of the arrows indicates the relative yield of each fragment. The yield of each fragment is proportional in the first approximation to the rate of cleavage of that particular phosphodiester bond (from Zachau et al., 182).

position. A very similar cleavage into 2 halves has been observed with the other tRNA's. Both T_1-RNase and pancreatic RNase have been used in the case of tRNA^{ser} to obtain large fragments. A selective attack by both of these enzymes was demonstrated (47). The partial enzymic cleavages produced by these 2 enzymes are indicated in Fig. 12 (182). The lengths of the arrows are proportional to the yields of the fragments, which, in turn, may be proportional to the rates of cleavage.

The conclusions from these studies are consistent with the cloverleaf model. They suggest that the molecules of tRNA are in the monomeric form and that they contain, in common with all RNA molecules, helical regions with a conformation similar to that determined for double-stranded helical RNA.

VII. THE THREE-DIMENSIONAL FORM OF TRANSFER RNA

It is very likely that the functional transfer RNA molecule has a definite native three-dimensional conformation, and that the primary nucleotide sequence is sufficient to specify this configuration. A knowledge of the exact conformation of tRNA is obviously of great importance for the understanding of its function.

Generally complete and instant reversibility is observed for all the changes in secondary structure and general conformation produced by complete melting of helices and thermal unfolding of the chain. Recently, however, it has been demonstrated that the native conformation may not reform spontaneously in all cases after exposure of the tRNA to denaturing agents. Special conditions, such as high $(1\,M)$ concentration of monovalent cations or smaller amounts of divalent organic as well as inorganic cations are required to convert the denatured (D) form of a tRNA back into its native (N) configuration. This behavior seems to be restricted to particular tRNA species. The rate at which the $D \rightarrow N$ transition occurs varies among different tRNA's (1, 57, 63, 118). There also is some evidence that aminoacyl–tRNA may have a less helical conformation than uncharged tRNA (145).

While the elucidation of the primary structure is a necessary prerequisite, the information obtained thus far has not been sufficient to deduce, unambiguously, the exact three-dimensional form of a tRNA molecule. However, comparison of the known nucleotide sequences of tRNA's reveal certain common features which, together with information from physical and chemical measurements, permit some guesses as to the probable conformation of these molecules. The exact three-dimensional conformation of transfer RNA's can be revealed only by X-ray diffraction analysis. Fibers of tRNA have thus far been studied by this technique similar to the study of DNA and double stranded RNA's (58). Recently tRNA$_f^{Met}$ has been crystallized (39a). This was the first time that any tRNA molecule has been obtained in a crystalline form, and it is now possible to think of using X-ray crystallography to determine the tertiary structure of tRNA.

Diagrams like Fig. 11 do not, of course, indicate the three-dimensional appearance of such structures and their implications. Studies with atomic models suggest that the single-stranded regions of the molecule may be highly flexible. For example, it is possible, in the cloverleaf model, to fold 1 side-leaf on top of the other, or any of the leaves back over the stem of the molecule (82).

In addition to classic base-pairing and base-stacking, typical for all

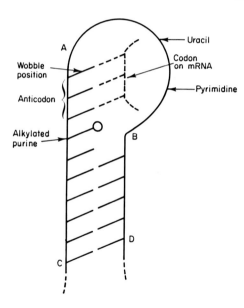

FIG. 13. Schematic diagram of the model for the tRNA anticodon arm illustrating its relationship to the codon. The helical regions are shown as straight in this diagram. CD is the first base pair in the double helical region of the anticodon arm and all the bases between A and C are stacked on one another and follow a regular helix. The companion set between B and D and the set of 3 bases in the codon follow the complementary helix. In space (see Fig. 14) A and B are quite close together because 5 nucleotide pairs is about half a turn of the helix (quoted from Fuller and Hodgson, 58).

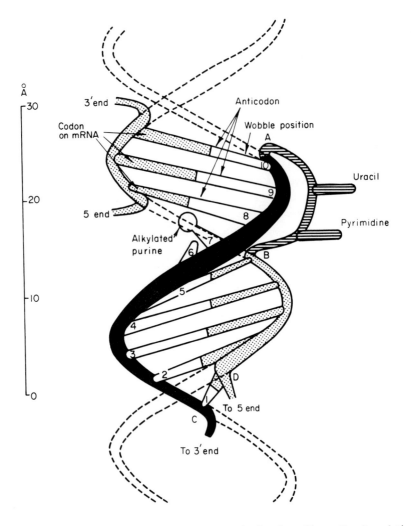

FIG. 14. Schematic diagram of the tRNA anticodon loop illustrating its relationship to the codon and the helical character of the structure. The letters A, B, C,
and D identify the same points on the structure as in Fig. 13. The bases in nucleotides 1 to 10 are stacked on one another and follow the regular helix which is
shown black. The chain of the anticodon double helix between D and B is shaded like
the codon to indicate that they follow the same helix. This helix is complementary
to the black one. The 2 nucleotides not in the standard conformation are represented by dark-line shading. The representation of their conformation is very
schematic because they lie behind nucleotides 8, 9, and 10 in the black chain. The
dotted lines indicate the generic helix from which the structure can be imagined to
be derived (quoted from Fuller and Hodgson, 58).

nucleic acids, other forces may contribute to the stabilization of the conformation of the native tRNA molecule. Metal-complex linkages, especially involving magnesium ions, and hydrogen bonding between substituents other than the bases may also contribute to tertiary structure.

It is interesting to note that tRNA, in contrast to other nucleic acids, contains a great number of non-hydrogen bond forming groups such as methyl, isopentenyl, and dihydrouracil. In this respect tRNA resembles a protein in which the secondary and tertiary structures are governed by secondary forces between the nonpolar side chains of amino acids in addition to electrostatic interactions and hydrogen bonding.

A three-dimensional model of the anticodon arm has been proposed (58). With chemical information about nucleotide sequence, X-ray evidence on the conformation of base-paired regions in RNA and the maintenance of reasonable stereochemical constraints, molecular model building leads to a unique solution for the conformation of the anticodon loop (Figs. 13 and 14). The proposed model allows codon–anticodon interaction through Watson-Crick base pairing. The codon and anticodon triplets have the conformation of the 2 strands in a regular RNA double helix (Fig. 14). The alternative or wobble pairings (see below) can be accommodated in the model by distortion of the anticodon conformation. It is not necessary to postulate distortion of the codon. The model allows adjacent mRNA codons to be recognized simultaneously by tRNA anticodon arms (Fig. 15). The 2 anticodon arms can interact with each other by intermolecular hydrogen bonds when they are recognizing simultaneously adjacent mRNA codons. The nucleotide sequences of the anticodon region do not suggest an intramolecular base-pairing scheme which would give a similar three-dimensional structure for all the anticodon loops. Therefore the principal assumption made was that the number of stacked bases should be a maximum. The model receives support from the base sequences in the anticodon loops (Fig. 10): The pyrimidines (thought to stack least well) are in the irregular part of the loop, the wobble base (see below) is at that position, in the stacked part of the structure which has the most conformational flexibility, and the purine with the modified hydrogen bonding donor group may increase hydrophobic stabilization of this stacked conformation thereby preventing ambiguity in the recognition of the first base of the codon (antiparallel recognition, see below). There is no evidence to conflict with this model. While model building does not prove the structure to be correct, its stereochemical neatness and the manner in which it accounts for what is known about the anticodon region suggest that it is essentially correct (58).

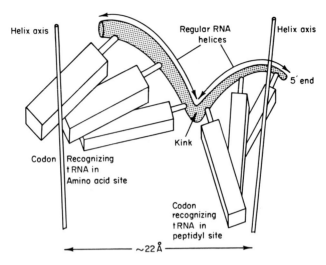

FIG. 15. Schematic diagram of successive codons in mRNA simultaneously recognizing anticodon arms. Each codon of the mRNA has the conformation illustrated in Figs. 13 and 14. The operation required to move the anticodon arm from the amino acid site to the peptidyl site involves a rotation θ about and a translation t along the anticodon helix axis and a translation d perpendicular to the helix axis along a line joining the 2 helix axes illustrated in this figure. If the anticodon helices are linked by a hydrogen bond and have a stereochemical relationship like 2 reovirus RNA helices in the crystalline fiber then the symmetry operation can be defined precisely as follows (otherwise it is an approximate description). $\theta = 87.3°$ (that is $120 - 32.7$), $t = -2.73$ Å, and $d = 22$ Å. Values of θ and t which would move nucleotide 9 into the position occupied by nucleotide 10 (Fig. 14) are taken as positive (quoted from Fuller and Hodgson, 58).

VIII. FUNCTIONAL SITES ON TRANSFER RNA

Some progress can be made in correlating chemical structure with biological function by comparing the nucleotide sequences of several tRNA's. However, without knowledge of exact three-dimensional conformation, these correlations must be considered highly speculative.

Four functional sites can be distinguished on the tRNA molecule: (1) the amino acid acceptor site, (2) the aminoacyl–tRNA synthetase recognition site, (3) the ribosome interaction site, and (4) the anticodon. There must be, in addition, regions in the molecule which are recognized by tRNA pyrophosphorylase and the various enzymes which generate the unusual bases, such as the tRNA methylases. Some tRNA's may be involved in regulatory processes which may require special structural features.

In earlier work aimed at defining the functional role of the various segments of the tRNA molecule, the biological activity of structurally altered tRNA's was studied. Structural alteration was achieved, more or less specifically, by various chemical modifications such as bromination, methylation, iodine–iodide oxidation; treatment with nitrous acid (*31*), semicarbazide (*120*), Girard-P reagent, hydroxylamine, ultraviolet light, and organic solvents; and base analog incorporation (such as 8-azaguanine).

Enzymically induced alterations produced either *in vivo* or *in vitro*, have also been useful in these considerations [e.g., methyl-deficient tRNA (Section IX,B,1)]. Other natural modifications include those responsible for the species specificity of the tRNA's, as for example in their reaction with heterologous amino-acyl–tRNA synthetases, the action of heterologous tRNA methylases, and the alteration of some tRNA species after phage infection, sporulation, etc. A promising variant of this approach is the isolation of mutationally altered tRNA's (Section XII,C).

Some progress has also been made by studying the inhibition of the function of a particular tRNA by chemically altered tRNA's, or by tRNA fragments obtained by partial degradation of tRNA with various nucleases. The biological activity of the fragments themselves, or of partially exonuclease-degraded tRNA, has also been studied (*75*).

The situation, however, is very similar to that in protein chemistry where modifications of specific amino acids in the active sites of enzymes have been studied for many years. Even if single identifiable residues were changed in a specific tRNA chain, the effects of such changes would be difficult to interpret without a knowledge of the three-dimensional conformation of the molecule. For all the functions the tertiary structure of the tRNA would be of major importance. Nevertheless, attempts will be made in the following to identify some of the functional sites on the tRNA molecule. It is unlikely that any of these sites could be formed by the ribose–phosphate backbone of the base-paired helical regions. It is more reasonable if the specific sites are formed by the bends and loops existing where unpaired nucleotide bases are exposed. The cloverleaf models will be used for orientation purposes.

There is very good evidence that the amino acid attachment site is the common CpCpA end group (Section X,C,1).

The position of the anticodon is less certain. However, a good possibility is the exposed sequence in the anticodon loop which, in all tRNA's studied so far, is the first to be cleaved by T_1-RNase. A three-dimen-

sional model for the anticodon arm has been discussed in Section VII.
The G–C change in the nucleotide sequence of the *su* tyrosine tRNA
(Section XII,C) formally designates this position as part of the anticodon (see also *31*). A comparison of the nucleotide sequences so far
determined makes it certain that the anticodons are IGC (alanine),
IGA (serine), GψA or GUA (tyrosine) (ψ can form the same base pairs
as uracil), IAC (valine), 2'-OMeGAA (phenylalanine), CAU (*N*-formyl-
methionine), and CUA (*amber* or chain termination). These assignments
fit the rule that the codon and anticodon pair in an antiparallel manner
and that the pairing in the first 2 positions is of the standard type (A
pairs with U and G pairs with C). The pairing in the third position of
the codon is more complicated. There is good experimental evidence that 1
tRNA can recognize several codons, provided that they differ only in the
third base (Section XI,B,1 and Part A, Chapter IV). F. H. C. Crick has
suggested that this is so because of a wobble in the pairing of the third
base (*41*). The theoretical model proposed by Crick, called the wobble
hypothesis, is consistent with many of the observed results (Part A,
Chapter IV). The likely codon–anticodon pairings for alanine, serine, tyrosine, valine, phenylalanine, and *su* tyrosine satisfy the standard base
pairings in the first 2 positions and the wobble hypothesis in the third,
whereas *N*-formyl-methionine tRNA exhibits ambiguous recognition of
the first base of the codon (Fig. 16). In most tRNA's the purine following
the last letter of the anticodon is alkylated. As discussed in the preceding
section this modification may increase hydrophobic stabilization of the
stacked conformation preventing wobble in the recognition of the first

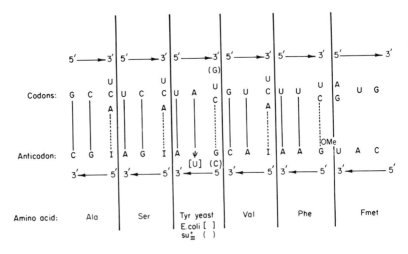

F1G. 16. The likely codon–anticodon pairings. Explanation in text.

base of the codon. In tRNA^{Val} (yeast) and in tRNA$_f^{Met}$ (*E. coli*) the base following the last letter in the anticodon is an unmodified A. The last letter in the anticodon of tRNA^{Val} is a C which would not be expected to wobble, recognizing the G in the first position of the GUX valine codons unambiguously. In tRNA$_f^{Met}$, however, the last letter is a U which can wobble by pairing with both A and G thus permitting recognition of both tRNA$_f^{Met}$ codons AUG and GUG.

The localization of the other functional sites is much more speculative, since they involve interactions of the tRNA with proteins. It is a reasonable assumption that common structures serve a common function. Several common binding sites that interact with proteins have to be present on all tRNA's. Such common sites may interact specifically with ribosomes, the tRNA-pyrophosphorylase, the various tRNA methylases, etc.

The aminoacyl–tRNA synthetase recognition site varies between different tRNA species. This site is specific for a given amino acid and synthetase. It also shows, in some cases, species specificity with regard to the source of the synthetase (Section XI). It is the most specific and presumably the most complex site on the tRNA molecule. A particular conformation involving the cooperation of several regions of the tRNA molecule, similar to the active site of an enzyme, may be required. At this stage it is not useful to speculate which parts of the tRNA structure serve as the aminoacyl–tRNA synthetase recognition site. The finding that multiple leucine tRNA species with different codon recognition and therefore presumably also different anticodons can be charged with leucine by the same leucyl-tRNA synthetase (Section XI,B,2,b) shows that the anticodon is not necessary for recognition of the enzymes. Similarly the G–C change in the nucleotide sequence of the *su* tyrosine tRNA changes its codon recognition but does not affect its charging with tyrosine. Thus at least the first position of the anticodon is not necessary for enzyme recognition. Similar conclusions can be drawn from studies of missense suppressors (see Section XII,C).

IX. BIOSYNTHESIS OF TRANSFER RNA

A. STRUCTURAL GENES FOR TRANSFER RNA

Structural cistrons for tRNA have been detected in several organisms by the DNA–RNA hybrid annealing procedure (Part A, Chapter II). Complex formation occurred only between an homologous DNA–tRNA pair *(64, 68, 98, 150)*. Thus, although the tRNA from *E. coli* can translate the genetic message of a rabbit into hemoglobin, this tRNA can still be uniquely identified with the genome of its origin.

In *E. coli* about 0.02% of the DNA is complementary to tRNA, corresponding to about 40 to 60 tRNA cistrons *(64, 68)*. In *Drosophila melanogaster* approximately 0.015% of the DNA is complementary to tRNA. This leads to the conclusion that there is about a 13-fold redundancy for each of the approximately 60 tRNA species in this organism *(140)*. Transfer RNA species with redundant coding properties will be described in Section XII,B. Unlike ribosomal RNA, tRNA does not seem to be synthesized in the nucleolus *(48, 140)*. In *Chironomus tentans* RNA with characteristics similar to tRNA (sedimentation constant of 4 S) is synthesized in all 4 chromosomes but not in the nucleolus *(48)*. *Herpes* virus-specific tRNA cistrons have been detected by specific hybridization of *Herpes* virus DNA and tRNA from infected

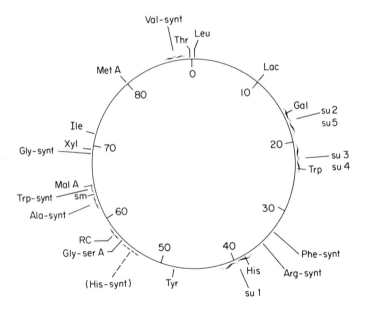

Fig. 17. Linkage map of *E. coli* (*161*) showing the position of the structural genes of some aminoacyl–tRNA synthetases: *his*-synt (in *Salmonella typhimurium*), *ala*-synt, *gly*-synt, *val*-synt, *phe*-synt, *arg*-synt, and *trp*-synt. The map location of 5 suppressor genes for nonsense mutations is also shown: su_1–su_5. The other gene symbols are as follows: *thr, leu, try, his, tyr, ile, metA*: requirement for threonine leucine, tryptophan, histidine, tyrosine, isoleucine–valine, methionine; *lac, gal, malA, xyl*: genes for metabolism of lactose, galactose, maltose, and xylose; RC: relaxed control locus for the regulation of RNA synthesis; SM: streptomycin-resistance locus.

cells. Chemical studies indicate that one of the virus-specified tRNA's may be specific for arginine (73, 156, 157).

Mutations in the structural genes for tRNA's and also for the enzymes involved in tRNA biosynthesis may be expressed as mutations in regulatory genes for certain cell functions. For example, in 1 class of histidine regulatory mutants (hisR) in Salmonella typhimurium the histidine-acceptor activity of tRNA^{His} is only about 50% of the wild type tRNA^{His}. It is uncertain at present whether the hisR strains have a mutation in the structural genes for tRNA^{His} or in an enzyme involved in tRNA biosynthesis (148).

Transfer RNA's with altered coding properties have been found in several bacterial strains carrying nonsense or missense suppressors (30, 32, 33, 51, 69). In one case it has been shown that the mutational event giving rise to the amber suppressor su_{III}^{+} allele in E. coli can be accounted for by a G–C change in the structural gene for a tyrosine tRNA (Section XII,C). It seems likely that the mutations leading to the su_{I}^{+} and su_{II}^{+} amber suppressor genes have similarly affected structural genes for serine and glutamine tRNA (30, 51, 61, 62). The map location of the suppressor genes (Fig. 17) (60, 62, 147, 161) indicates that the structural genes for tRNAs are not clustered in E. coli while the tRNA cistrons of Bacillus subtilis were found to be clustered together in 2 regions close to the rRNA loci on the proximal part of the chromosome. This result was obtained by combining the hybridization technique with the synchronization of chromosome replication in germinating spores (149a).

B. METHYLATION AND GENERATION OF OTHER UNUSUAL BASES

The transcription of the tRNA cistrons by the DNA-dependent RNA polymerase does not produce the complete primary structure of tRNA, nor indeed even the complete nucleotide sequence. It is probable that the RNA polymerase synthesizes a nucleotide chain lacking the CpCpA end group and that this end group is separately attached by the tRNA pyrophosphorylase (131). Turnover of the CpCpA end group can occur, for example, in reticulocytes in the absence of de novo synthesis of tRNA chains (86). The significance of this turnover is unknown. It does not seem to parallel peptide bond formation in these cells.

The unusual nucleotides are not incorporated as such during the polymerization. There is good evidence that the methylated bases, including ribothymidylic acid, are formed by enzymatic methylation of specific residues in the sequence after polymerization is complete (55, 56). The source of the methyl group is the amino acid methionine. The active agent, as in other methylation reactions involving methionine, is S-adenosylmethionine. In contrast, the source of the methyl group in deoxyribothymidylic acid, which is incorporated as dTTP into DNA, is methylene tetrahydrofolic acid.

The mutant strain of $E.$ $coli$ $K_{12}W6$ has 2 genetic defects; (1) it requires methionine for growth, and (2), due to a defect in its genetic control of RNA synthesis, it continues to synthesize RNA (unlike most $E.$ $coli$ amino acid auxotrophs) in the absence of the required amino acid. The genetic locus for the control of RNA synthesis has been designated as RC (Fig. 17), the mutant is called RC^{rel} (for relaxed control) and the wild type is called RC^{str} (for stringent control) (154).

The tRNA which accumulates in the absence of the methyl donor methionine is deficient in methylated nucleotides. This tRNA, lacking its normal complement of methyl groups, acts as methyl group acceptor in the enzymic methylation [Reaction (6)].

tRNA + $^{14}CH_3$-S-adenosylmethionine → ^{14}C-CH$_3$-tRNA + S-adenosylhomocysteine

$$(6)$$

At least 6 enzymes that methylate tRNA have been identified in $E.$ $coli$ (87, 88). The list may not be complete. Each enzyme has a definite base specificity. In addition these enzymes also show some tRNA specificity. As a general rule methyl-deficient tRNA can accept methyl groups in the presence of either homologous or heterologous enzymes. However, fully methylated tRNA cannot be hypermethylated by its homologous enzyme whereas a heterologous enzyme apparently can methylate additional sites on this tRNA. The tRNA methylases are widespread in nature. They have been found in many organisms including bacteria, plants, and animals.

By analogy with methylation it is likely that the other unusual nucleotides are produced at the polynucleotide level. There are probably enzymes that can cause the intramolecular rearrangement of uridine into pseudouridine, the thiolation of pyrimidines, the reduction of uracil or of cytosine followed by deamination of the latter to dihydrouracil, the deamination of adenosine to inosine, etc. These enzymes should act only on very specific bases in the completed polynucleotide chain, possibly recognizing specific base sequences and certain features of secondary and tertiary structure in the tRNA molecule.

1. Function of the Unusual Bases

The biological significance of the unusual nucleotides is unknown although many possibilities have been considered. The availability of methyl-deficient tRNA has lead to several studies aimed at elucidating the function of the methylated nucleotides. The tRNA isolated from the methionine-starved mutant $K_{12}W6$ contains, at best, about equal amounts of fully methylated and nonmethylated tRNA. Recently, separation of the 2 kinds of tRNA has been achieved, and differences in amino acid acceptor activity were revealed when unmethylated tRNA[Leu] was tested with a heterologous leucyl–RNA synthetase (130). A difference in the coding properties of unmethylated versus fully methylated tRNA[Phe] has also been reported (102, 138).

The results are consistent with the hypothesis that the methylated bases, and, for that matter, many of the other unusual nucleotides in tRNA, serve to stabilize the three-dimensional structure of the chain. The unusual bases may also enhance the ability of tRNA molecules to interact specifically with proteins. The concentration of the unusual nucleotides in the nonbase-paired loops and bends of the cloverleaf model is in agreement with this notion.

X. TRANSFER RNA AS THE AMINO ACID ACCEPTOR

The formation of aminoacyl–tRNA derivatives is accomplished by a 2-step reaction. In the first step the carboxyl group of the amino acid is activated by the formation of an aminoacyladenylate complex [Reaction (7)]. In the second step the amino acid is transferred to a specific tRNA molecule [Reaction (8)].

$$\text{Amino acid} + \text{Enzyme} + \text{ATP} \rightleftharpoons \text{Enzyme–aminoacyl-AMP} + \text{iPP} \qquad (7)$$
$$\text{Enzyme–aminoacyl-AMP} + \text{tRNA} \rightleftharpoons \text{Aminoacyl–tRNA} + \text{AMP} + \text{Enzyme} \qquad (8)$$

$$\text{Sum: Amino acid} + \text{tRNA} + \text{ATP} \rightleftharpoons \text{Aminoacyl-tRNA} + \text{AMP} + \text{iPP} \qquad (9)$$

Both steps are mediated by a single enzyme, the aminoacyl–tRNA synthetase. Each individual enzyme forms aminoacyl–tRNA derivatives of a single amino acid. Thus there are at least 20 different aminoacyl-tRNA synthetases, one for each of the 20 natural amino acids (17, 79).

A. THE AMINOACYL–tRNA SYNTHETASES

Aminoacyl–tRNA synthetases, like tRNA, are expected to be present in all living organisms. There is considerable evidence for the presence of these enzymes in a wide variety of species (115, 133). Several convenient assays are available. In addition to the formation of aminoacyl-tRNA, these enzymes catalyze an amino acid dependent ^{32}P-pyrophosphate–ATP exchange by reversal of Reaction (7), and an ATP dependent formation of amino acid hydroxamates in the presence of high concentrations of hydroxylamine. The latter assay has been made more sensitive and specific by the use of radioactive amino acids. The labeled amino acid hydroxamate can be separated from the free amino acid by chromatography on ion exchange paper. These enzymes also catalyze an amino acid and tRNA dependent exchange of AMP with ATP. This finding provided the first indication that RNA was involved in amino acid activation.

A number of aminoacyl–tRNA synthetases have been partially purified from such sources as *E. coli*, yeast, liver, and pancreas (*19*). Partial purification of all 20 enzymes from *E. coli* has been described (*117*). They are separable from one another by conventional chromatographic procedures. For several of these enzymes, apparent multiple peaks are obtained. In the case of the leucine enzyme the activity is distributed about equally between the 2 peaks, whereas a small satellite accompanies a major peak in the case of the valine, isoleucine, and glutamic acid enzymes. Three apparent peaks of about equal activity have been reported for lysyl–tRNA synthetase from *E. coli*. No functional differences have been found so far between the multiple peaks of these *E. coli* aminoacyl–tRNA synthetases; their multiplicity and their biological significance is a matter of speculation.

In *Neurospora crassa* 2 phenylalanyl– and 2 aspartyl–tRNA synthetases have been found. The major one is present in the cytoplasm and the minor one in the mitochondria. They are also functionally different. Each of the 2 enzymes aminoacylates a different species of *Neurospora* tRNA. The mitochondrial enzymes charge the mitochondrial tRNA's and the cytoplasmic ones are specific for the cytoplasmic tRNA's (*7–10*). They also differ in their specificity for aminoacylation of *E. coli* tRNA's (Section XI,B,2).

In another recent example, 2 distinguishable valyl–tRNA synthetases were found after phage T4 or T6 infection of *E. coli* KB. Phage infection of this strain results in at least 1 new chromatographically separable valyl–tRNA synthetase. Phage infection of a mutant of *E. coli* KB possessing a temperature-sensitive valyl–tRNA synthetase results in the appearance of a new relatively temperature-resistant enzyme activity. The possibility that this synthetase activity represents a new phage-induced enzyme has not yet been rigorously established (*38, 122*).

In spite of their importance only a few aminoacyl–tRNA synthetases have been isolated in pure homogeneous form (*133*).

The molecular weights of different aminoacyl–tRNA synthetases are very similar [e.g., isoleucyl–tRNA synthetase from *E. coli*, 112,000 (*127*); tyrosyl–tRNA synthetase from *E. coli*, 98,000 (*28*); lysyl–, valyl–, (*94*) and seryl–tRNA synthetases from yeast, 112,000, 116,000, and 89,000 (108), respectively]. Apart from amino acid compositions and preliminary peptide maps nothing is known about the detailed structures of the aminoacyl–tRNA synthetases. It will be of great interest to learn more about the macromolecular organization of these enzymes with respect to the presence of identical or nonidentical subunits and the conformation of monomer units. Although the crystallization of some aminoacyl–tRNA synthetases has been reported, crystals suitable for X-ray diffraction analysis have not been obtained so far.

The question of enzyme-bound oligonucleotide material has received some attention. In the purest enzyme preparations less than 1 equivalent of a mononucleotide per mole of enzyme could be detected. Thus, at present it seems unlikely that an oligonucleotide firmly bound to the enzyme takes part in the recognition between enzyme and tRNA.

In general, aminoacyl–tRNA synthetases are stable in crude extracts and in pure form under appropriate conditions. They are specific for ATP; other nucleoside triphosphates have been found to be inactive. Magnesium ions are required for activity and in several cases potassium ions are required for maximal activity. Many of the enzymes are inhibited by reagents, such as p-mercuribenzoate, which react with sulfhydryl groups. Thiol compounds, such as 2-mercaptoethanol and dithiothreitol, prevent and reverse loss of enzymic activity. The affinity of the enzymes for amino acids, ATP, and magnesium ions is relatively high. The K_m values in aminoacyl–tRNA synthesis are in the range of 10^{-4} to 10^{-6} M. However, the K_m values for different aminoacyl–tRNA synthetases, can vary by as much as 2 orders of magnitude. Therefore, no single reaction mixture will yield optimal activity for all enzymes. The K_m in aminoacyl–tRNA synthesis for $E.$ $coli$ tRNA^{Ile} with the $E.$ $coli$ isoleucyl–tRNA synthetase, for $E.$ $coli$ tRNA^{Tyr} with $E.$ $coli$ tyrosyl–tRNA synthetase, and for yeast tRNA^{Ser} with yeast seryl–tRNA synthetase is of the order of 10^{-7} M.

1. Structural Genes for Aminoacyl–tRNA Synthetases

a. Mutants. It had been assumed for a long time that the essential mechanisms of protein and nucleic acid synthesis would escape genetic analysis because the relevant mutants would be lethal. Recently, however, a new strategy involving the isolation of conditionally lethal mutants has been employed (123). Temperature-sensitive (ts) mutants have been the most useful. These mutants can be grown at permissive temperatures of 25°–30°C, but do not grow at nonpermissive temperatures of 37°–45°C. Conditionally lethal mutants are potentially of great value in studying the mechanism and the regulation of the synthesis of proteins and nucleic acids (Part A, Chapter I).

Several conditionally expressed mutants of *E. coli* with altered amino-acyl–tRNA synthetases have been isolated (*121*). Biochemical characterization has lead to the identification of three ts strains, having temperature-sensitive valyl–, phenylalanyl–, and alanyl–tRNA synthetases, respectively (*23, 178*).

Another class of mutants with altered aminoacyl–tRNA synthetases are resistant to specific amino acid analogs. For example, a mutant strain of *E. coli* has been isolated which is resistant to the phenylalanine analog *p*-fluorophenylalanine (*52, 53*). In this strain, the phenylalanyl–tRNA synthetase, in contrast to the wild-type enzyme, is unable to form a *p*-fluorophenylalanyl adenylate complex (Section XI,B,1,b).

An altered glycyl–tRNA synthetase was discovered in the laboratory strain *E. coli* KB. The K_m for glycine in ^{32}PP–ATP exchange and glycyl–tRNA formation in crude extracts of this strain was 50–100 times higher than in extracts from other common laboratory strains (*22, 121*). This observation illustrates the importance of selecting appropriate strains when working with aminoacyl–tRNA synthetases.

It is very likely that the already recognized structural genes for aminoacyl–tRNA synthetases are responsible for the major fraction of aminoacyl–tRNA forming capacity for the corresponding amino acids. However, the available data do not exclude the presence of quantitatively minor aminoacyl–tRNA synthetases with restricted specificities for minor tRNA species. It is also not known, as yet, how many cistrons are required to code for a single aminoacyl–tRNA synthetase.

Because of the involvement of the aminoacyl–tRNA complexes in several regulatory mechanisms in the cell, the structural genes for the aminoacyl–tRNA synthetases have occasionally been discovered as regulatory genes for a particular biosynthetic pathway. This was the case for the structural gene of the histidyl-tRNA synthetase of *Salmonella typhimurium* (*141*). Mutants in this gene (*hisS*) have an altered histidyl–tRNA synthetase. These mutants constitute one of the 4 classes of mutants which affect the regulation of the histidine operon (Section IX,A and Part A, Chapter I). One such mutant studied in detail was found to have a histidyl–tRNA synthetase with a 50-fold decreased affinity for histidine. The K_m was $8 \times 10^{-3} M$ as contrasted to $1.5 \times 10^{-4} M$ for the wild-type enzyme (*141*).

Another example of the regulatory function of an aminoacyl–tRNA synthetase involves the temperature-sensitive valyl–tRNA synthetase of *E. coli* mentioned above. From a study of this mutant it was concluded that valine must interact with the valyl–tRNA synthetase to be able to repress the enzymes of the valine–isoleucine biosynthetic pathway (*23, 49, 50*).

A simplified tactic to screen for aminoacyl–tRNA synthetase mutants has recently been used successfully for the isolation of many mutants of several aminoacyl–tRNA synthetases. Isoleucine and valine share a common biosynthetic pathway that branches at the terminal transamination step. By selecting for isoleucine auxotrophs that do not require valine, either the isoleucine-specific transaminase or the isoleucyl–tRNA synthetase could be affected. Indeed, about half the mutants isolated in this way had a mutationally altered isoleucyl–tRNA synthetase, most of which had a higher K_m for isoleucine. Similarly, the last step in the biosynthesis of glycine is the conversion of serine into glycine by the enzyme serine transhydroxymethylase. Again, selecting for glycine auxotrophs that do not require serine gave many mutants whose glycyl–tRNA synthetase had a higher K_m for glycine. Obviously, mutations in enzymes other than the aminoacyl–tRNA synthetases which are involved in the utilization of amino acids for regulatory or biosynthetic purposes other than protein synthesis would be picked up by this technique. The 2 examples above demonstrate that the aminoacyl–tRNA synthetases as expected are by far the most important enzymes for the utilization of amino acids in the cell. It should be possible to obtain mutants of any aminoacyl–tRNA synthetase in this way by an appropriate selection scheme. Another variant would be the selection of such auxotrophs as described above that, in addition, are temperature sensitive.

b. Map Location. The structural genes for aminoacyl-tRNA synthetases thus far located on the *E. coli* chromosome map are shown in Fig. 17 (*121, 161*). The structural gene for the histidyl-tRNA synthetase has been mapped in *Salmonella* (*141*) and its location may be compared to those of the other 6 enzymes, assuming that the genetic maps of *Salmonella* and *E. coli* are homologous. The structural genes for these enzymes are not clustered together. Their map locations are also quite distant from those of the known structural genes for the enzymes involved in the biosynthesis of these amino acids.

B. Formation and Structure of Aminoacyladenylate

The products of Reaction (7) are an acyl phosphate anhydride, the amino-acyladenylate (Fig. 18), and inorganic pyrophosphate. The amino-acyladenylate is a transient intermediate; it normally remains tightly bound to the enzyme. The combination of the aminoacyladenylate with the enzyme leads to considerable stabilization of the anhydride.

The evidence for the mechanism shown in Reaction (7) may be summarized as follows: (1) The enzymes catalyze an amino acid-dependent

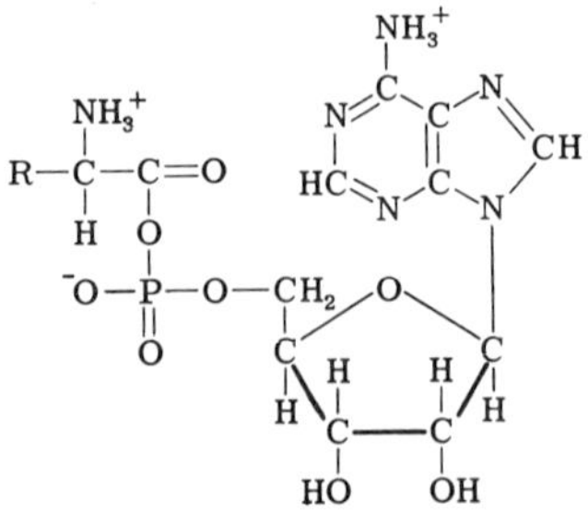

FIG. 18. Structure of the aminoacyladenylate.

ATP-P-^{32}P exchange and an ATP-dependent formation of amino acid hydroxamates. (2) Aminoacyladenylates have been isolated from several aminoacyl–tRNA synthetases, after deproteinization, in amounts approximately equivalent to the amount of enzyme. (3) The anhydride link between the carboxyl group of the amino acid and the 5′-phosphate group of AMP is indicated by the incorporation of adenosine triphosphate, labeled with ^{32}P in the innermost ester phosphate, into the enzyme-bound product. (4) Studies with the tryptophanyl–tRNA synthetase and ^{18}O-labeled tryptophan have shown that the formation of tryptophan hydroxamate is accompanied by the transfer of oxygen from the carboxyl group of tryptophan to the 5′-phosphate group of AMP. (5) In the absence of tRNA the enzymes do not catalyze exchange between AMP and ATP. (6) Enzyme-linked aminoacyladenylates have been separated from reaction mixtures by passage through a Sephadex column (3, 125). About 1 mole of isoleucyladenylate is bound per mole of pure isoleucyl-tRNA synthetase indicating a single binding site for the complex (125, 126). The natural enzyme-bound aminoacyl adenylates are utilized for ATP formation in the presence of pyrophosphate by reversal of Reaction (7), whereas amino-acyl–tRNA complexes are formed when tRNA is present (Section XI,B,1,a). Chemically synthesized aminoacyladenylates can be attached to the corresponding aminoacyl–tRNA synthetases. They react in a manner identical to that of the natural complexes. The amino acid specificity of the activation reaction will be discussed below, together with the specificity of the second step, that involving aminoacylation of tRNA (Section XI,B,1).

The mechanism of amino acid activation is analogous to the activation of acetate and other fatty acids. Acyladenylates are the intermediates in these reactions. In the case of fatty acid activation, the acceptor of the acyl group is coenzyme A. In amino acid activation the acceptor is tRNA. In both cases the same enzyme catalyses the activation and the transfer to the acceptor.

C. Aminoacyl–tRNA

1. *Structure of the Aminoacyl–tRNA Ester*

Amino acids are attached to tRNA molecules through an ester bond involving their carboxyl group and either the 2′- or the 3′-hydroxyl group of the ribose of the 3′-terminal adenosyl residue of the tRNA chain [Reaction (9)] (*76, 81*). The structure of the aminoacylated adenosyl end group of tRNA is shown in Fig. 19. The intact 3′-terminal -CpCpA sequence of the tRNA is required for full acceptor activity, although in certain cases chains terminating in -CpA are also able to accept amino acids. Removal of the -CpCpA terminus by snake venom phosphodiesterase, or by tRNA pyrophosphorylase [Reaction (4)], destroys the acceptor activity (*132*). Oxidation of the ribose of the terminal adenosyl residue with periodate (Fig. 7) also destroys the acceptor activity. The presence of an aminoacyl group on a tRNA molecule protects the end group from destruction by periodate (Section V,B,4) indicating the involvement of the 2′,3′-diol grouping of the terminal ribose (*18*). This fact, together with the observation that each amino acid is bound to unfractionated tRNA to a characteristic extent which is independent of the extent to which any other amino acid is bound (Table I), provided early evidence that bulk tRNA is a mixture of amino acid-specific acceptors. The -CpCpA attachment site is certainly the most ancient portion of the tRNA molecule participating in the basic functions of a proto-tRNA, the acceptance of the activated amino acids and their polymerization to form proteins.

Digestion of leucyl–tRNA with pancreatic RNase releases a compound that has been identified as the 2′(3′)-leucyl adenosine ester by degra-

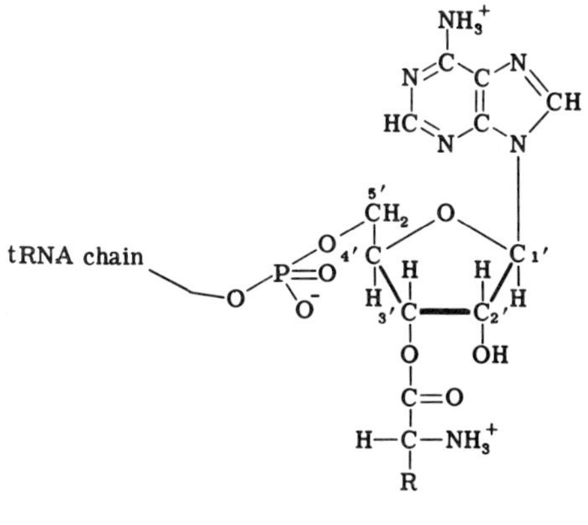

Fig. 19. Structure of the aminoacylated adenosyl end group of tRNA.

dative studies and by comparison with the chemically synthesized compound (*187*). Other aminoacyladenosine complexes have been isolated from pancreatic RNase digests of aminoacyl–tRNA's from various sources. Digestion of aminoacyl–tRNA with T_1-RNase yields aminoacyl oligonucleotides (Section V,B,4). Further evidence for the postulated ester linkage between amino acid and tRNA was obtained when aminoacyl–tRNA preparations were reduced with lithium borohydride to the corresponding amino alcohols.

The ester linkage is readily saponified with dilute alkali. At pH values between 7 and 10 the half times are of the order of minutes or seconds, depending on the identity of the esterified amino acid (*146, 175*). The ester linkage can also be split with neutral hydroxylamine, giving amino acid hydroxamates. Ammonolysis of glutamyl– and aspartyl–tRNA produces the α-amides isoglutamine and isoasparagine, indicating that the α-carboxyl group of these amino acids is esterified to tRNA (*40*). Aminoacyl–tRNA complexes are stable in dilute acid.

2. Reactivity of the Aminoacyl–tRNA Ester

The equilibrium constant for valyl– and threonyl–tRNA formation [Reaction (9)] ranges between 0.3 and 0.7. The free energy change in the hydrolysis of the aminoacyl–tRNA complex is therefore of the same order as that of a pyrophosphate bond in ATP.

The large free energy change associated with the hydrolysis of aminoacyl–tRNA esters is characteristic of amino acid esters in general. It is related to the presence of the positively charged α-amino group. N-Acylated aminoacyl–tRNA derivatives such as N-formylmethionyl–tRNA and peptidyl–tRNA seem to be considerably more stable to hydrolysis and less reactive than aminoacyl–tRNA esters with an unsubstituted amino group (*186*).

The cis-hydroxyl group adjacent to the ester linkage has a marked influence on the rate of hydrolysis and on the reactivity with hydroxylamine. Leucyl–tRNA is hydrolyzed 30 times faster than leucine ethyl ester and, in the reaction with hydroxylamine, the difference in reactivity is so large that the rates cannot be compared accurately under the same conditions.

Finally, the rate of hydrolysis of aminoacyl–tRNA esters depends on the nature of the amino acid. Valyl–tRNA is hydrolyzed much more slowly than the corresponding glycyl, arginyl, glutamyl, and aspartyl derivatives.

3. The Site of Esterification in Aminoacyl–tRNA

The determination of whether the amino acid is esterified to the 2'- or 3'-hydroxyl group of the terminal adenosyl residue of the tRNA has been difficult because of a very rapid base-catalyzed migration of the amino-acyl group to an equilibrium distribution between the 2'- and 3'-isomers (*186*).

Chemical studies aimed at directly identifying the site of esterification start with the terminal aminoacyladenosine ester obtained by pancreatic RNase treatment of aminoacyl–tRNA. In 1 set of experiments the hydroxyl groups not esterified by the amino acid were replaced by various stable chemical blocking groups such as tosyl, tetrahydropyranyl, and phosphoryl groups. The 2'- and 3'-isomers of the stable derivatives could then be separated, and, from their relative amounts, the ratio of the 2'- to 3'-amino acid ester in the starting material was inferred. In another approach the 2'- and 3'-isomers of valyladenosine were separated by thin-layer chromatography, without prior substitution.

Finally, the ratio of 2'- to 3'-ester has been determined directly by nuclear magnetic resonance spectroscopy. With this method more than 90% of natural aminoacyladenosine was shown to be the 3'-isomer, in agreement with the amount of 3'-isomer found using one of the blocking procedures. With other blocking agents, however, ratios of 2'- to 3'-isomer of 0.35 to 0.65 were obtained.

From studies of the acyl migration in model compounds it was concluded that the base-catalyzed migration is so rapid—the half time is much less than 1 second under physiological conditions—that the original site of esterification cannot be determined, and that, *in vivo*, the amino-acyl–tRNA exists as an equilibrium mixture of the 2 forms. This interpretation does not explain the finding by 2 different methods of more than 90% of the 3'-ester in the natural product.

Suggestive evidence that 3'-aminoacyl–tRNA is the biologically active isomer in peptide bond formation comes from the finding that the 3'-isomer of puromycin (an analog of aminoacyladenosine) inhibits protein synthesis whereas the 2'-isomer of the antibiotic is not inhibitory (*179*).

4. The in Vivo Esterification of tRNA

Two aspects of the intracellular esterification of tRNA have received considerable attention. One is the level of esterification of tRNA *in vivo* under a variety of growth conditions. The other is the presence on tRNA of esterified compounds other than the 20 natural amino acids, such as amino acid derivatives and peptides.

In exponentially growing *E. coli*, between 70 and 100% of the molecules of most of the different tRNA species are present in esterified form. In the stationary growth phase, these fractions are essentially the same. In a leucine auxotroph strain, starved for leucine, the degree of esterification of the tRNA^{Leu} fell to 20% in a strain carrying the RC^{str} gene, and to 13% in a strain carrying the RC^{rel} gene (*181*).

a. *Signals on tRNA.* Two developments have stimulated interest in other than the 20 natural amino acids, that are esterified to tRNA. One has been the accumulation of evidence that tRNA may be involved in a variety of regulatory systems. The other has been the realization that regulatory instructions coded on the template RNA may have to be read by tRNA adaptors. Thus, any compound, different from the natural amino acids that is esterified to tRNA is a potential signal in protein synthesis or part of a regulatory device.

One such signal is N-formylmethionine which has been found in tRNA from *E. coli* and yeast (*110, 111*). Two species of tRNA^{Met} can be separated in *E. coli*. Methionine bound to one of these can be formylated by enzymes from *E. coli* and yeast. The formyl donor is N^{10}-formyltetrahydrofolic acid. The formylatable species, methionyl–tRNA$_f^{Met}$ responds in the ribosomal binding assay to the triplets AUG and GUG, whereas the nonformylatable species, methionyl–tRNA$_m^{Met}$ responds to the triplet AUG only. N-Formylmethionyl–tRNA$_f^{Met}$ acts as a chain-initiating tRNA in bacterial systems by delivering N-formylmethionine to the NH$_2$-termini of starting polypeptide chains, whereas methionyl–tRNA$_m^{Met}$ places methionine into internal positions of proteins (*2, 39, 167*). The complete nucleotide sequence of tRNA$_f^{Met}$ has been worked out (Fig. 10). Its unusual features have been discussed above. The nucleotide sequence alone does not suggest which parts of the molecule are responsible for the recognition by the transformylase and methionyl–tRNA synthetase and which parts are of importance for the affinity of tRNA$_f^{Met}$ for the ribosomal binding site. Both tRNA$_f^{Met}$ and tRNA$_m^{Met}$ are charged with methionine by the same methionyl–tRNA synthetase.

N-Formylmethionine has been obtained in the form of N-formylmethionyladenosine after pancreatic RNase digestion of N-formylmethionyl–tRNA. Furthermore, the presence of N-formylmethionine protects the adenosyl end group of the tRNA moiety from destruction by periodate. It is very likely therefore that formylmethionine is bound, in the conventional way, as the aminoacyl–tRNA ester.

N-Formyl derivatives of at least 12 other amino acids (predominantly glycine, threonine, and valine) have been isolated from yeast tRNA (*72*). These compounds are also linked to an adenosine residue—probably not the terminal one—but to the N^6 position of the purine ring rather than the ribose. The general formula of these N^6-(N-formylaminoacyl) adenosines is shown in Fig. 5. The function of these derivatives is unknown. They may belong to the family of unusual bases having structural functions (Section IX,B,1). It is interesting to note that the alkali lability of the ester bound in N-formylmethionyladenosine is similar to that of the amide bond in the N^6-(N-formylaminoacyl) adenosines in contrast to the significantly weaker bond of the unsubstituted $2'(3')$-aminoacyladenosines.

Derivatives similar to N-formylmethionine, but involving other amino acids or other acyl groups, may exist on tRNA. However, none has so far been characterized.

Transfer RNA isolated from various sources contains short peptides, and, in some cases, other unidentified compounds that protect the adenosyl end group from periodate destruction. Peptide esters of tRNA have been shown to be intermediates in polypeptide chain assembly on the ribosomes, each 70 S ribosome carrying a nascent peptide chain (Part B, Chapter IV). Since there are about 10 tRNA molecules per each 70 S ribosome in the cell, only 10% of all tRNA molecules are expected to carry nascent peptide chains. The amount of peptidyl–tRNA in some instances was much greater than that which could be engaged in polypeptide chain assembly. Further characterization of these compounds is required.

XI. TRANSLATION OF AMINO ACID STRUCTURE INTO NUCLEOTIDE SEQUENCE

A. THE DICTIONARY

The attachment of amino acids to specific tRNA adaptors [Reaction (9)] is the crucial step in the translation of the 4-letter language of the nucleotides into the 20-letter language of the amino acids. In this reaction the chemical structures of the amino acids are translated into specific nucleotide sequences in the tRNA's, thus establishing the dictionary relating the 2 languages. Since the tRNA adaptors also contain the anticodons, the relationship between the amino acids and their codons is established at the same time, thus providing the key to the code. In the language analogy, translation and coding are 2 different processes. They might be likened to the transcription of clear text into Morse code with the help of the key to the code, and translation from one language into another requiring a dictionary. In protein synthesis the 2 processes are coupled in the aminoacyl–tRNA synthesis reaction.

It has been argued that the assignment of a particular amino acid to a particular codon could not have been the result of an historical accident but rather of the existence of an amino acid–codon logic (i.e., logic resulting from a stereospecific relationship between the codon and the amino acid). However, there is no plausible way in which an amino acid could have a significant stereochemical relationship to the codon representing it in the template nucleic acid. It was this very difficulty which led to the adaptor hypothesis. The data proving the adaptor hypothesis (Section I,B) instead show that the interaction of codon and tRNA at the ribosomal level is independent of the particular amino acid carried by the tRNA.

However, the symbols of the nucleic acid language still have to interact specifically at some stage with the symbols of the protein language. While the adaptor hypothesis does not eliminate the problem by moving it from the template RNA level to the aminoacyl–tRNA synthetase level, it has made the specificity of the interaction much more plausible. Instead of recognition between a monomer of one language (the amino acid) and an oligomer of the other language (the codon), polymers of both languages (enzyme and tRNA) recognize one another. With polymers, it is not difficult to visualize how specificity can be achieved. In addition to the exploitation of stereochemical differences by variations in the three-dimensional structure of recognition sites, selectivity can also be achieved by subtle differences in a variety of secondary forces operating at the macromolecular level, such as solubility in hydrophobic environments. It is therefore not surprising that the tRNA's are relatively large molecules with a definite three-dimensional structure and a great variety of unusual bases. Possessing features typical both for nucleic acids and for proteins, their design seems eminently suitable for their function as the link between the languages.

In general biochemical terms, the aminoacyl–tRNA synthetase is viewed as having 2 recognition sites, one for the amino acid and one for the tRNA. Similarly, 2 such sites (for recognition of the aminoacyl–tRNA synthetase and for the codon) are considered to be present on the tRNA (Section VIII). With the exception of the anticodon, the structure of these various sites, and therefore also the relationships between them, are completely unknown.

A knowledge of the exact three-dimensional structure of the aminoacyladenylate–enzyme–tRNA complex is obviously of central importance. This complex represents the most specific interaction known between a protein and a nucleic acid.

B. Specificity of Aminoacyl–tRNA Synthesis

Some hints about the mechanism and the general features of recognition between the amino acid, the tRNA, and the enzyme protein can be obtained by examining the amino acid and tRNA specificity of the aminoacyl–tRNA synthesis reaction.

1. *Specificity for Amino Acids*

Aminoacyl–tRNA synthesis is specific. An amino acid will only be linked to those tRNA species that are specific for this particular amino acid (with 1 exception; see below). From the adaptor function of tRNA it follows that aminoacyl–tRNA synthesis is the only step in protein synthesis at which discrimination between amino acids can occur. Thus, the degree of discrimination in this reaction must be at least as high as that in the overall process of protein synthesis.

a. Natural Amino Acids. Precise and well-controlled measurement of the overall specificity of protein synthesis have not been carried out. However, there is some suggestive evidence that the frequency at which natural amino acids are mistaken for each other in protein synthesis may be of the order of 1 in 3000 (*103*). Only few systematic studies have been carried out on the amino acid specificity of aminoacyl–tRNA synthesis. In one such study it was found that leucine, isoleucine, and valine, although structurally closely related, substitute for each other less than once in 10,000 times (*104, 127*). Thus it appears that the specificity of aminoacyl–tRNA synthesis is sufficient to account for the overall fidelity of protein synthesis.

Aminoacyladenylate formation [Reaction (7)] and aminoacyl–tRNA synthesis [Reaction (8)] have been uncoupled from each other by isolating the enzyme-bound aminoacyladenylate intermediate which accumulates in stochiometric amounts in the absence of tRNA acceptor (Section X,B) (*3, 126, 127*). Under these conditions the amino acid specificity of the second step is considerably greater than that of the first step.

Highly purified *E. coli* isoleucyl–tRNA synthetase converts valine as well as isoleucine to the corresponding aminoacyladenylate. Although considerable specificity is achieved in the first step at the level of substrate binding (K_m for isoleucine $= 5 \times 10^{-6} M$; K_m for valine $= 4 \times 10^{-4} M$) the selectivity in the overall synthesis of isoleucyl-tRNA^{Ile} is far greater. No transfer of valine occurs to tRNA^{Ile} from the Enz^{Ile}-valyladenylate complex under conditions in which virtually all the isoleucine in Enz^{Ile}-isoleucyladenylate is converted to isoleucyl-tRNA^{Ile}. Isoleucine-specific tRNA induces instead a breakdown of the Enz^{Ile}-valyladenylate to valine and AMP. This hydrolysis is dependent on intact tRNA^{Ile}. Any alteration which destroys the isoleucine acceptor activity of the tRNA also destroys its ability to induce the hydrolysis of the Enz^{Ile}-valyladenylate. This breakdown of the valyladenylate complex has been interpreted as being part of the mechanism which prevents the esterification of the wrong amino acid to tRNA (*125, 126*).

Evidence for a greater specificity in the second step of the reaction has also been obtained with purified tryptophanyl–tRNA synthetase (from pancreas). Furthermore, in the case of purified $E.$ $coli$ valyl–tRNA synthetase, both valine and threonine support ATP–PP exchange. Whether threonine was transferred to tRNA^{Val} was not tested, however.

Recently, the retention of aminoacyl–tRNA-synthetase–tRNA complexes on nitrocellulose filters was utilized to study the binding of tRNA to the enzyme without concomitant aminoacylation of the tRNA (180). $E.$ $coli$ isoleucyl-tRNA synthetase complexes only with tRNA^{Ile} and tyrosyl–tRNA synthetase complexes exclusively with tRNA^{Tyr}. ATP and amino acid are not required for binding. Also, the complex forms equally well whether the tRNA is aminoacylated or not. The apparent association constant for both charged and uncharged tRNA is approximately 10^8 liters/mole. There seems to be 1 binding site for tRNA per enzyme molecule in agreement with 1 binding site for ATP and amino acid (Section X,B). Aminoacyl–tRNA-synthetase–tRNA complexes have also been detected after filtration through Sephadex gel to separate enzyme bound from unbound tRNA (95). Similar conclusions were reached with this method which does not involve trapping of the enzyme and the complex on the solid-phase filter. The results from these studies are consistent with the hypothesis that a major contribution to the specificity of aminoacyl–tRNA synthetases is made by their selectivity in binding tRNA's.

$b.$ $Amino$ $Acid$ $Analogs.$ Aminoacyl–tRNA synthetases can also utilize amino acid analogs. Generally those analogs, which are esterified to tRNA adaptors, are also incorporated into protein, presumably in place of the corresponding natural amino acid. Analogs formed by modifications of the side chains of amino acids have been shown, with a variety of aminoacyl–tRNA synthetases, to serve alternatively as either substrates or inhibitors, or to be inert in these systems. In most cases only aminoacyladenylate formation was assayed.

A few systematic studies have been carried out on the substrate specificity of highly purified aminoacyl–tRNA synthetases toward analogs of the corresponding amino acids bearing modified carboxyl or amino groups. Such analogs of tyrosine, for example, have been examined for their effect on aminoacyladenylate formation with the purified tyrosyl–tRNA synthetases from $E.$ $coli$ and $B.$ $subtilis$ (29). Removal of the carboxyl group, its reduction, amidation, or esterification yielded nonutilizable substrates, although each of the analogs was a competitive inhibitor. Analogs having an acylated α-amino group, or containing a methyl group in place of the α-hydrogen, were neither substrates nor inhibitors. Surprisingly, the D-tyrosine isomer was activated and esterified to tRNA^{Tyr} by both enzymes. In each case the V_{max} was lower and the K_m was higher than with L-tyrosine. This may be an exceptional case, since other aminoacyl–tRNA synthetases seem to have a high specificity for the L-isomers of amino acids.

Amino acid analogs have been used successfully as selective agents for mutants that produce altered aminoacyl–tRNA synthetases. The altered enzymes appear to exclude the unnatural amino acid from esterification to tRNA by failing to form an aminoacyladenylate derivative of the analog (Section X,A,1,a) (52, 53). So far no case has been reported where an amino acid analog is activated by conversion into an aminoacyl-adenylate, but then not transferred to tRNA. This variety of mutants may include some that produce an altered tRNA.

2. The Specificity for tRNA

a. Heterologous Systems, Universality of the Dictionary. The dictionary and the key to the code as defined above are established in the specific interaction of a tRNA molecule with an enzyme–aminoacyl-adenylate complex, resulting in the attachment of the amino acid to the tRNA.

The code is universal and degenerate. In addition, the frequency of utilization of alternate code words for a particular amino acid may vary from species to species (Section XII,B,1). Similar, general features in the dictionary may become apparent by studying the tRNA specificity of the aminoacyl–tRNA synthetase reaction. It should be noted, however, that the dictionary and the code are not dependent on one another. Thus, although the code is universal it does not follow that the dictionary must also be universal. The mechanism responsible for universality of the code may be quite different from the mechanism that underlies the universality of the dictionary (see Part A, Chapter IV).

The base sequence of tRNA is species specific as demonstrated, for example, by specific hybridization to homologous DNA (Section IX,A). Compare, also, the nucleotide sequences of the tyrosine tRNA's from yeast and E. coli. Most of the sequence similarities are also shared by the other tRNA's, so that when these are taken into account the sequence differences between the 2 tyrosine tRNA's are very striking. The greatest similarity, as expected, is in the anticodon arm (Fig. 10). The amino acid sequences of aminoacyl–tRNA synthetases are probably also species specific. This is not surprising since the structures of both the tRNA's and the enzymes are under the genetic control of the organism and therefore presumably subject to heritable modifications. In spite of the extensive species variation that has occurred during evolution, the dictionary has remained largely universal. In numerous cases a heterologous aminoacyl–tRNA-synthetase–tRNA system will promote the attachment of amino acids to the corresponding tRNA's (177). Some apparent ex-

ceptions to this rule have been reported. Thus, a tRNA from one species
is not always charged by an enzyme from a different species (15, 177).
The cross reaction between enzymes and tRNA's from 2 organisms
depends not only upon the organisms in question but also the particular
amino acid (15). The inability of a particular aminoacyl–tRNA synthetase to attach the corresponding amino acid to a tRNA from a different organism has been interpreted as evidence for nonuniversality of
the dictionary. However, detailed examination of the rates and, in some
cases, of the products of the reaction led to the conclusion that, while
the rate of attachment of the amino acid is reduced by the lack of
homology between the enzyme and the tRNA, the specificity of attachment seems to be preserved.

Only a single case has been found to date where the specificity of
attachment has not been preserved in a heterologous enzyme–tRNA
system. Two phenylalanyl–tRNA synthetases from *Neurospora crassa*,
one from the cytoplasm, the other from mitochondria, can be separated
from each other (Section X,A). Each enzyme aminoacylates a different
species of (*Neurospora*) tRNA^{Phe}. The cytoplasmic enzyme attaches
phenylalanine to the major cytoplasmic tRNA^{Phe} from *Neurospora* and
also to tRNA^{Phe} from *E. coli*. The mitochondrial enzyme attaches phenylalanine to a minor mitochondrial (*Neurospora*) tRNA^{Phe}. With *E. coli*
tRNA, however, it attaches phenylalanine to tRNA^{Val} and tRNA^{Ala} to
form the hybrids phenylalanyl–tRNA^{Val} (*E. coli*) and phenylalanyl–
tRNA^{Ala} (*E. coli*) (7–10). Further study of this special case of "infidel-
ity" in the dictionary may help to elucidate the mechanism of enzyme–
tRNA recognition.

b. Multiple tRNA Species versus Multiple Enzyme Species. In all
systems studied so far, a single enzyme seems to be able to recognize
the multiple tRNA's which are specific for a particular amino acid.
The best evidence that the same enzyme molecule can recognize 2
different tRNA's is the demonstration of a rapid exchange of leucine
between *E. coli* tRNA$_1^{Leu}$ and *E. coli* tRNA$_2^{Leu}$ catalyzed by *E. coli*
leucyl–tRNA synthetase in the presence of AMP by reversal of Reaction
(8) (176).

$$^{14}C\text{-leucyl–tRNA}_1^{Leu} + AMP + Enz^{Leu} \rightleftharpoons Enz^{Leu}\text{–}^{14}C\text{-leucyl–AMP} + tRNA_1^{Leu} \quad (10)$$

$$Enz^{Leu}\text{–}^{14}C\text{-leucyl–AMP} + tRNA_2^{Leu} \rightleftharpoons {}^{14}C\text{-leucyl–tRNA}_2^{Leu} + Enz^{Leu} + AMP \quad (11)$$

$$\text{Sum: } {}^{14}C\text{-leucyl–tRNA}_1^{Leu} + tRNA_2^{Leu} \rightleftharpoons {}^{14}C\text{-leucyl–tRNA}_2^{Leu} + tRNA_1^{Leu} \quad (12)$$

This exchange has not been demonstrated for other amino acids. However, the highly purified leucyl–, valyl–, and isoleucyl–tRNA synthetases
from *E. coli* can attach leucine, valine, or isoleucine, respectively, to all

of the corresponding multiple *E. coli* tRNA species for these amino acids.

Heterologous aminoacyl–tRNA synthetases are, however, occasionally capable of recognizing only some of the multiple tRNA's specific for a particular amino acid. For example, the 2 minor leucine tRNA's from *E. coli* cannot be charged by the enzymes from rabbit, rat, or yeast *(14)*. A similar discrimination between multiple species of *E. coli* tRNA^{Met} and tRNA^{Ser} by heterologous enzymes has been reported.

These observations, although still scanty, can be interpreted as an indication of the degeneracy of the dictionary and of species variation in the frequency of utilization of alternate recognition sites on tRNA. It is intriguing that in the case of *E. coli* tRNA^{Leu}, and probably also of *E. coli* tRNA^{Ser}, those molecules that are not recognized by the heterologous enzymes also have different coding properties. A generalization on the connection between the dictionary and the code is not possible on the basis of data now available.

XII. SPECIFICITY OF TRANSFER RNA AS THE AMINO ACID ADAPTOR

A. SPECIES SPECIFICITY

The evidence that tRNA acts as amino acid adaptor in the assembly of polypeptide chains on the ribosomes has been described in Section I,B.

No species specificity is apparent in the interaction of tRNA from one species with ribosomes and template RNA from another species. Hemoglobin can be synthesized in a cellfree system using polysomes from rabbit reticulocytes combined with ^{14}C-aminoacyl–tRNA's from *E. coli*, yeast, or *Micrococcus lysodeikticus* (the latter organism has a DNA with a GC content of 72%). Radioautographs of fingerprints of tryptic digests, or analysis of tryptic peptides after column chromatography, show that normal hemoglobin is apparently synthesized in all cases *(166, 172)*. It should be noted that tRNA from *E. coli* functions as a true amino acid adaptor in this heterologous system *(165)* (Section I,B). These experiments are consistent with the universality of the code.

B. CODE WORD SPECIFICITY OF tRNA; DEGENERACY

The degeneracy of the code, i.e., the assignment of more than one codon to a particular amino acid, manifests itself at 2 stages in protein synthesis. (1) At the level of tRNA multiplicity, more than one tRNA with different coding properties exists for many amino acids. (2) At the level of tRNA–codon interaction, 1 tRNA molecule can recognize more than one codon. The way in which a single tRNA molecule recognizes more than one codon is explained by the wobble theory (41) (Section VIII).

1. The Coding Properties of Multiple tRNA Species

Early evidence that multiple forms of tRNA have different coding properties was obtained when the polynucleotide-stimulated ribosomal incorporation of leucine was tested using 2 separable leucine tRNA's. A 200 transfer countercurrent distribution separates (E. coli) leucine tRNA into 2 fractions, $tRNA_1^{Leu}$ and $tRNA_2^{Leu}$; ^{14}C-leucyl–$tRNA_1^{Leu}$ transfers leucine into polypeptide material in response to poly(U_5,C), whereas ^{14}C-leucyl–$tRNA_2^{Leu}$ transfers leucine preferentially in response to poly(U_5,G) (173). More extensive (400 transfer) countercurrent distribution reveals 3 species of tRNA in peak 1 designated as $tRNA_{1a}^{Leu}$, $tRNA_{1b}^{Leu}$, and $tRNA_{1c}^{Leu}$ and 2 species in peak 2, $tRNA_{2a}^{Leu}$ and $tRNA_{2b}^{Leu}$. With random copolymers, and in hemoglobin synthesis, the coding properties of peak 2a and peak 2b are clearly different from one another (164). Subsequently, $tRNA_{2b}^{Leu}$ has been shown to correspond to the triplet UUG and to recognize the rabbit hemoglobin α-chain codon α46 exclusively (172).

The coding specificity of many more tRNA's for different amino acids from yeast and E. coli has subsequently been studied in several laboratories (59, 91, 151–153, 171). The tRNA's were purified by countercurrent distribution or by column chromatography on DEAE–cellulose. The purified tRNA fractions were aminoacylated with radioactive amino acids and then tested for binding to ribosomes in the presence of the triplets assigned to the respective amino acids. The coding properties of purified tRNA's known to date are summarized in Fig. 20. The pattern of multiple codon recognition by single tRNA's are in striking agreement with predictions of the wobble theory. The wobble hypothesis not only accounts for the recognition of more than one codon by a single tRNA molecule but also predicts the patterns of multiple codon recognition by separate tRNA species and the minimum number of tRNA's required to recognize the multiple codons. The following generalizations can be deduced from the data in Fig. 20.

(1) Separate tRNA's are required when 2 codons specifying 1 amino acid begin with different bases (as in the sets of codons for arginine, leucine, and serine). An exception of this rule is $tRNA_f^{Met}$, which recognizes the codons AUG and GUG (Section VIII).

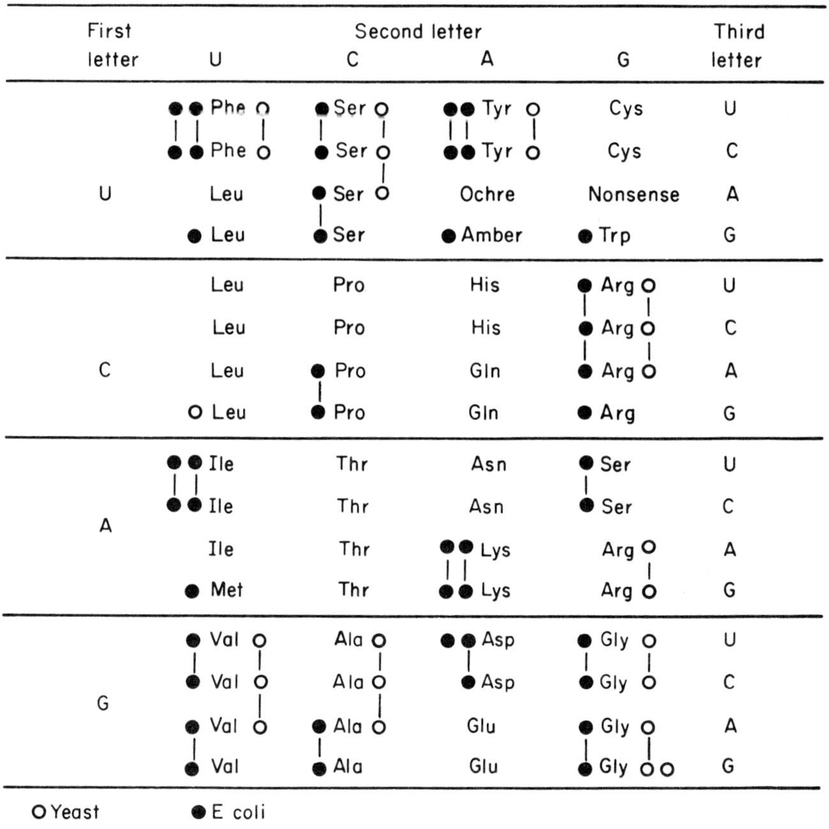

First letter	Second letter U	C	A	G	Third letter
U	•• Phe O	• Ser O	•• Tyr O	Cys	U
	•• Phe O	• Ser O	•• Tyr O	Cys	C
	Leu	• Ser O	Ochre	Nonsense	A
	• Leu	• Ser	• Amber	• Trp	G
C	Leu	Pro	His	• Arg O	U
	Leu	Pro	His	• Arg O	C
	Leu	• Pro	Gln	• Arg O	A
	O Leu	• Pro	Gln	• Arg	G
A	•• Ile	Thr	Asn	• Ser	U
	•• Ile	Thr	Asn	• Ser	C
	Ile	Thr	•• Lys	Arg O	A
	• Met	Thr	•• Lys	Arg O	G
G	• Val O	Ala O	•• Asp	• Gly O	U
	• Val O	Ala O	• Asp	• Gly O	C
	• Val O	• Ala O	Glu	• Gly O	A
	• Val	• Ala	Glu	• Gly O O	G

O Yeast   • E. coli

FIG. 20. The coding properties of multiple tRNA's from *E. coli* and yeast (modified and extended from Söll *et al.*, *151*).

(2) A single tRNA can recognize 3 codons, if these have the first 2 bases in common and U, C, or A as the third base (these pair with inosine in the anticodon). Two codons are recognized by 1 tRNA if the first 2 bases are the same and the third either U or C (pairing with G in the anticodon) or A or G (pairing with U in the anticodon), but not if 1 codon ends in a purine and the other in a pyrimidine.

(3) tRNA's specific for only 1 codon recognize codons ending in G or U but not in A or C.

(4) tRNA anticodons which recognize 3 codons contain inosine. In support of this 3 tRNA's from yeast but only one from *E. coli* recognize 3 codons (yeast tRNA is much richer in inosine than *E. coli* tRNA).

(5) The amino acid-codon assignments in *E. coli* and yeast are identical, consistent with the universality of the code. A comparison of

the recognition pattern of tRNA^{Ala} from yeast with that from *E. coli,* and the anticodons of tRNA^{Val} and tRNA^{Ser} from yeast with the recognition pattern of the corresponding tRNA's from *E. coli,* makes it seem very likely that the recognition set actually used can vary from species to species. Thus, a particular amino acid can be associated with different tRNA anticodons in different organisms. The relative amounts of certain adaptors responding to particular codons may be variable from organism to organism. However, no systematic data are available on the quantitative composition of the tRNA complement in different organisms. It is likely, that the wide span of GC content of DNA, ranging from 25% for *Tetrahymena pyriformis* to 72% for *Micrococcus lysodeikticus,* can be accounted for by specific variation in the frequency of ultilization of alternate code words. It is conceivable that the population of tRNA molecules is optimally tuned to the translation of template RNA's with a particular GC content and thus becomes species specific (*157*).

(6) There are several examples in Fig. 20 of redundant tRNA's for a particular codon. Thus, not all tRNA's appear to be indispensable for the correct translation of templates, providing safeguards against mutation. The redundant tRNA species may serve special purposes in the cell, such as in suppression and regulation. Three structural genes for tyrosine tRNA have been identified in *E. coli* (Section IV,B). The structural cistrons for tRNA in *Drosophila* show a 13-fold redundancy for the synthesis of 60 postulated tRNA species (Section IX,A).

C. Suppressor tRNA's

Transfer RNA's with altered coding properties have been shown to be responsible for the suppression of nonsense and missense mutations in the su^+ cell (*24, 61*). In 1 amber suppressor strain su_I^+, a tRNA^{Ser} was found to be able to recognize the *amber* triplet, UAG (*4, 30, 51*). In another *amber* suppressor strain su_{II}^+ a tRNA^{Tyr} was shown to bind to ribosomes in response to the *amber* triplet (*150*). The amino acid inserted into protein in response to the UAG *amber* triplet in 3 su^+ strains had previously been shown to be specific for each suppressor (su_I^+, serine; su_{II}^+, glutamine; su_{III}^+, tyrosine) (*168–170*). Similar results have been reported in *E. coli* strains carrying missense suppressors. In 1 strain a glycine (GGA) to arginine (AGA) mutation was suppressed by restoring the wild-type sequence in 5% of the protein molecules produced. Presumably a modified tRNA^{Gly} in the suppressor strain recognizes the AGA-arginine triplet (*33*). Similar studies of suppression of a glycine (GGU) to cysteine (UGU) mutation also led to the suggestion that an altered tRNA^{Gly} was responsible for the suppression by recognizing the UGU–cysteine triplet (69).

Although there is no doubt that the suppressor itself is a tRNA molecule these experiments do not show how the suppressor gene acts (25, 114). Most simply the suppressor could specify the nucleotide sequence of a tRNA. This has recently been shown for the amber suppressor su^-_{III} gene (67, 98, 150). It is the structural gene for one of the tyrosine tRNA's of E. coli. The mutation su^-_{III} to su^+_{III} results in a G to C change in the anticodon of this tRNA. The nucleotide sequence of the su^+_{III} tyrosine tRNA is shown in Fig. 10 with the differences that occur in su^-_{III}, and species I and II. The tRNA$^{Tyr}_{su^-_{III}}$ and tRNA$^{Tyr}_{I}$ have indentical sequences. They differ from tRNA$^{Tyr}_{su^+_{III}}$ in having a modified G* in the anticodon (the nature of the modification is unknown). tRNA$^{Tyr}_{II}$ has the same anticodon sequence containing G* but is different in 2 nucleotides in the S-region.

The change of the anticodon from GUA to CUA enables this minor tRNA$^{Tyr}_{su^+_{III}}$ to recognize the amber codon UAG instead of the tyrosine codons UAU and UAC. This type of suppressor gene—where a new codon recognition occurs at the expense of the original one—can only arise in strains with redundant genes for a particular tRNA. Multiple tRNA genes are not required in another type of suppressor gene that has been postulated. In this case the altered tRNA recognizes the new codon in addition to its original codon in an ambiguous fashion.

There is evidence for at least 3 structural genes for tyrosine RNA (tRNA$^{Tyr}_{I}$, tRNA$^{Tyr}_{II}$, tRNA$^{Tyr}_{su}$). These genes are expressed at different rates. The 3 tyrosine tRNA's I, II, and su are present in the approximate ratios 25:60:15. The need for this differential control is obscure because the 3 tRNA's have nearly identical sequences. Regulatory mutants affecting the differential transcription of tRNA cistrons could be assayed in such suppressor strains since the efficiency of suppression seems to be correlated with the amount of su tRNA.

The G to C change in the nucleotide sequence of the su tyrosine tRNA specifically changes its codon recognition but apparently does not affect any of the other functional properties of the tRNA. To identify the function of other parts of the molecule, different mutants of this gene are being studied. These have been obtained by isolating su⁻ revertants of the su⁺ gene (62, 67). Any mutation which gives rise to a nonfunctional tRNA will have the su⁻ phenotype. Characterizing the structural and functional changes in such su⁻ tyrosine tRNA should provide insight into the structural basis for function in tRNA molecules (1a).

In strains carrying nonsense or missense suppressors, ambiguities are introduced into the readout of all templates containing that particular triplet. It would certainly not be without consequence to the organism if common triplets were misread frequently. However rare triplets occurring only in a few proteins could perhaps be misread at a higher frequency without serious effect to the organism. Therefore, the level of suppression should be low when common triplets are involved, but could be higher when a rare triplet is misread.

REFERENCES

1a. I. Abelson, L. Barnett, S. Brenner, M. Gefter, A. Landy, R. Russell, and J. D. Smith, Mutant tyrosine transfer ribonucleic acids. *FEBS Letters* **3**, 1–4 (1969).

1. A. Adams, T. Lindahl, and J. R. Fresco, Conformational differences between the biologically active and inactive forms of a transfer ribonucleic acid. *Proc. Natl. Acad. Sci. U. S.* **57**, 1684–1691 (1967).

2. J. M. Adams and M. R. Capecchi, N-formylmethionyl-sRNA as the initiator of protein synthesis. *Proc. Natl. Acad. Sci. U. S.* **55**, 147–155 (1966).

3. C. C. Allende, J. E. Allende, M. Gatica, J. Celis, G. Mora, and M. Matamala, The amino acyl ribonucleic acid synthetases. I. Properties of the threonyl-adenylate-enzyme complex. *J. Biol. Chem.* **241**, 2245–2251 (1966).

4. T. Andoh and A. Garen, Fractionation of a serine transfer RNA containing suppressor activity. *J. Mol. Biol.* **24**, 129–132 (1967).

5. J. Apgar, G. A. Everett, and R. W. Holley, Analyses of large oligonucleotide fragments obtained from a yeast alanine transfer ribonucleic acid by partial digestion with ribonuclease T₁. *J. Biol. Chem.* **241**, 1206–1211 (1966).

6. J. Apgar, R. W. Holley, and S. H. Merrill, Purification of alanine-, valine-, histidine-, and tyrosine-acceptor ribonucleic acids from yeast. *J. Biol. Chem.* **237**, 796–802 (1962).

7. W. E. Barnett and D. H. Brown, Mitochondrial transfer ribonucleic acids. *Proc. Natl. Acad. Sci. U. S.* **57**, 452–458 (1967).

8. W. E. Barnett, D. H. Brown, and J. L. Epler, Mitochondrial-specific aminoacyl-RNA synthetases. *Proc. Natl. Acad. Sci. U. S.* **57**, 1775–1781 (1967).

9. W. E. Barnett and J. L. Epler, Fractionation and specificities of two aspartyl-ribonucleic acid and two phenylalanyl-ribonucleic acid synthetases. *Proc. Natl. Acad. Sci. U. S.* **55**, 184–189 (1966).

10. W. E. Barnett and J. L. Epler, Multiple aminoacyl-RNA synthetase systems and the genetic code in *Neurospora. Cold Spring Harbor Symp. Quant. Biol.* **31**, 549–555 (1966).

10a. B. G. Barrell and F. Sanger, The sequence of phenylalanine tRNA from *E. coli. FEBS Letters* **3**, 275–278 (1969).

11. A. A. Bayev, T. V. Venkstern, A. D. Mirsabekov, A. I. Krutilina, L. Li, and V. D. Axelrod, The primary structure of valine tRNA of baker's yeast. *Mol. Biol.* **1**, 754–766 (1967) (in Russian).

12. A. A. Bayev, T. V. Venkstern, A. D. Mirsabekov, A. I. Krutilina, L. Li, and V. D. Axelrod, Oligonucleotide composition of enzymatic hydrolysates of yeast valine-specific transfer RNA. *Biochim. Biophys. Acta* **108**, 162–164 (1965).

13. D. Bell, R. V. Tomlinson, and G. M. Tener, Chemical studies on mixed soluble ribonucleic acids from yeast. *Biochemistry* **3**, 317–326 (1964).

14. T. P. Bennett, J. Goldstein, and F. Lipmann, Coding properties of *E. coli* leucyl-sRNA's charged with homologous or yeast activating enzymes. *Proc. Natl. Acad. Sci. U. S.* **49**, 850–857 (1963).

15. S. Benzer and B. Weisblum, On the species specificity of acceptor RNA and attachment enzymes. *Proc. Natl. Acad. Sci. U. S.* **47**, 1149–1154 (1961).
16. P. Berg, U. Lagerquist, and M. Dieckmann, The enzymic synthesis of aminoacyl derivatives of ribonucleic acid. VI, Nucleotide sequences adjacent to the . . . .pCpCpA end groups of isoleucine- and leucine-specific chains. *J. Mol. Biol.* **5**, 159–171 (1962).
17. P. Berg, Specificity in protein synthesis. *Ann. Rev. Biochem.* **30**, 293–324 (1961).
18. P. Berg, F. H. Bergmann, E. J. Ofengand, and M. Dieckmann, The enzymatic synthesis of aminoacyl derivatives of ribonucleic acid. I. The mechanism of leucyl-, valyl-, isoleucyl-, and methionyl ribonucleic acid formation. *J. Biol. Chem.* **236**, 1726–1734 (1961).
19. F. H. Bergmann, P. Berg, and M. Dieckmann, The enzymic synthesis of aminoacyl derivatives of ribonucleic acid. II. The preparation of leucyl-, valyl-, isoleucyl- and methionyl ribonucleic acid synthetases from *Escherichia coli*. *J. Biol. Chem.* **236**, 1735–1740 (1961).
20. P. L. Bergquist, Degenerate transfer RNA's from Brewer's yeast. *Cold Spring Harbor Symp. Quant. Biol.* **31**, 435–447 (1966).
21. K. Biemann, S. Tsunakawa, J. Sonnenbichler, H. Feldmann, D. Dütting, and H. G. Zachau, Struktur eines ungewöhnlichen Nucleotids aus serinspesifischer transfer-Ribonucleinsäure. *Angew. Chem.* **78**, 600 (1966).
22. A. Böck and F. C. Neidhardt, Location of the structural gene for glycyl ribonucleic acid synthetase by means of a strain of *Escherichia coli* possessing an unusual enzyme. *Z. Vererbungslehre* **98**, 187–192 (1966).
23. A. Böck, L. E. Faiman, and F. C. Neidhardt, Biochemical and genetic characterization of a mutant of *Escherichia coli* with a temperature-sensitive valyl ribonucleic acid synthetase. *J. Bacteriol.* **92**, 1076–1082 (1966).
24. S. Brenner, A. O. W. Stretton, and S. Kaplan, Genetic code: The "nonsense" triplets for chain termination and their suppression. *Nature* **206**, 994–998 (1965).
25. S. Brody and C. Yanofsky, Mechanism studies of suppressor-gene action. *J. Bacteriol.* **90**, 687 (1965).
26. G. L. Brown, Preparation, fractionation and properties of sRNA. *Progr. Nucleic Acid. Res.* **2**, 259–310 (1963).
27. G. G. Brownlee and F. Sanger, Nucleotide sequences from the low molecular weight ribosomal RNA of *Escherichia coli*. *J. Mol. Biol.* **23**, 337–353 (1967).
28. R. Calendar and P. Berg, Purification and physical characterization of tyrosyl ribonucleic acid synthetase from *Escherichia coli* and *Bacillus subtilis*. *Biochemistry* **5**, 1681–1690 (1966).
29. R. Calendar and P. Berg, The catalytic properties of tyrosyl ribonucleic acid synthetases from *Escherichia coli* and *Bacillus subtilis*. *Biochemistry* **5**, 1690–1695 (1966).
30. M. R. Capecchi and G. N. Gussin, Suppression *in vitro*: Identification of a serine-s-RNA as a "nonsense" suppressor. *Science* **149**, 417–422 (1965).
31. J. Carbon and J. B. Curry, A change in the specificity of transfer RNA after partial deamination with nitrous acid. *Proc. Natl. Acad. Sci. U. S.* **59**, 467–474 (1968).
32. J. Carbon, P. Berg, and C. Yanofsky, Missense suppression due to a genetically altered tRNA. *Cold Spring Harbor Symp. Quant. Biol.* **31**, 487–497 (1966).
33. J. Carbon, P. Berg, and C. Yanofsky, Studies of missense suppression of the tryptophan synthetase A-protein mutant, A36. *Proc. Natl. Acad. Sci. U. S.* **56**, 764–771 (1966).

34. J. Carbon, L. Hung, and D. S. Jones, Reversible oxidative inactivation of specific transfer RNA species. *Proc. Natl. Acad. Sci. U. S.* **53**, 979–986 (1965).

35. R. W. Chambers, The chemistry of pseudouridine. *Progr. Nucleic Acid Res. Mol. Biol.* **5**, 349–398 (1966).

36. F. Chapeville, F. Lipmann, G. von Ehrenstein, B. Weisblum, W. Y. Ray, Jr., and S. Benzer, On the role of soluble ribonucleic acid in coding for amino acids. *Proc. Natl. Acad. Sci. U. S.* **48**, 1086–1092 (1962).

37. J. D. Cherayil and R. M. Bock, A column chromatographic procedure for the fractionation of sRNA. *Biochemistry* **4**, 1174–1183 (1965).

38. M. J. Chrispeels, R. F. Boyd, L. S. Williams, and F. C. Neidhardt, Modification of valyl tRNA synthetase by bacteriophage in *Escherichia coli. J. Mol. Biol.* **31**, 463–475 (1968).

39. B. F. C. Clark and K. A. Marcker, The role of N-formylmethionyl-sRNA in protein biosynthesis. *J. Mol. Biol.* **17**, 394–406 (1966).

39a. B. F. C. Clark, B. P. Doctor, K. C. Holmes, A. Klug, K. A. Marcker, S. J. Morris, and H. H. Paradies, Crystallization of transfer RNA. *Nature* **219**, 1222–1226 (1968).

40. N. Coles, M. W. Bukenberger, and A. Meister, Incorporation of dicarboxylic amino acids into soluble ribonucleic acid. *Biochemistry* **1**, 317–322 (1962).

40a. Suzanne Cory, K. A. Marcker, S. K. Dube, and B. F. C. Clark, Primary structure of a methionine transfer RNA from *Escherichia coli. Nature* **220**, 1039–1040 (1968).

41. F. H. C. Crick, Codon-anticodon pairing: The wobble hypothesis. *J. Mol. Biol.* **19**, 548–555 (1966).

42. F. H. C. Crick, The recent excitement in the coding problem. *Progr. Nucleic Acid Res.* **1**, 163–217 (1963).

43. F. H. C. Crick, On protein synthesis. *Symp. Soc. Exptl. Biol.* **12**, 138–163 (1958).

44. B. P. Doctor, J. Apgar, and R. W. Holley, Fractionation of yeast amino acid-acceptor ribonucleic acids by countercurrent distribution. *J. Biol. Chem.* **236**, 1117–1120 (1961).

45. S. K. Dube, K. A. Marcker, B. F. C. Clark, and S. Cory, Nucleotide sequence of N-formyl-methionyl-transfer RNA. *Nature* **218**, 232–233 (1968).

45a. B. S. Dudock and G. Katz, Large oligonucleotide sequences in wheat germ phenylalanine transfer ribonucleic acid. *J. Biol. Chem.* **244**, 3069–3074 (1969).

46. D. B. Dunn, J. D. Smith, and P. F. Spahr, Nucleotide composition of soluble ribonucleic acid from *Escherichia coli. J. Mol. Biol.* **2**, 113–117 (1960).

47. D. Dütting, H. Feldmann, and H. G. Zachau, Partial digestions with pancreatic and T_1-ribonucleases of serine transfer ribonucleic acid. *Z. Physiol. Chem.* **347**, 249–267 (1966).

48. J. E. Edström and B. Daneholt, Sedimentation properties of the newly synthesized RNA from isolated nuclear components of *Chironomus tentans* salivary gland cells. *J. Mol. Biol.* **28**, 331–343 (1967).

49. L. Eidlic and F. C. Neidhardt, Protein and nucleic acid synthesis in two mutants of *Escherichia coli* with temperature-sensitive aminoacyl ribonucleic acid synthetases. *J. Bacteriol.* **89**, 706–711 (1965).

50. L. Eidlic and F. C. Neidhardt, Role of valyl-sRNA synthetase in enzyme repression. *Proc. Natl. Acad. Sci. U. S.* **53**, 539–543 (1965).

51. D. L. Engelhardt, R. E. Webster, R. C. Wilhelm, and N. D. Zinder, *In vitro* studies on the mechanism of suppression of a nonsense mutation. *Proc. Natl. Acad. Sci. U. S.* **54**, 1791–1797 (1965).

52. W. L. Fangman, G. Nass, and F. C. Neidhardt, Immunological and chemical studies of phenylalanyl sRNA from *Escherichia coli*. *J. Mol. Biol.* 13, 202–219 (1965).

53. W. L. Fangman and F. C. Neidhardt, Demonstration of an altered aminoacyl ribonucleic acid synthetase in a mutant of *Escherichia coli*. *J. Biol. Chem.* 239, 1839–1843 (1964).

54. G. D. Fasman, C. Lindblow, and E. Seaman, Optical rotatory dispersion studies on the conformational stabilization forces of yeast soluble ribonucleic acid. *J. Mol. Biol.* 12, 630–640 (1965).

55. E. Fleissner and E. Borek, Studies on the enzymatic methylation of soluble RNA. I. Methylation of the sRNA polymer. *Biochemistry* 2, 1093–1100 (1963).

56. E. Fleissner and E. Borek, A new enzyme of RNA synthesis: RNA methylase. *Proc. Natl. Acad. Sci. U. S.* 48, 1199–1203 (1962).

56a. B. G. Forget and S. M. Weissman, Nucleotide sequence of KB cell 5 S RNA. *Science* 158, 1695–1699 (1967).

57. J. R. Fresco, A. Adams, R. Ascione, D. Henley, and T. Lindahl, Tertiary structure in transfer ribonucleic acids. *Cold Spring Harbor Symp. Quant. Biol.* 31, 527–537 (1966).

58. W. Fuller and A. Hodgson, Conformation of the anticodon loop in tRNA. *Nature* 215, 817–821 (1967).

59. A. Galizzi, Different coding properties of two aspartyl-tRNA species. *J. Mol. Biol.* 27, 619–621 (1967).

60. E. Gallucci and A. Garen, Suppressor genes for nonsense mutations. II. The *su*-4 and *su*-5 suppressor genes of *Escherichia coli*. *J. Mol. Biol.* 15, 193–200 (1966).

61. A. Garen, Sense and nonsense in the genetic code. *Science* 160, 149–159 (1968).

62. A. Garen, S. Garen, and R. C. Wilhelm, Suppressor genes for nonsense mutations. I. The *su*-1, *su*-2 and *su*-3 genes of *Escherichia coli*. *J. Mol. Biol.* 14, 167–178 (1965).

63. W. J. Gartland and N. Sueoka, Two interconvertible forms of tryptophanyl-sRNA in *E. coli*. *Proc. Natl. Acad. Sci. U. S.* 55, 948–956 (1966).

64. D. Giacomoni and S. Spiegelman, Origin and biologic individuality of the genetic dictionary. *Science* 138, 1328–1331 (1962).

65. J. Gillam, S. Millward, D. Blew, M. von Tigerstrom, E. Wimmer, and G. M. Tener, The separation of soluble ribonucleic acids on benzoylated diethylaminoethyl cellulose. *Biochemistry* 6, 3043–3056 (1967).

66. J. Goldstein, T. P. Bennett, and L. C. Craig, Countercurrent distribution studies of *E. coli* B sRNA. *Proc. Natl. Acad. Sci. U. S.* 51, 119–125 (1964).

67. H. M. Goodman, J. Abelson, A. Landy, S. Brenner, and J. D. Smith, Amber suppression: A nucleotide change in the anticodon of a tyrosine transfer RNA. *Nature* 217, 1019–1024 (1968).

68. H. M. Goodman and A. Rich, Formation of a DNA-soluble RNA hybrid and its relation to the origin, evolution and degeneracy of soluble RNA. *Proc. Natl. Acad. Sci. U. S.* 48, 2101–2109 (1962).

69. N. K. Gupta and H. G. Khorana, Missense suppression of the tryptophan synthetase A-protein mutant A78. *Proc. Natl. Acad. Sci. U. S.* 56, 772–779 (1966).

70. R. H. Hall, M. J. Robins, L. Stasiuk, and R. Thedford, Isolation of N(6)-(γ,γ-dimethylallyl) adenosine from soluble ribonucleic acid. *J. Am. Chem. Soc.* 88, 2614–2615 (1966).

71. R. H. Hall, A general procedure for the isolation of "minor" nucleosides from ribonucleic acid hydrolysates. *Biochemistry* **4**, 661–670 (1965).

72. R. H. Hall and G. B. Cheda, The presence of N^6-(N-formyl-α-aminoacyl) adenosines in an enzymatic hydrolysate of yeast soluble ribonucleic acid. *J. Biol. Chem.* **240**, 2754–2755 (1965).

72a. F. Harada, H. J. Gross, F. Kimuza, S. H. Chang, S. Nishimura, and U. L. RajBhandary, 2-Methylthio N^6-(Δ^2-isopentenyl) adenosine: A component of *E. coli* tyrosine transfer RNA. *Biochem. Biophys. Res. Comm.* **33**, 299–306 (1968).

73. J. Hay, G. J. Köteles, H. M. Keir, and H. Suback-Sharpe, Herpes virus specified ribonucleic acids. *Nature* **210**, 387–390 (1966).

74. H. Hayashi, Species specificity of tyrosine transfer RNA's from *Escherichia coli* and yeast. *J. Mol. Biol.* **19**, 161–173 (1966).

75. H. Hayashi and K. I. Miura, Anticodon sequence as a possible site for the activity of transfer RNA. *Cold Spring Harbor Symp. Quant. Biol.* **31**, 63–70 (1966).

76. L. I. Hecht, M. L. Stephenson, and P. C. Zamecnik, Binding of amino acids to the end group of a soluble ribonucleic acid. *Proc. Natl. Acad. Sci. U. S.* **45**, 505–518 (1959).

77. E. Herbert and C. J. Smith, Effect of limiting concentrations of T₁ ribonuclease on the release of aminoacyl-oligonucleotides from unfractionated yeast acceptor RNA. *J. Mol. Biol.* **28**, 281–294 (1967).

78. J. Hindley, Fractionation of ³²P-labelled ribonucleic acids on polyacrylamide gel and their characterization by fingerprinting. *J. Mol. Biol.* **30**, 125–136 (1967).

79. M. Hoagland, The relationship of nucleic acid and protein synthesis as revealed by studies in cell-free systems. *In* "The Nucleic Acids" (E. Chargaff and J. N. Davidson, eds.), Vol. 3, pp. 349–408. Academic Press, New York, 1960.

80. M. B. Hoagland, The present status of the adaptor hypothesis. *Brookhaven Symp. Biol.* **12**, 40–46 (1959).

81. M. B. Hoagland, M. L. Stephenson, J. F. Scott, L. I. Hecht, and P. C. Zamecnik, A soluble ribonucleic acid intermediate in protein synthesis. *J. Biol. Chem.* **231**, 241–257 (1958).

82. R. W. Holley, The nucleotide sequence of a nucleic acid. *Sci. Am.* **214**, No. 2, 30–39 (1966).

83. R. W. Holley, G. A. Everett, J. T. Madison, and A. Zamir, Nucleotide sequences in the yeast alanine transfer RNA. *J. Biol. Chem.* **240**, 2122–2128 (1965).

84. R. W. Holley, J. Apgar, G. A. Everett, J. T. Madison, M. Marquisee, S. H. Merrill, J. R. Penswick, and A. Zamir, Structure of a ribonucleic acid. *Science* **147**, 1462–1465 (1965).

85. R. W. Holley, J. T. Madison, and A. Zamir, A new method for sequence determination of large oligonucleotides. *Biochem. Biophys. Res. Commun.* **17**, 389–394 (1964).

86. C. E. Holt, P. B. Joel, and E. Herbert, Turnover of terminal nucleotides of soluble ribonucleic acid in intact reticulocytes. *J. Biol. Chem.* **241**, 1819–1829 (1966).

87. J. Hurwitz, M. Gold, and M. Anders, The enzymatic methylation of ribonucleic acid and deoxyribonucleic acid. IV. The properties of the soluble ribonucleic acid-methylating enzymes. *J. Biol. Chem.* **239**, 3474–3482 (1964).

88. J. Hurwitz, M. Gold, and M. Anders, The enzymatic methylation of ribonucleic acid and deoxyribonucleic acid. III. Purification of soluble ribonucleic acid-methylating enzymes. *J. Biol. Chem.* **239**, 3462–3473 (1964).

89. T. Ishida and K. Miura, Heterogeneity in the nucleotide sequence near the amino acid accepting terminal of transfer RNA. *J. Mol. Biol.* **11**, 341–357 (1965).

90. C. M. Kay and K. Oikawa, Hydrodynamic and optical rotatory dispersion studies on wheat germ soluble ribonucleic acid. *Biochemistry* **5**, 213–223 (1966).

91. D. A. Kellogg, B. P. Doctor, G. E. Loebel, and M. W. Nirenberg, RNA codons and protein synthesis. IX. Synonym codon recognition by multiple species of valine-, alanine-, and methionine-sRNA. *Proc. Natl. Acad. Sci. U. S.* **55**, 912–919 (1966).

92. A. D. Kelmers, G. D. Novelli, and M. P. Stulberg, Separation of transfer ribonucleic acid by reverse phase chromatography. *J. Biol. Chem.* **240**, 3979–3983 (1965).

93. J. X. Khym, Partition of transfer ribonucleic acid between aqueous salt solutions and trichlorotrifluoroethane containing a quaternary amine. *J. Biol. Chem.* **241**, 4529–4533 (1966).

94. U. Lagerquist and J. Waldenström, Purification and some properties of valyl ribonucleic acid synthetase from yeast. *J. Biol. Chem.* **242**, 3021–3025 (1967).

95. U. Lagerquist, L. Rymo, and J. Waldenström, Structure and function of transfer ribonucleic acid. II. Enzyme-substrate complexes with valyl ribonucleic acid synthetase from yeast. *J. Biol. Chem.* **241**, 5391–5400 (1966).

96. U. Lagerquist and P. Berg, The enzymatic synthesis of aminoacyl derivatives of ribonucleic acid. V. Nucleotide sequences adjacent to the. . . .pCpCpA endgroup. *J. Mol. Biol.* **5**, 139–158 (1962).

97. J. A. Lake and W. W. Beeman, On the conformation of yeast transfer RNA. *J. Mol. Biol.* **31**, 115–125 (1968).

98. A. Landy, J. Abelson, H. M. Goodman, and J. D. Smith, Specific hybridization of tyrosine transfer ribonucleic acids with DNA from a transducing bacteriophage ϕ 80 carrying the amber suppressor gene *su* III. *J. Mol. Biol.* **29**, 457–471 (1967).

99. P. Lebowitz, P. L. Ipata, M. H. Makman, H. H. Richards, and G. L. Cantoni, Resolution of cytidine- and adenosine-terminal transfer ribonucleic acids. *Biochemistry* **5**, 3617–3625 (1966).

100. M. N. Lipsett, Disulfide bonds in sRNA. *Cold Spring Harbor Symp. Quant. Biol.* **31**, 449–455 (1966).

101. M. N. Lipsett, The isolation of 4-thiouridylic acid from the soluble ribonucleic acid of *Escherichia coli. J. Biol. Chem.* **240**, 3975–3978 (1965).

102. U. Z. Littauer, M. Revel, and R. Stern, Coding properties of methyl-deficient phenylalanine transfer RNA. *Cold Spring Harbor Symp. Quant. Biol.* **31**, 501–514 (1966).

103. R. B. Loftfield, The frequency of errors in protein biosynthesis. *Biochem. J.* **89**, 82–92 (1963).

104. R. B. Loftfield and E. A. Eigner, Species specificity of transfer RNA. *Acta Chem. Scand.* **17**, Suppl., 1–117 (1963).

105. J. T. Madison and H. K. Kung, Large oligonucleotides isolated from yeast tyrosine transfer ribonucleic acid after partial digestion with ribonuclease T_1. *J. Biol. Chem.* **242**, 1324–1330 (1967).

106. J. T. Madison, G. A. Everett, and H. Kung, Nucleotide sequence of a yeast tyrosine transfer RNA. *Science* **153**, 531–534 (1966).

106a. J. T. Madison, G. A. Everett, and H. K. Kung, On the nucleotide sequence of yeast tyrosine transfer RNA. *Cold Spring Harbor Symp. Quant. Biol.* **31,** 409–416 (1966).

107. J. T. Madison and R. W. Holley, The presence of 5,6-dihydrouridylic acid in yeast "soluble" ribonucleic acid. *Biochem. Biophys. Res. Commun.* **18**, 153–157 (1965).

108. M. H. Makman and G. L. Cantoni, Isolation of seryl and phenylalanyl ribonucleic acid synthetases from baker's yeast. *Biochemistry* **4**, 1434–1442 (1965).

109. J. D. Mandell and A. D. Hershey, A fractionating column for analysis of nucleic acids. *Anal. Biochem.* **1**, 66–77 (1960).

110. K. Marcker, The Formation of *N*-formylmethionyl-sRNA. *J. Mol. Biol.* **14**, 63–70 (1965).

111. K. Marcker and F. Sanger, *N*-formyl-methionyl-s-RNA. *J. Mol. Biol.* **8**, 835–840 (1964).

112. R. Markham, R. E. F. Matthews, and J. D. Smith, Evidence for the existence of two types of ribonucleic acid. *Nature* **173**, 537–539 (1954).

113. R. Markham and J. D. Smith, The structure of ribonucleic acids. I. Cyclic nucleotides produced by ribonucleases and by alkaline hydrolysis. *Biochem. J.* **52**, 552–557 (1952).

114. J. R. Menninger, "Amber" suppression and activating enzyme. *J. Mol. Biol.* **16**, 556–561 (1966).

115. "Methods in Enzymology." (L. Grossman and K. Moldave, eds.), Vol. 12, Part A. Academic Press, New York, 1967.

116. K. Miura, Specificity in the structure of transfer RNA. *Progr. Nucleic Acid Res. Mol. Biol.* **6**, 39–82 (1967).

117. K. H. Muench and P. Berg, Preparation of aminoacyl ribonucleic acid synthetases from *Escherichia coli.* In "Procedures in Nucleic Acid Research" (G. L. Cantoni and D. R. Davies, eds.), pp. 375–383. Harper, New York, 1966.

118. K. H. Muench, Chloroquine-mediated conversion of transfer ribonucleic acid of *Escherichia coli* from an inactive to an active state. *Cold Spring Harbor Symp. Quant. Biol.* **31**, 539–542 (1966).

119. K. H. Muench and P. Berg, Fractionation of transfer ribonucleic acid by gradient partition chromatography on sephadex columns. *Biochemistry* **5**, 970–981 (1966).

120. A. Muto, K. Miura, H. Hayatsu, and T. Ukita, Effect of semicarbazide modification on the aminoacylation of transfer RNA. *Biochim. Biophys. Acta* **95**, 669–671 (1965).

121. F. C. Neidhardt, Roles of amino acid activating enzymes in cellular physiology. *Bacteriol. Rev.* **30**, 701–719 (1966).

122. F. C. Neidhardt and C. F. Earhart, Phage-induced appearance of a valyl sRNA synthetase activity in *Escherichia coli. Cold Spring Harbor Symp. Quant. Biol.* **31**, 557–563 (1966).

123. F. C. Neidhardt, The regulation of ribonucleic acid synthesis in bacteria. *Progr. Nucleic Acid Res. Mol. Biol.* **3**, 145–181 (1964).

124. S. Nishimura, F. Harada, V. Narushima, and T. Seno, Purification of methionine-, valine-, phenylalanine- and tyrosine-specific tRNA from *Escherichia coli. Biochim. Biophys. Acta* **142**, 133–148 (1967).

125. A. T. Norris and P. Berg, Mechanism of aminoacyl RNA synthesis: Studies with isolated aminoacyl adenylate complexes of isoleucyl RNA synthetase. *Proc. Natl. Acad. Sci. U. S.* **52**, 330–337 (1964).

126. A. Norris-Baldwin and P. Berg, Transfer ribonucleic acid-induced hydrolysis of valyladenylate bound to isoleucyl ribonucleic acid synthetase. *J. Biol. Chem.* **241**, 839–845 (1966).

127. A. Norris-Baldwin and P. Berg, Purification and properties of isoleucyl ribonucleic acid synthetase from *Escherichia coli. J. Biol. Chem.* **241**, 831–838 (1966).

128. R. L. Pearson and A. D. Kelmers, Separation of transfer ribonucleic acids by hydroxyapatite columns. *J. Biol. Chem.* **241**, 767–769 (1966).

129. J. R. Penswick and R. W. Holley, Specific cleavage of the yeast alanine RNA into two large fragments. *Proc. Natl. Acad. Sci. U. S.* **53**, 543–546 (1965).

130. A. Peterkofsky, C. Jesensky, and J. D. Capra, The role of methylated bases in the biological activity of *E. coli* leucine tRNA. *Cold Spring Harbor Symp. Quant. Biol.* **31**, 515–524 (1966).

131. J. Preiss, M. Dieckmann, and P. Berg, The enzymic synthesis of aminoacyl derivatives of ribonucleic acid. IV. The formation of the 3'-hydroxyl terminal trinucleotide sequence of amino acid-acceptor ribonucleic acid. *J. Biol. Chem.* **236**, 1748–1757 (1961).

132. J. Preiss, P. Berg, E. J. Ofengand, F. H. Bergmann, and M. Dieckmann, The chemical nature of the RNA-amino acid compound formed by amino acid-activating enzymes. *Proc. Natl. Acad. Sci. U. S.* **45**, 319–328 (1959).

133. "Procedures in Nucleic Acid Research" (G. L. Cantoni and D. R. Davies, eds.). Harper, New York, 1966.

134. U. L. RajBhandary, S. H. Chang, A. Stuart, R. D. Faulkner, R. M. Hoskinson, and H. G. Khorana, Studies on polynucleotides. LXVIII. The primary structure of yeast phenylalanine transfer RNA. *Proc. Natl. Acad. Sci.* **57**, 751–758 (1967).

135. U. L. RajBhandary, A. Stuart, R. D. Faulkner, S. H. Chang, and H. G. Khorana, Nucleotide sequence studies on yeast phenylalanine sRNA. *Cold Spring Harbor Symp. Quant. Biol.* **31**, 425–434 (1966).

136. U. L. RajBhandary and A. Stuart, Nucleic acids-sequence analysis. *Ann. Rev. Biochem.* **35**, 759–788 (1966).

137. U. L. RajBhandary, R. J. Young, and H. G. Khorana, The labeling of end groups in polynucleotide chains: The selective phosphorylation of phosphomonoester groups in amino acid acceptor ribonucleic acids. *J. Biol. Chem.* **239**, 3875–3884 (1964).

138. M. Revel and U. Z. Littauer: The coding properties of methyl-deficient phenylalanine transfer RNA from *Escherichia coli. J. Mol. Biol.* **15**, 389–394 (1966).

139. C. C. Richardson, Phosphorylation of nucleic acids by an enzyme from T4 bacteriophage-infected *Escherichia coli. Proc. Natl. Acad. Sci. U. S.* **54**, 158–165 (1965).

140. F. M. Ritossa, K. C. Atwood, and S. Spiegelman, On the redundancy of DNA complementary to amino acid transfer RNA and its absence from the nucleolar organizer region of *Drosophila melanogaster. Genetics* **54**, 663–676 (1966).

141. J. R. Roth and B. N. Ames, Histidine regulatory mutants in *Salmonella typhimurium*. II. Histidine regulatory mutants having altered histidyl-tRNA synthetase. *J. Mol. Biol.* **22**, 325–334 (1966).

142. G. W. Rushizky, E. M. Bartos, and H. A. Sober, Chromatography of mixed oligonucleotides on DEAE Sephadex. *Biochemistry* **3**, 626–629 (1964).

143. F. Sanger, G. G. Brownlee, and B. G. Barrell, A two-dimensional fractionation procedure for radioactive nucleotides. *J. Mol. Biol.* **13**, 373–398 (1965).

144. P. S. Sarin, P. C. Zamecnik, P. L. Bergquist, and J. F. Scott, Conformational differences among purified samples of transfer RNA from yeast. *Proc. Natl. Acad. Sci.* **55**, 579–585 (1966).

145. P. S. Sarin and P. C. Zamecnik, Conformational differences between sRNA and amino-acyl-sRNA. *Biochem. Biophys. Res. Commun.* **20**, 400–405 (1965).

146. P. S. Sarin and P. C. Zamecnik, On the stability of aminoacyl sRNA to nucleophilic catalysis. *Biochim. Biophys. Acta* **91**, 653–655 (1964).

147. E. R. Signer, J. R. Beckwith, and S. Brenner, Mapping of suppressor loci in *Escherichia coli. J. Mol. Biol.* **14**, 153–166 (1965).

148. D. F. Silbert, G. R. Fink, and B. N. Ames, Histidine regulatory mutants in *Salmonella typhimurium.* III. A class of mutants deficient in tRNA for histidine. *J. Mol. Biol.* **22**, 335–347 (1966).

149. C. J. Smith and E. Herbert, Terminal nucleotide sequences of yeast transfer RNA's specific for serine, tyrosine glycine, threonine, phenylalanine and alanine. *Science* **150**, 384 (1965).

149a. I. Smith, D. Dubnau, P. Morell, and J. Marmur, Chromosomal location of DNA base sequences complementary to transfer RNA and to 5 S, 16 S and 23 S ribosomal RNA in *Bacillus subtilis. J. Mol. Biol.* **33**, 123–140 (1968).

150. J. D. Smith, J. N. Abelson, B. F. C. Clark, H. M. Goodman, and S. Brenner, Studies on *amber* suppressor tRNA. *Cold Spring Harbor Symp. Quant. Biol.* **31**, 479–485 (1966).

151. D. Söll, J. D. Cherayil, and R. M. Bock, Studies on polynucleotides LXXV. Specificity of tRNA for codon recognition as studied by the ribosomal binding technique. *J. Mol. Biol.* **29**, 97–112 (1967).

152. D. Söll and U. L. RajBhandary, Studies on polynucleotides. LXXVI. Specificity of transfer RNA for codon recognition as studied by amino acid incorporation. *J. Mol. Biol.* **29**, 113–124 (1967).

153. D. Söll, D. S. Jones, E. Ohtsuka, R. D. Faulkner, R. Lohrmann, H. Hayatsu, H. G. Khorana, J. D. Cherayil, A. Hampel, and R. M. Bock, Specificity of sRNA for recognition of codons as studied by the ribosomal binding technique. *J. Mol. Biol.,* **19**, 556–573 (1966).

153a. M. Staehelin, H. Rogg, B. C. Baguley, T. Ginsberg, and W. Wehrli, Structure of a mammalian serine tRNA. *Nature* **219**, 1363–1365 (1968).

154. G. S. Stent and S. Brenner, A genetic locus for the regulation of ribonucleic acid synthesis. *Proc. Natl. Acad. Sci. U. S.* **47**, 2005–2014 (1961).

155. G. Streisinger, Y. Okada, J. Emrich, J. Newton, A. Tsugita, E. Terzaghi, and M. Inouye, Frameshift mutations and the genetic code. *Cold Spring Harbor Symp. Quant. Biol.* **31**, 77–84 (1966).

156. H. Suback-Sharpe, W. M. Shepherd, and J. Hay, Studies on sRNA coded by Herpes virus. *Cold Spring Harbor Symp. Quant. Biol.* **31**, 583–594 (1966).

157. H. Suback-Sharpe and J. Hay, An animal virus with DNA of high guanine + cytosine content which codes for sRNA. *J. Mol. Biol.* **12**, 924–928 (1965).

158. N. Sueoka and T. Y. Cheng, Fractionation of nucleic acids with the methylated albumin column. *J. Mol. Biol.* **4**, 161–172 (1962).

159. N. Sueoka and T. Yamane, Fractionation of aminoacyl-acceptor RNA on a methylated albumin column. *Proc. Natl. Acad. Sci. U. S.* **48**, 1454–1461 (1962).

160. N. Sueoka and T. Kano-Sueoka, A specific modification of leucyl-sRNA of

Escherichia coli after phage T₂ infection. *Proc. Natl. Acad. Sci. U. S.* **52**, 1535–1540 (1964).

160a. S. Takemura, T. Mizutani, and M. Miyazaki, The primary structure of valine-I transfer ribonucleic acid from *Torulopsis utilis. J. Biochem. (Tokyo)* **63**, 277–278 (1968).

160b. S. Takemura, M. Murakami, and M. Miyazaki, Nucleotide sequence of isoleucine transfer RNA from *Torulopsis utilis. J. Biochem. (Tokyo)* **65**, 489–491 (1969).

161. A. L. Taylor and C. Dunham-Trotter, Revised linkage map of *Escherichia coli. Bacteriol. Rev.* **31**, 332–353 (1967).

162. R. Thiebe and H. G. Zachau, Zur Fraktioniering der löslichen ribonukleinsäure. *Biochim. Biophys. Acta* **103**, 568–578 (1965).

163. R. V. Tomlinson and G. M. Tener, The effect of urea, formamide and glycols on the secondary binding forces in the ion-exchange chromatography of polynucleotides on DEAE cellulose. *Biochemistry* **2**, 697–702 (1963).

164. G. von Ehrenstein and D. Dais, A leucine acceptor sRNA with ambiguous coding properties in polynucleotide-stimulated polypeptide synthesis. *Proc. Natl. Acad. Sci. U. S.* **50**, 81–86 (1963).

165. G. von Ehrenstein, B. Weisblum, and S. Benzer, The function of sRNA as amino acid adaptor in the synthesis of hemoglobin. *Proc. Natl. Acad. Sci. U. S.* **49**, 669–675 (1963).

166. G. von Ehrenstein and F. Lipmann, Experiments on hemoglobin biosynthesis. *Proc. Natl. Acad. Sci. U. S.* **47**, 941–950 (1961).

167. R. E. Webster, D. L. Engelhardt, and N. D. Zinder, *In vitro* protein synthesis: Chain initiation. *Proc. Natl. Acad. Sci. U. S.* **55**, 155–161 (1966).

168. M. G. Weigert, E. Lanka, and A. Garen, Amino acid substitutions resulting from suppression of nonsense mutations. III. Tyrosine insertion by the *Su*-4 gene. *J. Mol. Biol.* **23**, 401–404 (1967).

169. M. G. Weigert, E. Lanka, and A. Garen, Amino acid substitutions resulting from suppression of nonsense mutations. II. Glutamine insertion by the *Su*-2 gene; tyrosine insertion by the *Su*-3 gene. *J. Mol. Biol.* **14**, 522–527 (1965).

170. M. G. Weigert and A. Garen, Amino acid substitutions resulting from suppression of nonsense mutations. I. Serine insertion by the *Su*-1 suppressor gene. *J. Mol. Biol.* **12**, 448–455 (1965).

171. B. Weisblum, J. D. Cherayil, R. M. Bock, and D. Söll, An analysis of arginine codon multiplicity in rabbit hemoglobin. *J. Mol. Biol.* **28**, 275–280 (1967).

172. B. Weisblum, F. Gonano, G. von Ehrenstein, and S. Benzer, A demonstration of coding degeneracy for leucine in the synthesis of a protein. *Proc. Natl. Acad. Sci. U. S.* **53**, 328–334 (1965).

173. B. Weisblum, S. Benzer, and R. W. Holley, A physical basis for degeneracy in the amino acid code. *Proc. Natl. Acad. Sci. U. S.* **48**, 1449–1454 (1962).

174. B. Weiss and C. C. Richardson, End-group labeling of nucleic acids by enzymatic phosphorylation. *Cold Spring Harbor Symp. Quant. Biol.* **31**, 471–478 (1966).

175. R. Wolfenden, The mechanism of hydrolysis of aminoacyl RNA. *Biochemistry* **2**, 1090–1092 (1963).

176. T. Yamane and N. Sueoka, Enzymic exchange of leucine between different components of leucine acceptor RNA in *Escherichia coli. Proc. Natl. Acad. Sci. U. S.* **51**, 1178–1184 (1964).

177. T. Yamane and N. Sueoka, Conservation of specificity between amino acid

acceptor RNA and aminoacyl-sRNA synthetase. *Proc. Natl. Acad. Sci. U. S.*
50, 1093–1100 (1963).

177a. M. Yanif and B. G. Barrell, Nucleotide sequence of *E. coli* B tRNA$_I^{Val}$.
Nature **222**, 278–279 (1969).

178. M. Yanif, F. Jacob, and F. Gros, Mutations thermosensibles des systems
activant la valine chez *E. coli. Bull. Soc. Chim. Biol.* **47**, 1609–1626 (1965).

179. M. B. Yarmolinsky and G. De la Haba, Inhibition by puromycin of amino acid
incorporation into protein. *Proc. Natl. Acad. Sci. U. S.* **45**, 1721–1729 (1959).

180. M. Yarus and P. Berg, Recognition of tRNA by aminoacyl tRNA synthetases.
J. Mol. Biol. **28**, 479–490 (1967).

181. C. D. Yegian, G. S. Stent, and E. M. Martin, Intracellular condition of
Escherichia coli transfer RNA. *Proc. Natl. Acad. Sci. U. S.* **55**, 839–846 (1966).

182. H. G. Zachau, D. Dütting, H. Feldmann, F. Melchers, and W. Karau, Serine
specific transfer ribonucleic acids. XIV. Comparison of nucleotide sequences
and secondary structure models. *Cold Spring Harbor Symp. Quant. Biol.* **31**,
417–424 (1966).

183. H. G. Zachau, D. Dütting, and H. Feldmann, Nucleotidsequenzen zweier
serinspezifischer transfer-Ribonucleinsäuren. *Angew. Chem.* **78**, 392–393 (1966).

184. H. G. Zachau, D. Dütting, and H. Feldmann, The structures of two serine
transfer ribonucleic acids. *Z. Physiol. Chem.* **347**, 212–235 (1966).

185. H. G. Zachau, D. Dütting, and H. Feldmann, On the primary structure of
transfer ribonucleic acids. *In* "Properties and Function of Genetic Elements"
(D. Shugar, ed.), pp. 271–285. Academic Press, New York, 1966.

186. H. G. Zachau and H. Feldmann, Amino acid esters of RNA, nucleosides and
related compounds. *Progr. Nucleic Acid Res. Mol. Biol.* **4**, 217–230 (1965).

187. H. G. Zachau, G. Acs, and F. Lipmann, Isolation of adenosine amino acid esters
from a ribonuclease digest of soluble, liver ribonucleic acid. *Proc. Natl. Acad.
Sci. U. S.* **44**, 885–889 (1958).

188. A. Zamir, R. W. Holley, and M. Marquisee, Evidence for the occurrence of a
common pentanucleotide sequence in the structures of transfer ribonucleic
acids. *J. Biol. Chem.* **240**, 1267–1273 (1965).

IV

The Flow of Information from Gene to Protein

Marshall Nirenberg

The salient features of the code and the deciphering experiments will be described in this article; relevant topics described in other chapters in this volume and extensively reviewed elsewhere are discussed only briefly here. Additional information can be found in the comprehensive reviews in the *Cold Spring Harbor Symposium on Quantitative Biology* of 1966 (*24*), and in recent books by Woese (*99*) and by Jukes (*49*).

I. THE CONCEPT OF A GENE-PROTEIN CODE

The idea of a genetic code for protein structure originated on three independent occasions during the early 1950's. Intuition was spectacular in the light of the knowledge then available. By 1950, genetic and biochemical studies, particularly of the phenomenon of transformation, had revealed that DNA is the genetic material that directs the synthesis of protein. It was known that DNA is a linear polymer of four nucleotides, that RNA is required for protein synthesis, and that proteins are polymers of twenty or more amino acids. In addition, the finding by Astbury and Bell (*3*) that the spacing of amino acids in proteins is approximately the same as the spacing of bases in DNA suggested a direct correspondence between base sequence in nucleic acids and amino acid sequence in proteins.

In 1950, Caldwell and Hinshelwood (*19*) suggested that each amino acid incorporated into protein is determined by two adjacent nucleotide units in nucleic acid. The assumption was made that RNA is composed of five kinds of units, the four bases and ribose phosphate. Therefore, the suggestion was advanced that 25 doublets in RNA might serve as unique determinants for amino acids in protein.

In 1952, Dounce (*34*) proposed that three adjacent bases in RNA correspond to one amino acid in protein. A completely overlapping triplet code was suggested. The concepts of an RNA template for the synthesis of protein, polarity of translation of the template, and ATP-dependent formation of activated amino acid intermediates were formulated in considerable detail. Although the hypothesized steps were not those found in later studies, the ideas relating to the nature of the code were remarkably accurate. Interestingly, Dounce's conviction that templates are necessary for protein synthesis originated during his oral doctoral examination when he was asked by James Sumner to consider the problem of how proteins synthesize other proteins.

Concurrently, Gamow (*37*) predicted the existence of a simple code relating the structures of DNA and protein. Gamow conceived the idea after reading the paper by Watson and Crick (*92*) on the pairing of bases in DNA. Gamow initially suggested that double-stranded DNA contains specific diamond-shaped cavities for amino acids, and that amino acids are assembled and are converted to protein directly on double-stranded DNA templates. Each site for an amino acid was delineated by three bases (one member of a pair of bases and adjacent noncomplementary bases on opposite DNA strands). Only twenty unique sets of cavities can be constructed if one assumes that right-handed and left-handed cavities are equivalent. The two bases delineating the end of one amino acid site could also specify part of the site for the next amino acid. Hence a completely overlapping, fully degenerate code was suggested.

Gamow later proposed another code, the triangle code, wherein a triplet consisting of three sequential bases in a single-strand of RNA corresponds to one amino acid in protein (*38*). Triplets with the same base composition but differing in base sequence were assumed to be equivalent. The 64 triplets were reduced to 20 sets of triplets.

Every triplet was assumed to correspond to an amino acid. Amino acids again were thought to interact directly with the bases in the nucleic acid template. Triplets were overlapping so that the distance between two amino acids in protein approximated the distance between two bases in RNA. However, Gamow emphasized that an overlapping code imposes restrictions on the kinds of dipeptide sequences that may be synthesized. Brenner (*13*) later ruled out a fully overlapping triplet code on the basis of known dipeptide sequences of protein.

The triangle code was soon replaced by the comma-less code of Crick,

Griffith, and Orgel (28). The 64 triplets were reduced to the magic number of 20 and the phase of reading was limited to 1 of 3 possible phases by assuming that 20 "sense" triplets correspond to the 20 amino acids, whereas 44 triplets were "nonsense" triplets that did not correspond to amino acids. The essential feature was that adjacent "sense" triplets were restricted to only one of the three reading phases. Overlapping triplets formed by shifting the phase of reading then would be "nonsense" triplets. A comma-less code (one of 288 solutions) is shown below:

$$\mathrm{AB}^{\mathrm{A}}_{\mathrm{B}} \qquad \mathrm{^A_B CB}^{\mathrm{A}}_{\mathrm{C}} \qquad \mathrm{BD}^{\mathrm{A~A}}_{\mathrm{C}}{}^{\mathrm{B}}_{\mathrm{C~D}}$$

Sinsheimer (76) suggested a two-letter code based on the chemical resemblance between G and U and between A and C. Both G and U have a keto group at position 6 of the ring, whereas A and C have an amino group at this position. Hence G was assumed to be equivalent to U and A equivalent to C.

Many other theories concerning the code were proposed during the latter part of the 1950's. Also, attempts were made to deduce the nature of the code by correlating base compositions of nucleic acids with amino acid compositions of proteins. These pioneering speculations have been described recently by Woese (99).

II. DECIPHERING THE BASE COMPOSITIONS OF CODONS

Although the concept that RNA is a template for protein was well established, direct biochemical evidence was lacking. However, Hershey's finding that a fraction of RNA is rapidly synthesized and then degraded in *Escherichia coli* infected with T2 bacteriophage, and the demonstration by Volkin and Astrachan (90) that the composition of this RNA fraction resembles phage DNA rather than *E. coli* DNA suggested that the unstable RNA fraction might function as templates for the synthesis of phage proteins.

Advances in several areas paved the way for the experimental translation of codons. First, the puzzle of protein synthesis was unraveled bit by bit. Many of the reactions, enzymes, and other components that are required for protein synthesis were identified, and *in vitro* systems for protein synthesis were slowly refined. Tissieres, Schlessinger, and Gros (89), Kameyama and Novelli (50), and Nisman and Fukuhara (67) found that DNase inhibits amino acid incorporation into protein in

E. coli extracts. Next, RNA was shown to function *in vitro* as a template for the synthesis of protein (*66*). RNA from yeast, ribosomes, and to-bacco mosaic virus were found to be highly active in stimulating the in-corporation into polypeptide material of every species of amino acid tested. In contrast, poly U served as a template for phenylalanine polym-erization, yielding polyphenylalanine. Hence, residues of U in poly U serve as an RNA codon for phenylalanine. Single-stranded poly U is an active template for phenylalanine incorporation, but double- or triple-stranded poly U·poly A helices do not serve as templates for polypeptide synthesis (*61*).

Synthetic mRNA preparations and cell-free protein synthesizing sys-tems provided a means of exploring the base compositions of codons and the nature of the RNA code. A summary of the minimum base composi-tions of polynucleotides that serve as codons for amino acids is shown in Table I. Only polynucleotides containing the minimum species of bases required to stimulate an amino acid into protein are shown. Poly U, poly C, and poly A stimulate the incorporation into protein of phenylalanine, proline, and lysine, respectively. No template activity was detected with poly G. The available evidence indicates that RNA preparations with much secondary structure in solution (such as poly G) do not serve as templates for protein synthesis (*75*).

Poly (U,C), poly (C,G), and poly (A,G) are templates for two amino acids per polynucleotide in addition to those mentioned above, whereas

TABLE I

MINIMUM SPECIES OF BASES REQUIRED FOR mRNA CODONS[a]

Polynucleotides	Amino acids			
U	Phe			
C	Pro			
A	Lys			
G	—			
UC	Leu	Ser		
UA	Leu	Tyr	Ile	Asn
UG	Leu	Val	Cys	Trp
CA	His	Thr	Gln	Asn
CG	Arg	Ala		
AG	Arg	Glu		
UAG	Asp	Met		
CAG	Asp	Ser		

[a] The specificity of randomly ordered polynucleotide templates in stimulating amino acid incorporation into protein in *E. coli* extracts is shown. As only the minimum species of bases necessary for template activity are shown, many amino acids responding to polymers composed of three kinds of bases are omitted.

poly (U,A), poly (U,G), and poly (C,A) each serve as templates for four additional amino acids. Polynucleotides composed of three species of bases are templates for ten or more amino acids.

Randomly ordered polynucleotides composed of 1, 2, 3, or 4 kinds of bases contain 1, 8, 27, and 64 kinds of triplets, respectively. The relative abundance of each kind of triplet can be calculated easily if the base ratio of a randomly ordered polynucleotide is known. One can derive both the *kinds* of bases that correspond to an amino acid and the *number* of bases of each kind, since the amount of each species of amino acid that is incorporated into protein due to the addition of a polynucleotide preparation and the base ratio of the polynucleotide can be determined experimentally. In this manner the base compositions of approximately 50 codons were assigned to amino acids. The results showed that multiple codons can correspond to the same amino acid; hence, the code is highly degenerate. In most cases synonym codons differ by only one base; therefore, it was assumed that the nonvariable bases occupy the same relative positions within each synonym word.

Analysis of the coat protein of mutant strains of tobacco mosaic virus provided evidence that triplets in mRNA are translated in a nonoverlapping fashion because the replacement of one base by another in mRNA usually results in the replacement of one amino acid, rather than several, in protein. By means of genetic studies, Crick, Barnett, Brenner, and Watts-Tobin (*27*) showed that the code is a triplet code, and biochemical studies confirmed this conclusion (see also Chapters I and II).

III. BASE SEQUENCES OF CODONS

The order of nucleotides within codons was deciphered by determining the attachment of aminoacyl-tRNA (AA-tRNA) to ribosomes in response to trinucleotide templates of known sequence (*64*) and by stimulating polypeptide synthesis with polynucleotides containing known repeating sequences of nucleotides (*53*).

The genetic code is shown in Fig. 1. Most triplets correspond to amino acids. Codons for the same amino acid usually differ only in the base occupying the third position of the triplet. Therefore, synonym codons are *systematically* related to one another. Five patterns of codon degeneracy for amino acids are found, each pattern determined by the kinds of bases that occupy the third positions of synonym triplets. The third base of each degenerate triplet is shown below; the dashes correspond to the first and second bases of each triplet.

(1) — — G
(2) — — U
     — — C
(3) — — A
     — — G

(4) — — U
     — — C
     — — A
(5) — — U
     — — C
     — — A
     — — G

Codon base sequence studies confirm 43 of the 50 base compositions of codons that were estimated previously in studies with randomly ordered polynucleotides and the cell-free protein synthesizing system.

The number of codons per amino acid varies considerably. Six degenerate codons correspond to serine, five or six to arginine, five or six to leucine, and from one to four codons for each of the remaining amino acids. It should be noted that only one codon is found for tryptophan and one codon for methionine residues within protein chains. One consequence of systematic degeneracy is that the replacement of one base by another in DNA often does not result in the replacement of one amino acid by another in protein. Many mutations, therefore, are silent ones.

Sets of codons for certain amino acids may be arranged in an orderly, rather than a random, manner because amino acids that are structurally or metabolically related are often coded by RNA codons of similar struc-

UUU △ ○ PHE UUC △ ○	UCU △ ○ UCC △ ○ SER	UAU △ ○ TYR UAC △ ○	UGU △ ○ CYS UGC △
UUA ○ LEU UUG △ ○	UCA △ UCG △ ○	UAA △ TERM UAG △	UGA △ TERM UGG △ ○ TRP
CUU △ ○ CUC △ ○ LEU CUA △ CUG △	CCU △ ○ CCC △ ○ PRO CCA △ ○ CCG △ ○	CAU △ ○ HIS CAC △ ○ CAA △ ○ GLN CAG △	CGU △ ○ CGC △ ○ ARG CGA △ ○ CGG △
AUU △ ○ AUC △ ○ ILE AUA △ ○ AUG △ ○ MET	ACU △ ○ ACC △ ○ THR ACA △ ○ ACG △	AAU △ ○ ASN AAC △ ○ AAA △ ○ LYS AAG △	AGU △ SER AGC △ ○ AGA △ ○ ARG AGG △
GUU △ ○ GUC △ GUA △ ○ VAL GUG △	GCU △ ○ GCC △ ○ ALA GCA △ ○ GCG △	GAU △ ○ ASP GAC △ ○ GAA △ ○ GLU GAG △	GGU △ ○ GGC △ ○ GLY GGA △ ○ GGG △

FIG. 1. The symbols represent the following: △, base sequences of mRNA codons determined by stimulating binding of E. coli AA-tRNA to E. coli ribosomes with trinucleotide templates; ○, base compositions of mRNA codons determined by stimulating the incorporation of amino acids into protein with randomly ordered polynucleotide templates in extracts of E. coli. TERM corresponds to terminator codons.

ture. For example, Asp-codons, GAU and GAC, are similar to Glu-codons, GAA and GAG. Similarly, Ser-codons are related to Thr-codons. Codons for the aromatic amino acids, phenylalanine (UUU, UUC), tyrosine (UAU, UAC), and tryptophan (UGG) appear to be related. The code appears to be arranged so that the effects of base replacements in DNA, or erroneous translations of bases in mRNA, may be minimized. Amino acid replacements in protein that occur due to the replacement of one base by another in nucleic acid can be read in Fig. 1 by moving horizontally or vertically from the amino acid in question, but not diagonally. Woese (99) (see also Chapter VI) has discussed the nonrandom relationship in detail, particularly the problem of why the code is arranged as it is.

EXPERIMENTAL APPROACHES

The enzymology of AA-tRNA binding to ribosomes and the nature of the ribosomal binding sites will be described in detail elsewhere in these volumes and was summarized recently in an elegant fashion (57a). It should be noted, however, that binding of AA-tRNA to ribosomes in response to codons is not dependent on peptide bond formation. However, certain protein factors and GTP participate in the binding of f-Met-tRNA, an initiator of protein synthesis, to the 30 S ribosome, and "T factors" and GTP participate in the binding of AA-tRNA to the 70 S ribosome. The available evidence suggests that [f-Met-tRNA·protein] and [AA-tRNA·protein] complexes interact with ribosomes. The formation of relatively stable [AA-tRNA·codon·ribosome] complexes usually is dependent on the formation of specific base pairs between tRNA anticodon and mRNA codon in the presence of Mg^{2+}. It should also be emphasized that ribosomal sites for f-Met-tRNA and for peptidyl-tRNA differ in various ways from sites for AA-tRNA (perhaps also from tRNA sites).

The assay for ribosomal bound AA-tRNA is based on the retention of ribosomes and ribosome-bound AA-tRNA by cellulose nitrate filters (65). Unbound AA-tRNA is removed by washing. The assay does not depend on filtration since the pore size of the discs (0.45 μ) is considerably larger than the diameter of a 70 S ribosome. Also, cellulose *acetate* discs with 0.45 μ pores do not retain AA-tRNA-ribosome-codon complexes. Hence, ribosomes must interact with cellulose nitrate filters that are impregnated with a detergent. However, the nature of the interaction has not been elucidated.

The trinucleotide codons can be synthesized by chemical methods devised by Khorana and associates (53), by enzymic methods with polynucleotide phosphorylase and dinucleoside monophosphate primers as reported by Leder, Singer and Brimacombe (*56*) and Thach and Doty (*88*), or with pancreatic RNase A, which catalyzes the synthesis of oligonucleotides from pyrimidine-2′,3′-cyclic phosphates in the presence of mono- or oligonucleotide acceptors (*7, 8, 10*).

The abilities of unfractionated AA-tRNA preparations to bind to ribosomes in response to codons were studied as well as those of purified AA-tRNA fractions. Both approaches were employed because species of AA-tRNA with some similarity in anticodon structure compete with one another for codon-specified sites on ribosomes. Hence the fidelity of codon recognition sometimes depends on the relative concentrations of species of AA-tRNA with similar anticodon structure.

The relation between oligonucleotide chain-length and template activity is shown in Table II. The dinucleotide, ApA, does not stimulate Lys-tRNA binding to ribosomes, whereas, the trinucleotide, ApApA, stimulates Lys-tRNA binding to ribosomes as actively as do ApApApA or ApApApApA. Hence at least three *sequential* A residues are required for binding of Lys-tRNA to ribosomes.

Hatfield (*45*) has shown that the binding of ^{3}H-trinucleotide codons to ribosomes is dependent upon AA-tRNA, just as AA-tRNA binding to ribosomes is dependent on trinucleotides (Fig. 2). Both the rate and the extent of binding of UpUpC-^{3}H to ribosomes are similar to Phe-tRNA-^{14}C binding. Trinucleotide-^{3}H can be used to study codon recognition by deacylated tRNA or by tRNA that might be acylated with compounds other than common amino acids.

At 0.01 *M* Mg^{2+}, the specificity of AA-tRNA binding to ribosomes in response to trinucleotides is quite high, probably higher than the specificity of amino acid incorporation into protein in response to polynucleotide templates. Errors in codon recognition may be detected more readily by

TABLE II
Lys-tRNA-^{14}C Binding to Ribosomes[a]

Addition	Lys-tRNA-^{14}C bound due to oligo A ($\mu\mu$moles)
ApA	0.01
ApApA	1.92
ApApApA	1.92
ApApApApA	1.92
ApApApApApA	2.71

[a] The effect of oligo A preparations upon the binding of *E. coli* Lys-tRNA to ribosomes. In the absence of oligo A, 0.49 $\mu\mu$moles of Lys-tRNA-^{14}C bound to ribosomes; this amount has been subtracted from each value shown above.

assaying protein synthesis than by the estimation of AA-tRNA binding because the sensitivity of the protein synthesis assay is higher than the binding assay. It is also possible that incorrect [AA-tRNA-ribosome-codon] complexes are not stable enough to detect via the binding assay, but the interactions may be sufficiently stable to permit peptide bond formation.

Some trinucleotides do not cause detectable stimulation of AA-tRNA binding to ribosomes, or have only a slight effect upon such binding. As discussed in a later section, little or no AA-tRNA may be present for certain codons. *Escherichia coli* B AA-tRNA does not respond appreciably to UAA, UAG, UGA, UUA, or AUA and AA-tRNA corresponding to AGA or AGG, and CUU or CUC is found only in relatively low concentrations.

Khorana and his associates have synthesized polynucleotides containing known oligonucleotide sequences in repeating order and have used such preparations as templates in the cell-free protein synthesizing system (*53*). First, chemical methods were developed for the synthesis of short oligodeoxynucleotides of known sequence which then were linked together to form longer oligodeoxynucleotides containing several repeating nucleotide sequences. Some oligodeoxynucleotides serve as templates for RNA polymerase so that polyribonucleotides can be synthesized directly. However, the use of DNA polymerase to synthesize polydeoxynucleotides with repeating sequences that correspond to the oligonucleotide template is a more satisfactory method. The polydeoxynucleotide preparations then serve as templates for polyribonucleotide synthesis that is catalyzed by RNA polymerase. Single-stranded oligodeoxynucleotides do not serve as templates for DNA polymerase; complementary oligonucleotide strands that form antiparallel Watson-Crick base pairs are required. However, each oligodeoxynucleotide need be synthesized only once since it can be copied by DNA polymerase, and part of each DNA product can be removed and used again as a template for the synthesis of more DNA.

The specificity of the polynucleotide preparations in stimulating protein synthesis in *E. coli* extracts is shown in Table III. Polynucleotides with repeating di-, tri-, and tetranucleotide sequences direct a maximum of two, three, and four amino acids into protein, respectively. Most of the polynucleotides are translated in every phase possible. For example, poly UUC directs the synthesis of polyphenylalanine, polyserine, and polyleucine, and hence is functionally equivalent to a mixture of poly UUC, poly UCU, and poly CUU. Some phases are translated much more frequently than others; however, preferred phases usually cannot be predicted from the base sequence of the polynucleotide. Both poly GUA and poly GAU are translated to protein in only two of the three possible

TABLE III

POLYNUCLEOTIDES WITH REPEATING BASE SEQUENCES
AS TEMPLATES FOR POLYPEPTIDE SYNTHESIS[a]

Polynucleotide template	Polypeptide products
RNA with Repeating Dinucleotides	
Poly UC	$[\text{Ser-Leu}]_n + [\text{Leu-Ser}]_n$
Poly AG	$[\text{Arg-Glu}]_n + [\text{Glu-Arg}]_n$
Poly UG	$[\text{Val-Cys}]_n + [\text{Cys-Val}]_n$
Poly AC	$[\text{Thr-His}]_n + [\text{His-Thr}]_n$
RNA with Repeating Trinucleotides	
Poly UUC	$[\text{Phe}]_n + [\text{Ser}]_n + [\text{Leu}]_n$
Poly AAG	$[\text{Lys}]_n + [\text{Glu}]_n + [\text{Arg}]_n$
Poly UUG	$[\text{Cys}]_n + [\text{Leu}]_n + [\text{Val}]_n$
Poly CAA	$[\text{Gln}]_n + [\text{Thr}]_n + [\text{Asn}]_n$
Poly GUA	$[\text{Val}]_n + [\text{Ser}]_n$
Poly UAC	$[\text{Tyr}]_n + [\text{Thr}]_n + [\text{Leu}]_n$
Poly AUC	$[\text{Ile}]_n + [\text{Ser}]_n + [\text{His}]_n$
Poly GAU	$[\text{Met}]_n + [\text{Asp}]_n$
RNA with Repeating Trinucleotides	
Poly UAUC	$[\text{Tyr-Leu-Ser-Ile}]_n + [\text{Leu-Ser-Ile-Tyr}]_n$ $+ [\text{Ser-Ile-Tyr-Leu}]_n + [\text{Ile-Tyr-Leu-Ser}]$
Poly UUAC	$[\text{Leu-Leu-Thr-Tyr}]_n + [\text{Leu-Thr-Tyr-Leu}]_n$ $+ [\text{Thr-Tyr-Leu-Leu}]_n + [\text{Tyr-Leu-Leu-Thr}]_n$
Poly GUAA	Di- and tripeptides
Poly AUAG	Di- and tripeptides

[a] Summary of Khorana and co-workers studies on polypeptide products synthesized in *E. coli* B extracts in the presence of polynucleotide templates with repeating base sequences (adapted from Khorana, *52*).

phases and poly GUAA and poly AUAG are templates only for di- and tripeptides. These results agree well with findings from other laboratories that show that UAA, UAG, and UGA correspond to peptide chain termination (discussed in Section V). The data provide additional evidence that triplets are read in nonoverlapping fashion. Also, it is most encouraging that the codon base sequence data do not conflict with the sequence data found with trinucleotides and AA-tRNA binding studies.

IV. CODON DEGENERACY

The intuitive suggestions of Gamow (*38*) and Dounce (*34*) that the code is composed of triplet words led naturally to the question of how the 64 triplets are allocated to the 20 amino acids. Gamow assumed that the code is completely degenerate and that all triplets correspond to amino acids. Later, Crick *et al.* (*28*) suggested that some, but not all, triplets correspond to amino acids.

The results obtained with randomly ordered mRNA and the cell-free protein synthesizing system showed that the code is extensively degenerate and that synonym codons often differ by only one base. Therefore, the bases common to each synonym codon were assumed to occupy the same positions within each triplet.

A systematic form of degeneracy seemed probable because U was often equivalent to C, and A was equivalent to G. Attempts were made to deduce the rules governing degeneracy from the available data on base compositions of codons and amino acid replacements in protein (*35, 98*).

Two species of Leu-tRNA were found that respond to different mRNA codons (*97*). However, further work was required to determine whether one species of tRNA responds only to one codon, or to two or more codons.

As the order of bases within codons was established, it became abundantly clear that synonym codons are systematically related to one another. As discussed earlier, alternate bases occupy the third position of synonym triplets. Since only a few kinds of degeneracy patterns were found for the 20 amino acids, it seemed likely that correspondingly few codon recognition mechanisms were operative.

Evidence that one molecule of AA-tRNA can respond to two kinds of codons was provided by the demonstration that most molecules of Phe-tRNA respond both to UUU and to UUC (*9*). Further evidence was obtained by determining the specificity of purified tRNA fractions for trinucleotide codons.

A summary of results (*21, 23, 33, 45, 51, 64, 70, 77–80*) with purified fractions of *E. coli* AA-tRNA is shown in Table IV. Responses of AA-tRNA from other organisms are discussed in Section VI. Most species of AA-tRNA respond to alternate codons. Four, possibly five, kinds of synonym sets are found with *E. coli* AA-tRNA as shown below. The dashes represent the first and second base, and the letter, the third base of each triplet.

(1)	— — G	(4)	— — U
(2)	— — U		— — C
	— — C		— — A
(3)	— — A	(5)	— — A
	— — G		— — G
			— — U

The fifth pattern of degeneracy is found with *E. coli* Ser-tRNA (*80*) but has not been found thus far with AA-tRNA from other organisms.

The number of words in the code depends on the number of tRNA

TABLE IV

CODONS RECOGNIZED BY SPECIES OF *E. coli* AA-tRNA[a]

Amino acid	AA-tRNA Species				
	1	2	3	4	5
	mRNA Codons				
Leu	CUU CUC	CUA CUG	CUG	UUG	—(2)
Ser	UCU UCC	UCA UCG (2)	UCG	AGU AGC	
Arg	CGU CGC CGA	AGA (2) AGG	CGG		
Gly	GGU GGC	GGA GGG			
Ala	GCU GCC	GCA GCG			
Val	GUU GUC	GUA GUG			
Trp	UGG				
Met	AUG				
Ile	AUU AUC (2)				
Phe	UUU UUC (2)				
Tyr	UAU UAC (2)				
Cys	UGU UGC (3)				
His	CAU CAC				
Lys	AAA AAG (3)				
Glu	GAA GAG				
Pro	CCA CCG				

[a] Aminoacyl-tRNA preparations from *E. coli* were fractionated by reverse phase column chromatography and the effect of trinucleotide templates upon the binding of purified AA-tRNA fractions to ribosomes was determined (21). The dash, —, represents Leu-tRNA fractions that do not repond to trinucleotide codons. Numerals within parentheses indicate the number of redundant peaks of AA-tRNA found.

anticodons rather than the number of amino acids. It is clear that the mRNA-AA-tRNA code differs from the mRNA-amino acid code.

Several varieties of tRNA are found for most amino acids. In some cases, AA-tRNA from different fractions responds with identical specificity to codons. In other cases, multiple species of tRNA differ in specificity of response to codons. Therefore, patterns of degeneracy for *amino acids* are due to two factors: the ability of many species of tRNA to respond to alternate codons, and the presence of two or more species of tRNA corresponding to the same amino acid that respond to different codons.

Multiple varieties of tRNA that have the same specificity for codons may be products of the same gene that have been altered in different ways by enzymes *in vivo* or perhaps have been altered *in vitro* during the fractionation procedure. Alternatively, redundant AA-tRNA fractions may be products of different genes.

Crick suggested that codon degeneracy is due to the formation of alternate base pairs between a base in a tRNA anticodon and alternate bases occupying the third positions of synonym mRNA anticodons (*26*). The first and second bases of mRNA codons form antiparallel, Watson-Crick base pairs with corresponding bases in the tRNA anticodon. Alternate base pairs proposed by Crick are shown in Table V; U in the tRNA anticodon pairs alternately with A or G occupying the third position of synonym mRNA codons; C pairs with G; G pairs with C or U; and I pairs with U, C, or A.

The elucidation of base sequences of various species of tRNA provides a means of relating base sequences of tRNA anticodons with base sequences of mRNA codons. Some anticodon-codon interactions, defined with tRNA and codons of known sequence, are shown in Fig. 2. In every case that has been examined, codon-anticodon relations are in accord with wobble base pairing.

The available evidence suggests that a newly synthesized molecule of tRNA contains the four common bases and that trace bases in tRNA such as inosine and ψ are formed after tRNA is released from DNA templates in the presence of specific enzymes. For example, deamination of A in an appropriate position in tRNA might result in the formation

TABLE V
ALTERNATE BASE PAIRING[a]

tRNA Anticodon	mRNA Codon (Base 3)
G	C or U
U	A or G
G	C
I	U or C or A

[a] Alternate base pairing between a base in a tRNA anticodon, shown in the left-hand column, and the base(s) in the third position of synonym mRNA codons. Relationships are antiparallel, "wobble" hydrogen bonds suggested by Crick (*26*).

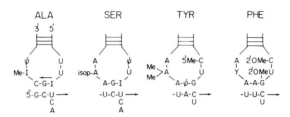

F<small>IG</small>. 2. Codon-anticodon base pairing is antiparallel in direction. The mRNA codons are written in the conventional manner; tRNA anticodon base sequences are written backward.

of inosine. The point that deserves emphasis is that the conversion of "embryonic" to mature tRNA may depend on many enzymically catalyzed reactions; hence, cells probably contain many intermediate forms of tRNA for each tRNA gene. It is also possible that certain modifications affect the interaction of the tRNA with codons, ribosomes, or tRNA synthetases.

V. PUNCTUATION

Punctuation of transcription and translation is illustrated schematically in Figs. 3 and 4. RNA polymerase attaches to specific site(s) on DNA and selects the strand of DNA to be transcribed, the direction of transcription, and the first base to be transcribed. Many questions remain to be answered about the structure of the attachment sites.

The direction of mRNA synthesis is opposite to that of the DNA strand being read. The first base to be incorporated into the nascent mRNA chain is the 5′-terminus of the mRNA, the last base is the 3′-terminus. Similarly, the RNA template is translated during protein synthesis starting at or near the 5′-terminus of the RNA and proceeding three bases at a time, sequentially, toward the 3′-terminus of the RNA. Therefore, mRNA is synthesized and then translated with the same polarity. The first amino acid corresponds to the NH_2-terminus of the peptide chain; the COOH-terminal amino acid is the last amino acid incorporated.

A. I<small>NITIATION</small>

Nascent chains of mRNA, synthesized *in vitro*, do not spontaneously detach from the *E. coli* DNA templates (*11, 12, 18, 82*). A 30 S ribosomal

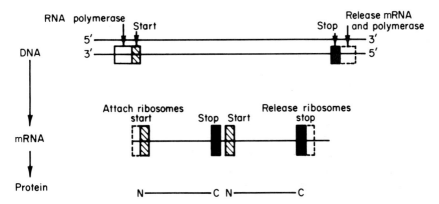

FIG. 3. The punctuation of transcription and translation is illustrated diagrammatically. Ribosomal subunits attach to mRNA near the 5′-terminus of the mRNA and are released near the 3′-terminus of the mRNA. Speculations are indicated by the dotted lines. N and C represent the amino and carboxy terminal amino acid residues of protein, respectively (redrawn from Nirenberg, *64*).

particle attaches to the nascent chain of mRNA near its 5′-terminus before the mRNA detaches from the DNA template. At least three nondialyzable factors and GTP are required for the initiation of protein synthesis. The reactions have not been clarified fully; however, the available evidence suggests that one factor participates in the attachment of a 30 S ribosomal subunit to a nascent chain of mRNA and that other factors, together with GTP, are required for the binding of a unique species of tRNA, *N*-formyl-Met-tRNA (*58*), to the 30 S ribosomal complex in response to the initiator codon, AUG or GUG (*23, 51*). The 50 S

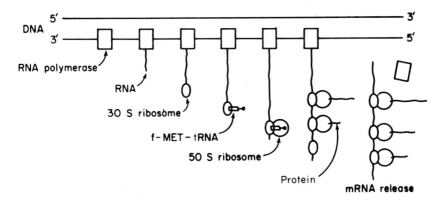

FIG. 4. Diagrammatic illustration of early steps of protein synthesis (redrawn from Nirenberg, *64*).

ribosomal subunit then attaches to the 30 S ribosomal complex before
the next codon is recognized by AA-tRNA. N-formyl-Met-tRNA thus
selects the first codon to be translated and phases the translation of subsequent codons.

It is possible that protein synthesis in $E.\ coli$ extracts is initially physically coupled to the corresponding genome by means of transient
[ribosome-mRNA-DNA] intermediates $(18,\ 74,\ 82)$.

The pattern of degeneracy observed with N-formyl-Met-tRNA$_f$ differs
from the patterns observed with other species of AA-tRNA because
initiator codons have alternate first bases rather than alternate third
bases. Another species of tRNA from $E.\ coli$, Met-tRNA$_m$, does not ac-

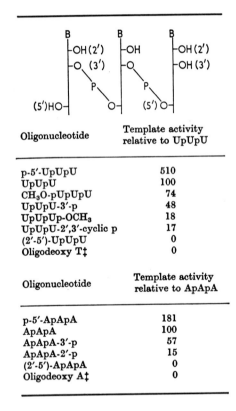

Oligonucleotide	Template activity relative to UpUpU
p-5'-UpUpU	510
UpUpU	100
CH$_3$O-pUpUpU	74
UpUpU-3'-p	48
UpUpUp-OCH$_3$	18
UpUpU-2',3'-cyclic p	17
(2'-5')-UpUpU	0
Oligodeoxy T‡	0

Oligonucleotide	Template activity relative to ApApA
p-5'-ApApA	181
ApApA	100
ApApA-3'-p	57
ApApA-2'-p	15
(2'-5')-ApApA	0
Oligodeoxy A‡	0

Fig. 5. Relative template activities (‡) of substituted oligonucleotides are ap-
proximations obtained by comparing the amount of AA-tRNA bound to ribosomes
in the presence of limiting concentrations of oligonucleotides compared to either
UpUpU for Phe-tRNA-^{14}C or ApApA for Lys-tRNA-^{14}C (each designated at 100%).
The data are from Rottman and Nirenberg (71).

cept formyl moieties, responds only to AUG, and corresponds to methionine at internal positions in protein.

Each triplet can occur in three structural forms: as 5'-terminal, 3'-terminal, or internal codons. Substituents attached to ribose hydroxyl groups of codons can influence profoundly the ability of codons to stimulate AA-tRNA binding to ribosomes (Fig. 5). Relative template activities of oligo U preparations, at limiting oligonucleotide concentrations, are as follows: p-5'-UpUpU > UpUpU > CH_3O-p-5'-UpUpU > UpUpU-3'-p > UpUpU-3'-p-OCH_3 > UpUpU-2',3'-cyclic phosphate (71). Trimers with (2'–5') phosphodiester linkages, (2'–5')-UpUpU and (2'–5')-ApApA, do not serve as templates for Phe- or Lys-tRNA, respectively. The relative template efficiencies of oligo A preparations are as follows: p-5'-ApApA > ApApA > ApApA-3'-p > ApApA-2'-p. Ikehara and Ohtsuka (47) have shown that N^6-DiMeApApA does not stimulate Lys-tRNA binding to ribosomes, whereas, the tubercidin analog (7-deazaadenosine), TupApA, is a template for Lys-tRNA.

RNA polymerase catalyzes the synthesis of mRNA with 5'-terminal triphosphate. Also, many enzymes have been described that catalyze the transfer of molecules to or from hydroxyl groups of nucleic acids. It is possible, therefore, that certain modifications of ribose or deoxyribose hydroxyl groups of nucleic acids provide a means of regulating the rate of transcription or translation.

B. TERMINATION

The first evidence for "nonsense" codons was provided by Benzer and Champe in 1962 (6). They obtained a mutant of bacteriophage T4 with a deletion in the rII region spanning part of the A gene and part of the contiguous B gene. Presumably, the remaining segments of gene A and B are joined and thus form one gene. Nevertheless, a functional B gene product was found. However, a second mutation that mapped in the A gene resulted in the loss of a functional B gene product. These results suggested that a "sense" codon is converted by mutation to a "nonsense" codon that cannot be read; hence subsequent regions of the gene also are not read. Sarabhai, Stretton, Brenner, and Bolle (72) then found that "nonsense" mutations at various sites within the gene for the head protein of bacteriophage T4 determine the chain length of the corresponding polypeptide. These dramatic results showed that "nonsense" codons correspond to the termination of protein synthesis. Additional evidence obtained by Brenner $(14, 15)$ and by Garen $(93, 94)$ and their colleagues showed that three codons, UAA, UAG, and UGA, correspond to the termination of protein synthesis (see also the recent review by Garen, *40*).

The mechanism of peptide-chain termination was investigated by stimulating cell-free protein synthesis with randomly ordered polynucleotides (*17, 39, 86*), oligonucleotides (*55*), and polynucleotides (*63*) of known sequence, and viral RNA (*16, 20*). Capecchi showed that the release of peptides from ribosomes is dependent on both a release factor and a terminator codon (*20*). The codons, UAA, UAG, and UGA, do not stimulate binding of AA-tRNA to ribosomes (although mutant strains of bacteria have been found that contain species of AA-tRNA that respond to terminator codons).

The process of termination also can be studied with trinucleotide templates (*22*). Incubation of terminator trinucleotides and the release factor with the [N-formyl-Met-tRNA-AUG-ribosome] complex results in the release of free N-formyl-methionine from the ribosomal intermediate. The release factor of *E. coli* then was separated into two components that correspond to different sets of codons: R_1, active with UAA or UAG, and R_2, active with UAA or UGA (*73*). It is clear, therefore, that terminator codons are recognized by specific molecules. The simplest hypothesis is that R_1 and R_2 interact with terminator codons on ribosomes; however, the codon recognition step and the mechanism of termination have not been clarified thus far. Another factor, the S factor, has been found that stimulates the rate of release of N-formyl methionine from the ribosomal complex (*62*). Evidence that a factor is required for the dissociation of 70 S ribosomes to 30 and 50 S subunits has also been reported (*83*).

As shown in Table V, the pattern of codon degeneracy found with R_1 (UAA and UAG) resembles that found with some species of AA-tRNA; i.e., A is equivalent to G at the third position of codons. However, the degeneracy pattern found with R_2 (UAA and UGA) differs from those found with AA-tRNA because A and G are equivalent at the second, but not the third, position of triplets.

VI. UNIVERSALITY

The results of many studies suggest that most, if not all, forms of life on this planet employ essentially the same code. One of the earliest indications came from the demonstration that the bacterium *Serratia marcescens* synthesizes alkaline phosphatase resembling the *E. coli* enzyme (57). Also, the *in vitro* synthesis of mammalian protein was reportedly obtained with mammalian ribosomes (presumably mRNA also) and *E. coli* AA-tRNA (91). In addition, base compositions of RNA codons defined with randomly ordered polynucleotides and extracts from many organisms, including plant and mammalian tissues, were shown to be similar to codons defined with extracts from *E. coli*.

However, the specificity of codon translation can be altered *in vivo* by mutations affecting cellular machinery required for nucleic acid and protein synthesis, or *in vitro* by altering components of reactions or incubation conditions. It is clear that cells sometimes differ in the specificity of codon translation. Comparative studies on the binding of AA-tRNA from *E. coli*, yeast, amphibian (*Xenopus laevis*), and guinea pig tissues to ribosomes in response to trinucleotides were undertaken to provide some information on the extent of variation in the fine structure of the code. Almost identical translations of nucleotide sequences to amino acids were found with bacterial, yeast, amphibian, and mammalian AA-tRNA (59).

Codon recognition by unfractionated AA-tRNA from *Xenopus* skeletal muscle and liver were also compared. No differences were detected either in base sequences recognized or in relative responses to degenerate codons (60).

In some cases, however, AA-tRNA from one organism was found to respond to fewer synonym codons than AA-tRNA from another organism (59). Such differences provide potential mechanisms for selectively regulating the rate of protein synthesis and also resemble certain suppressors (see Chapter I) that alter codon translation. To clarify the reason for the nonuniversal responses, guinea pig liver, yeast, and *E. coli* AA-tRNA preparations were fractionated by reverse phase column chromatography, and the binding of pooled AA-tRNA fractions to ribosomes in response to trinucleotide codons was determined (21).

The results are summarized in Table VI. Although some variation in

TABLE VI
PUNCTUATION CODONS[a]

Initiation f-MET-tRNA	AUG
	GUG
Termination (Release Factor 1)	UAA
	UAG
Termination (Release Factor 2)	UAA
	UGA

[a] Codons corresponding to the initiation or termination of protein synthesis in *E. coli* are shown. Release factors 1 and 2 are required for termination with the codons indicated, but it is not known whether they interact directly with terminator codons.

codon translation clearly does occur, codon base sequences recognized by bacterial, yeast, amphibian, and mammalian AA-tRNA are remarkably similar which suggests that most, perhaps all, forms of life on this planet use essentially the same genetic language. However, AA-tRNA from bacterial, yeast, and mammalian sources were found to differ in the number and relative abundance of AA-tRNA species and also in codon-anticodon relationships corresponding to certain amino acids. It should be emphasized that anticodons can vary in certain ways without altering amino acid-codon relationships. For example, one species of liver Ser-tRNA responds to UCU, UCC, and UCA, another species only to UCG, whereas one species of *E. coli* Ser-tRNA responds to UCU and UCC, and another responds to UCA, UCG, and UCU (see Chapter III).

The dissimilar responses found earlier with unfractionated AA-tRNA are due to marked differences in abundance of AA-tRNA for some codons. Perhaps the most striking observations relate to codons that actively stimulate binding of AA-tRNA from one organism but not from another. For example, liver Ile-tRNA responds well to three codons, AUU, AUC, and AUA, whereas, *E. coli* Ile-tRNA responds detectably only to AUU and AUC (AUA-deficient). Also, *E. coli* Arg-tRNA responds well to four codons, CGU, CGC, CGA, and CGG, but is relatively deficient in response to AGA and AGG, whereas, liver Arg-tRNA responds well to five codons and is deficient only in response to AGA. These observations suggest that some organisms have little or no AA-tRNA for certain codons.

The possibility that the rates of synthesis of certain proteins may be regulated by the concentration of certain species of AA-tRNA and the relative frequencies of the corresponding codons in mRNA has been considered on many occasions (*1, 21, 48, 64a, 82, 84*). This hypothesis has been confirmed recently, for the rate of protein synthesis directed by poly (A,G) in *E. coli* S 30 preparations is limited by the concentration of Arg-tRNA responding to AGA or AGG (*2*).

Relationships between tRNA anticodon and mRNA codons are apparently universal, at least in regard to the following patterns: C in a tRNA anticodon corresponds to G in the third position of an mRNA codon, G in the anticodon pairs with U or C in the third position of mRNA codons, U in an anticodon pairs either with A or G in the third position of mRNA codons, I in tRNA anticodon pairs with U, C, or A in mRNA codons (*26*). However, another pattern of degeneracy, in which A, G, or U may occupy the third position of synonym triplets, is found with two fractions of *E. coli* Ser-tRNA (*21*), but not with yeast or guinea pig AA-tRNA. Whether this unusual redundancy pattern is due to the presence of an unusual base in the tRNA anticodon or to incomplete separation of two or more species of Ser-tRNA has not been resolved

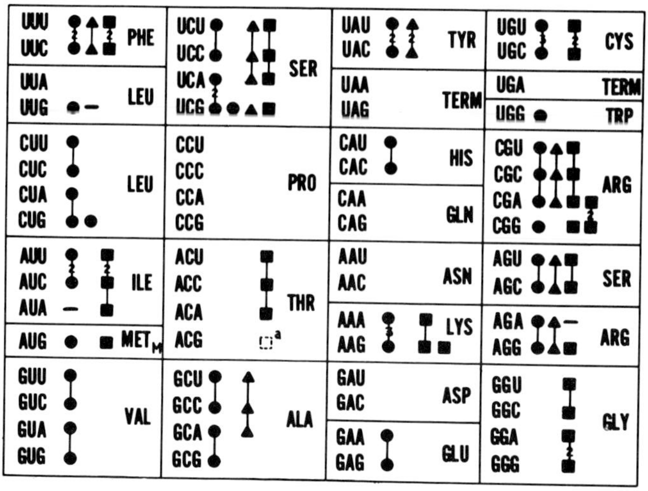

FIG. 6. "Universality" of the genetic code. Synonym codon sets were determined by stimulating the binding of purified aminoacyl-tRNA fractions of E. coli, yeast, or guinea pig liver to E. coli ribosomes with trinucleotide codons (21). The joined symbols adjacent to the codons represent synonym codons recognized by one purified aminoacyl-tRNA fraction from ●, E. coli; ▲, yeast; or ■, guinea pig liver. The numeral between symbols represents the number of redundant peaks of aminoacyl-tRNA found that respond similarly to codons.

thus far. The possibility has been suggested that the A, G, U pattern of degeneracy is due to the pairing of a base such as ψ, 4-thio U, or U in the appropriate position of a tRNA anticodon with A, G, or U in mRNA codons (21).

The code is not a universal code in the absolute sense because the ensemble of molecules that encode and decode genetic information occasionally makes mistakes. Also, DNA templates may undergo mutation. From a statistical point of view, codon translation probably fluctuates about a universal mean.

The code may have originated in conjunction with an early cellular form of life, perhaps $1–3 \times 10^9$ years ago. The oldest fossils known are thought to be microorganisms 3.1×10^9 years of age (5). However, fossil organisms appeared in abundance for the first time approximately 0.6×10^9 years ago. Virtually all of the invertebrate phyla and the first vertebrates had evolved 0.5×10^9 years ago; amphibians and mammals first appeared 0.35×10^9 and 0.18×10^9, respectively. Thus, the genetic code may have originated $0.6–3.0 \times 10^9$ years ago. Hinegardner and Engelberg (46) and Sonneborn (81) have suggested that the evolution of the code stopped after organisms as complex as bacteria had evolved because major alterations in the code probably are limited to those that do not greatly restrict the retrieval of the information that had been acquired.

One wonders whether the genetic code is the *only* code that could have evolved under the circumstances. In this connection it is necessary to distinguish between the *origin* and the subsequent *evolution* of the code. If the code arose only once due to an exceedingly improbable combination of events, then the evolution of the code may have been restricted in various ways by the nature of the original code. Alternatively, a large population of precursor codes may have originated independently of one another, and one code may have been selected from many due to evolutionary pressures.

The affinity of mono-, oligo-, and polynucleotides for one another may have favored their polymerization and their role as carriers of amino acids that similarly might favor the polymerization of amino acids forming peptides and peptidylnucleic acids. Some of the products may have functioned as primative enzymelike catalysts of various kinds depending on certain amino acid–base relationships. This would provide a means of selecting preferred amino acid-anticodon-codon relationships one at a time, in stepwise fashion and evolving ultimately into the genetic code. Woese has considered this problem in depth and suggested that amino acids may have specific affinities for anticodons and that such interactions may have influenced the sorting out of amino acid-anticodon relationship (*99*). A more extensive discussion of the possible mechanisms of evolution of the code is presented in Chapter VI of this volume.

VII. PRECISION AND RELIABILITY

A. SERIAL VERSUS PARALLEL STEPS

The overall accuracy of a multistep process depends on the number of steps required, the precision of each step, and whether steps are performed serially or in parallel. Precision rapidly deteriorates when an error at one step affects the accuracy of subsequent steps.

It is clear that many serial steps are required for the synthesis of DNA, mRNA, or protein. To synthesize one molecule of protein composed of 400 amino acid residues correctly, 400 molecules of AA-tRNA must be selected, one after another in the proper sequence. For the synthesis of the corresponding molecule of mRNA, a minimum of 1206 sequential selections for ribonucleoside triphosphate molecules of appropriate species are required.

Three serial selective steps are required for the synthesis of one molecule of AA-tRNA: selection by an amino acid activating enzyme of species of amino acid, nucleotide triphosphate, and tRNA. In Section VII,B the serial steps required for the synthesis of at least one species of AA-tRNA are shown to *increase*, rather than decrease, the accuracy of matching amino acid with tRNA.

The strategy of protein and nucleic acid synthesis is that *each unit usually is selected independently of other units.* Thus, one error usually does not affect the accuracy of selection for subsequent units. However, if an error in translation alters the phase of reading, subsequent codon translations will also be erroneous; furthermore, if the translation error results in premature termination of polypeptide synthesis, subsequent codons obviously will not be translated. Also, it should be emphasized that many molecules of AA-tRNA, protein, and nucleic acid are synthesized in parallel. In *E. coli*, for example, protein is synthesized simultaneously at 1000–50,000 sites per cell. Such parallel operations greatly enhance the efficiency of protein and nucleic acid syntheses within the cell.

B. Error Correction

Aminoacyl-tRNA synthetase enzymes catalyze the synthesis of aminoacyladenylates and then the transfer of the amino acid moiety to an appropriate species of tRNA, as shown below for Ile-tRNA synthesis:

$$\text{Isoleucine} + \text{ATP} + \text{enzyme}^{\text{Ile}} \rightleftharpoons [\text{Ile-AMP·enzyme}^{\text{Ile}} + \text{PP} \tag{1}$$

$$[\text{Ile-AMP·enzyme}^{\text{Ile}}] + \text{tRNA}^{\text{Ile}} \rightleftharpoons \text{Ile-tRNA}^{\text{Ile}} + \text{AMP} + \text{enzyme}^{\text{Ile}} \tag{2}$$

Baldwin and Berg (4) showed that highly purified isoleucyl-tRNA synthetase from *E. coli* erroneously catalyzes the synthesis of valyladenylate in addition to isoleucyladenylate. However, in the presence of tRNA^{Ile}, the enzyme corrects the error by catalyzing the hydrolysis of valyladenylate, as shown below.

$$\text{Valine} + \text{ATP} + \text{enzyme}^{\text{Ile}} \rightarrow [\text{Val-AMP·enzyme}^{\text{Ile}}] + \text{PP} \tag{3}$$

$$[\text{Val-AMP·enzyme}^{\text{Ile}}] + \text{tRNA}^{\text{Ile}} \rightarrow \text{Valine} + \text{AMP} + \text{enzyme}^{\text{Ile}} \tag{4}$$

Hydrolysis of the valyladenylate bound to enzyme^{Ile} is dependent on *E. coli* tRNA^{Ile}; other species of tRNA are without effect.

It is clear that the enzyme checks the accuracy of both the species of aminoacyladenylate synthesized and the species of tRNA selected. If an error has been made in forming the aminoacyladenylate, the reaction is terminated by hydrolysis of the erroneous aminoacyladenylate. The essential point is that serial steps enhance, rather than decrease, the precision of Ile-tRNA synthesis.

Incorrectly phased translations probably are terminated rapidly in most cases because one terminator triplet would be encountered per 22 triplets, assuming that triplet frequency is random when translations are not phased correctly.

C. SYSTEMATIC REDUNDANCY

As discussed earlier, synonym codons are systematically related to one another since synonyms usually differ only in the base at the third position of the triplets. Therefore, the replacement of one base pair by another in DNA often does not result in the replacement of one amino acid by another in protein. Similarly, many errors in translation are "silent" ones.

It should be emphasized that most amino acid replacements in protein that are due to errors in codon translation are not random errors because two out of three bases per codon usually are recognized correctly. For example, when the precision of translation deteriorates, the codon, UUU, may be translated as phenylalanine 80% of the time, as isoleucine 15%, and as leucine 5% of the time. The total frequency of error with 17 other amino acids may be less than 0.2%. The specificity of codon-amino acid translation is low, but amino acid incorporation into protein is far from random.

D. FACTORS AFFECTING THE FIDELITY OF PROTEIN SYNTHESIS

In 1960 Yanofsky and St. Lawrence (100) suggested that certain mutations might result in the production of structurally modified tRNA or AA-tRNA synthetases with altered specificity for amino acid incorporation into protein. Much information is now available concerning suppressor mutations that affect components required for protein synthesis. Also, much information has been obtained in studies with synthetic polynucleotide templates and in *in vitro* protein-synthesizing systems and in AA-tRNA binding studies. The results show that the fidelity of protein synthesis can be altered in many ways. Most codons probably are translated with relatively little error (0.1–0.01% error or less); however, the level of error can be as high as 50% with certain codons. Hence, the precision of translation can vary from one codon to another by at least 5000-fold.

A comprehensive review of suppression is available (41), so this topic will not be covered here. *In vitro* studies show that the specificity of codon recognition is affected by the temperature of incubation, pH, concentration of various species of tRNA, concentration of Mg^{2+}, aliphatic amines such as putrescine, spermidine, and spermine, streptomycin and related antibiotics, and other compounds.

For example, studies with synthetic polynucleotide templates and cell-free protein synthesizing systems have shown that the accuracy of codon translation often is reduced at Mg^{2+} concentrations greater than 0.02 M (*85*). Usually only one base is translated erroneously per codon. However, Weinstein *et al.* (*96*) have shown that translation errors due to Mg^{2+} occur less frequently with mammalian extracts than bacterial extracts.

Both pH and temperature of incubation markedly affect the accuracy of translation. Grunberg-Manago and Michelson have shown that the level of error in translation is considerably higher at pH 7.8 than at 6.5–7.0 (*44*). The effect of incubation temperature upon the accuracy of translation was studied with protein synthesizing systems derived from *E. coli* (*85*) and from a thermophilic bacterium, *B. stearothermophilus*, that grows optimally at 65°C (*36*). In both cases, the most accurate translations were obtained at the optimal temperature for growth; the accuracy decreased at lower temperatures. The accuracy of codon translation is also dependent on the concentration of *free* AA-tRNA (*96*, or *43, 69*). For example, in the absence of Phe-tRNA, poly U stimulates the binding of Ile-tRNA to ribosomes. Essentially all molecules of Ile-tRNA are capable of binding to ribosomes in response to poly U. Species of tRNA probably compete with one another during the process of codon recognition. The accuracy of codon translation therefore may depend upon the relative affinities of species of tRNA with similar anticodons for mRNA codons and ribosomes and the concentration of each species of AA-tRNA. Phe-tRNA and Ile-tRNA probably compete for ribosomal binding sites in response to poly U; but the affinity of Phe-tRNA for UUU codons would be expected to be higher than that of Ile-tRNA assuming that Phe- and Ile-tRNA anticodons are GAA and GAU, respectively. In the presence of poly U the concentration of free Phe-tRNA may be greatly reduced due to the attachment of this species of tRNA to ribosomes. Hence the ratio of free Phe-tRNA to free Ile-tRNA can change dramatically during the course of incubation. The magnitude of the shift in the ratio of free to ribosome-bound AA-tRNA and the consequences of a shift in ratio are not widely appreciated.

Some species of tRNA are far more abundant than others. *Escherichia coli* W3100 contains little detectable Ile-tRNA corresponding to AUA and Leu-tRNA corresponding to UUA, whereas guinea pig liver is deficient in Arg-tRNA corresponding to AGA (*21*). Recently, the rate of protein synthesis directed by poly (A,G) in *E. coli* extracts was shown to be limited by the concentration of Arg-tRNA corresponding to AGA and AGG (*2*). Streptomycin-induced miscoding has been studied extensively (*42*). The available data show that streptomycin binds to the 30 S portion of streptomycin-sensitive ribosomes (*25, 29*) and thereby alters interactions between tRNA anticodons and mRNA codons. Hence codons are misread frequently. Uridine, in the first or second positions of an mRNA codon, often is read as either U, C, or A in the presence of streptomycin (*31*). Usually, only one base per triplet is misread.

Streptomycin inhibits the binding of Phe-tRNA to ribosomes in response to UpUpU or the trinucleotide with the 3′-terminal phosphate, UpUpUp, but stimulates binding of Phe-tRNA in response to the trinucleotide with 5′-terminal phosphate, pUpUpU (*68*).

Streptomycin can affect protein synthesis in different ways, depending on the codon recognized. In the presence of streptomycin, the frequency of translational error may increase and the rate of protein synthesis may either increase or decrease. For example, streptomycin stimulates both the rate and extent of polysine synthesis with poly A templates but inhibits protein synthesis directed by poly (A,C) or poly (A,G) (*32*). In the presence of streptomycin, little misreading is detected with poly A, poly (A,C), or poly (A,G), but with polynucleotides such as poly U, poly (U,C), or poly (U,G), translational errors are induced. Although the level of error in translation of poly A, poly (A,C), and poly (A,G) is low in the presence of streptomycin, alterations in the rate of protein synthesis may be due to streptomycin-induced changes in recognition of initiator, terminator, or other rate-limiting codons. Streptomycin may also occasionally alter the stability of [codon-ribosome-AA-tRNA] complexes.

Ribosomes from streptomycin resistant strains of *E. coli* are affected only slightly by streptomycin. Weinstein and his colleagues (*95*) have shown that 80 S mammalian ribosomes translate codons more accurately in the presence of streptomycin than sensitive 70 S *E. coli* ribosomes.

Aminoglycoside antibiotics related to streptomycin, such as neomycin, kanamycin, paramycin, and gentamycin, also affect the accuracy of codon translation; however, the antibiotics differ in the specificity of error that they induce. Presumably the ribosomal site or sites affected are changed in different ways by different aminoglycoside antibiotics.

The frequency of error induced by neomycin is ten times that induced by streptomycin. In the presence of neomycin, single-stranded DNA serves directly as a template for protein synthesis without an mRNA intermediate. Similarly, ribosomal RNA in the presence of neomycin serves as a template for protein synthesis. Spectinamycin inhibits protein synthesis but does not affect the accuracy of codon translation (*31*).

Kasugamycin, another aminoglycosydic antibiotic, inhibits protein synthesis and binding of AA-tRNA to ribosomes. Kasugamycin also inhibits protein synthesis directed by single-stranded DNA in the presence of neomycin or kanamycin. Kasugamycin, like spectinamycin, does not induce miscoding. Tanaka and his colleagues (*87*) have suggested that aminoglycosydic antibiotics that induce miscoding contain either deoxystreptamine or streptomine moieties (present in kanamycin, neomycin, paramycin, gentamycin, hygromycin B, and streptomycin). Kasugamycin and spectinamycin do not induce errors in translation and contain an N,N'-methylactinamine moiety.

Certain strains of bacteria are known to contain a gene that converts the cell to a form resistant to a variety of antibiotics. Resistance to multiple antibiotics is due, in certain cases at least, to the production of enzymes catalyzing the conversion of the antibiotics to inactive products.

REFERENCES

1. B. N. Ames and P. E. Hartman, The Histidine Operon. *Cold Spring Harbor Symp. Quant. Biol.* **28**, 349 (1963).

2. W. F. Anderson, The effect of tRNA concentration on the rate of protein synthesis. *Proc. Natl. Acad. Sci. U. S.* **62**, 566 (1969).

3. W. T. Astbury and F. O. Bell, X-ray study of thymonucleic acid. *Nature* **141**, 747 (1938).

4. A. N. Baldwin and P. Berg, Transfer ribonucleic acid-induced hydrolysis of valyladenylate bound to isoleucyl ribonucleic acid synthetase. *J. Biol. Chem.* **241**, 839 (1966).

5. E. S. Barghoorn and J. W. Schopf, Microorganisms three billion years old from the precambrian of South Africa. *Science* **152**, 758 (1966).

6. S. Benzer and S. P. Champe, A change from nonsense to sense in the genetic code. *Proc. Natl. Acad. Sci. U. S.* **48**, 1114 (1962).

7. M. R. Bernfield, Ribonuclease and oligoribonucleotide synthesis. I. Synthetic activity of bovine pancreatic ribonuclease derivatives. *J. Biol. Chem.* **240**, 4753 (1965).

8. M. R. Bernfield, Ribonuclease and oligoribonucleotide synthesis. II. Synthesis of oligonucleotides of specific sequence. *J. Biol. Chem.* **241**, 2014 (1966).

9. M. R. Bernfield and M. W. Nirenberg, RNA codewords and protein synthesis: The nucleotide sequences of multiple codewords for phenylalanine, serine, leucine, and proline. *Science* **147**, 479 (1965).

10. M. R. Bernfield and F. M. Rottman, Ribonuclease and oligoribonucleotide synthesis. III. Oligonucleotide synthesis with 5'-substituted uridine 2',3'-cyclic phosphates. *J. Biol. Chem.* **242**, 4134 (1967).

11. H. A. Bladen, R. Byrne, J. G. Levin, and M. W. Nirenberg, An electron microscopic study of a DNA-ribosome complex formed *in vitro*. *J. Mol. Biol.* **11**, 78 (1965).

12. H. Bremer and M. W. Konrad. A complex of enzymatically synthesized RNA and template DNA. *Proc. Natl. Acad. Sci. U. S.* **51**, 801 (1964).

13. S. Brenner, On the impossibility of all overlapping triplet codes in information transfer from nucleic acid to proteins. *Proc. Natl. Acad. Sci. U. S.* **43**, 687 (1957).

14. S. Brenner, L. Barnett, E. R. Katz, and F. H. C. Crick, UGA: A third nonsense triplet in the genetic code. *Nature* **213**, 449 (1967).

15. S. Brenner, A. O. W. Stretton, and S. Kaplan, Genetic code: The nonsense triplets for chain termination and their suppression. *Nature* **206**, 994 (1965).

16. M. S. Bretscher, Polypeptide chain termination: An active process. *J. Mol. Biol.* **34**, 131 (1968).

17. M. S. Bretscher, H. M. Goodman, J. R. Menninger, and J. D. Smith, Polypeptide chain termination using synthetic polynucleotides. *J. Mol. Biol.* **14**, 634 (1965).

18. R. Byrne, J. G. Levin, H. A. Bladen, and M. W. Nirenberg, The *in vitro* formation of a DNA-ribosome complex. *Proc. Natl. Acad. Sci. U. S.* **52**, 140 (1964).

19. P. C. Caldwell and C. Hinshelwood, Some considerations on autosynthesis in bacteria. *J. Chem. Soc. Part 4*, p. 3156 (1950).

20. M. R. Capecchi, Polypeptide chain termination *in vitro*: Isolation of a release factor. *Proc. Natl. Acad. Sci. U. S.* **58**, 1144 (1967).

21. C. T. Caskey, A. Beaudet, and M. Nirenberg, RNA codons and protein synthesis. 15. Dissimilar responses of mammalian and bacterial transfer RNA fractions to messenger RNA codons. *J. Mol. Biol.* **37**, 99 (1968).

22. C. T. Caskey, R. Tompkins, E. Scolnick, T. Caryk, and M. Nirenberg, Sequential translation of trinucleotide codons for the initiation and termination of protein synthesis. *Science* **162**, 135 (1968).

23. B. F. C. Clark and K. A. Marcker, The role of *N*-formyl-methionyl-sRNA in protein biosynthesis. *J. Mol. Biol.* **17**, 394 (1966).

24. "Cold Spring Harbor Symposium on Quantitative Biology–1966," Vol. 31. Cold Spring Harbor Lab. Quant. Biol., New York, 1967.

25. E. C. Cox, J. R. White, and J. G. Flaks, Streptomycin action and the ribosome. *Proc. Natl. Acad. Sci. U. S.* **51**, 703 (1964).

26. F. H. C. Crick, Codon-anticodon pairing. The wobble hypothesis. *J. Mol. Biol.* **19**, 548 (1966).

27. F. H. C. Crick, L. Barnett, S. Brenner, and R. J. Watts-Tobin, General nature of the genetic code for proteins. *Nature* **192**, 1227 (1961).

28. F. H. C. Crick, J. S. Griffith, and L. E. Orgel, Codes without commas. *Proc. Natl. Acad. Sci. U. S.* **43**, 416 (1957).

29. J. E. Davies, Studies on the ribosomes of streptomycin-sensitive and resistant strains of *Escherichia coli. Proc. Natl. Acad. Sci. U. S.* **51**, 659 (1964).

30. J. E. Davies, Streptomycin and the genetic code. *Cold Spring Harbor Symp. Quant. Biol.* **31**, 665 (1966).

31. J. E. Davies, L. Gorini, and B. D. Davis, Misreading of RNA codewords introduced by aminoglycoside antibiotics. *J. Mol. Pharmacol.* **1**, 93 (1965).

32. J. E. Davies, D. S. Jones, and H. G. Khorana, A further study of misreading of codons induced by streptomycin and neomycin using ribopolynucleotides containing nucleotides in alternating sequence as templates. *J. Mol. Biol.* **18**, 48 (1966).

33. B. P. Doctor, J. E. Loebel, and D. A. Kellogg, Studies on the species specificity

of yeast and *E. coli*. Tyrosine tRNAs. *Cold Spring Harbor Symp. Quant. Biol.* **31**, 543 (1966).

34. A. L. Dounce, Duplicating mechanism for peptide chain and nucleic acid synthesis. *Enzymologia* **15**, 251 (1952).

35. R. V. Eck, Genetic code: Emergence of a symmetrical pattern. *Science* **140**, 477 (1963).

36. S. M. Friedman and I. B. Weinstein, Protein synthesis in a subcellular system from *Bacillus Stearothermophilus. Biochim. Biophys. Acta* **114**, 593 (1966).

37. G. Gamow, Possible relation between deoxyribonucleic acid and protein structures. *Nature* **173**, 318 (1954).

38. G. Gamow, A. Rich, and M. Ycas, In: The problem of information transfer from the nucleic acids to proteins. *Advan. Biol. Med. Phys.* **4**, 23 (1956).

39. M. C. Ganoza and T. Nakamoto, Studies on the mechanism of polypeptide chain termination in cell-free extracts of *E. coli. Proc. Natl. Acad. Sci. U. S.* **55**, 162 (1966).

40. A. Garen, Sense and nonsense in the genetic code. *Science* **160**, 149 (1968).

41. L. Gorini and J. R. Beckwith, Suppression. *Ann. Rev. Microbiol.* **20**, 401 (1966).

42. L. Gorini, A. Jacoby, and L. Breckenridge, Ribosomal ambiguity. *Cold Spring Harbor Symp. Quant. Biol.* **31**, 657 (1966).

43. M. Grunberg-Manago and J. Dondon, Influence of pH and sRNA concentration on coding ambiguities. *Biochem. Biophys. Res. Commun.* **18**, 517 (1965).

44. M. Grunberg-Manago and A. M. Michelson, Polynucleotide analogues. II. Stimulation of amino acid incorporation by polynucleotide analogues. *Biochim. Biophys. Acta* **80**, 431 (1964).

45. D. Hatfield, Oligonucleotide-ribosome-AA-sRNA interactions. *Cold Spring Harbor Symp. Quant. Biol.* **31**, 619 (1966).

46. R. T. Hinegardner and J. Engelberg, Comment on: Universality in the genetic code by C. R. Woese. *Science* **144**, 1031 (1964).

47. M. Ikehara and E. Ohtsuka, Stimulation of the binding of aminoacyl-sRNA to ribosomes by tubercidin (7-deazoadenosine) and N^6-dimethyladenosine containing trinucleoside diphosphate analogs. *Biochem. Biophys. Res. Commun.* **21**, 257 (1965).

48. H. A. Itano, The synthesis and structure of normal and abnormal hemoglobins. *In* "Abnormal Haemoglobins in Africa" (J. H. P. Jonxis, ed.), pp. 3–16. Blackwell, Oxford, 1965.

49. T. H. Jukes, "Molecules and Evolution." Columbia Univ. Press, New York, 1966.

50. T. Kameyama and G. D. Novelli, The cell-free synthesis of β-galactosidase by *Escherichia coli. Biochem. Biophys. Res. Commun.* **2**, 393 (1960).

51. D. A. Kellogg, B. P. Doctor, J. E. Loebel, and M. W. Nirenberg, RNA codons and protein synthesis. IX. Synonym codon recognition by multiple species of valine-, alanine-, and methionine-sRNA. *Proc. Natl. Acad. Sci. U. S.* **55**, 912 (1966).

52. H. G. Khorana, Synthesis in the study of nucleic acids. *Biochem. J.* **109**, 709 (1968).

53. H. G. Khorana, H. Büchi, H. Ghosh, N. Gupta, T. M. Jacob, H. Kössel, R. Morgan, S. A. Narang, E. Ohtsuka, and R. D. Wells, Polynucleotide synthesis and the genetic code. *Cold Spring Harbor Symp. Quant. Biol.* **31**, 39 (1966).

54. H. Kössel, Studies on polynucleotides. LXXXIII. Synthesis *in vitro* of the

244   M. NIRENBERG

tripeptide valyl-seryl-lysine directed by poly r(G-U-A-A). *Biochim. Biophys. Acta* **157**, 91 (1968).

55. J. A. Last, W. M. Stanley, Jr., M. Salas, M. B. Hille, A. J. Wahba, and S. Ochoa, Translation of the genetic message. IV. UAA as a chain termination codon. *Proc. Natl. Acad. Sci. U. S.* **57**, 1062 (1967).
56. P. Leder, M. F. Singer, and R. L. C. Brimacombe, Synthesis of trinucleoside diphosphates with polynucleotide phosphorylase. *Biochemistry* **4**, 1561 (1965).
57. C. Levinthal, E. R. Signer, and K. Fetherolf, Reactivation and hybridization of reduced alkaline phosphatase. *Proc. Natl. Acad. Sci. U. S.* **48**, 1230 (1962).
57a. F. Lipmann, Polypeptide chain elongation in protein biosynthesis. *Science* **164**, 1024 (1969).
58. K. Marcker and F. Sanger, N-formyl-methionyl-sRNA. *J. Mol. Biol.* **8**, 835 (1964).
59. R. E. Marshall, C. T. Caskey, and M. Nirenberg, Fine structure of RNA codewords recognized by bacterial, amphibian, and mammalian transfer RNA. *Science* **155**, 820 (1967).
60. R. E. Marshall and M. Nirenberg, RNA codons recognized by transfer RNA from amphibian embryos and adults. *Develop. Biol.* **19**, 1 (1969).
61. J. H. Matthaei and M. W. Nirenberg, Characteristics and stabilization of DNAase-sensitive protein synthesis in *E. coli* extracts. *Proc. Natl. Acad. Sci. U. S.* **47**, 1580 (1961).
62. G. Milman, J. Goldstein, E. Scolnick, and T. Caskey, Peptide chain termination. III. Stimulation of *in vitro* termination. *Proc. Natl. Acad. Sci. U. S.* **63**, 183 (1969).
63. A. R. Morgan, R. D. Wells, and H. G. Khorana, Studies on polynucleotides. LIX. Further codon assignments from amino acid incorporations directed by ribopolynucleotides containing repeating trinucleotide sequences. *Proc. Natl. Acad. Sci. U. S.* **56**, 1899 (1966).
64. M. Nirenberg, "The Genetic Code." Prix Nobel, 1969 (in press).
64a. M. Nirenberg, T. Caskey, R. Marshall, R. Brimacombe, D. Kellogg, B. Doctor, D. Hatfield, J. Levin, F. Rottman, S. Pestka, M. Wilcox, and F. Anderson, The RNA code and protein synthesis. *Cold Spring Harbor Symp. Quant. Biol.* **31**, 11 (1966).
65. M. Nirenberg and P. Leder, RNA codewords and protein synthesis. I. The effect of trinucleotides upon the binding of sRNA to ribosomes. *Science* **145**, 1399 (1964).
66. M. W. Nirenberg and J. H. Matthaei, The dependence of cell-free protein synthesis in *E. coli* upon naturally occurring or synthetic polyribonucleotides. *Proc. Natl. Acad. Sci. U. S.* **47**, 1588 (1961).
67. B. Nisman and H. Fukuhara, Incorporation des acides amines et synthèse de la β-galactosidase parles fractions enzymatiques de *Escherichia coli. Compt. Rend.* **249**, 2240 (1959).
68. S. Pestka, The action of streptomycin on protein synthesis *in vitro. Bull. N. Y. Acad. Sci.* [2] **43**, 126 (1967).
69. S. Pestka, R. Marshall, and M. Nirenberg, RNA codewords and protein synthesis. V. Effect of streptomycin on the formation of ribosome-sRNA complexes. *Proc. Natl. Acad. Sci. U. S.* **53**, 639 (1965).
70. S. Pestka and M. Nirenberg, Codeword recognition of 30 S ribosomes. *Cold Spring Harbor Symp. Quant. Biol.* **31**, 641 (1966).
71. F. Rottman and M. Nirenberg, RNA codons and protein synthesis. XI. Template activity of modified RNA codons. *J. Mol. Biol.* **21**, 555 (1966).

72. A. S. Sarabhai, A. O. W. Stretton, S. Brenner, and A. Bolle, Co-linearity of the gene with the polypeptide chain. *Nature* **201**, 13 (1964).

73. E. Scolnick, R. Tompkins, T. Caskey, and M. Nirenberg, Release factors differing in specificity for terminator codons. *Proc. Natl. Acad. Sci. U. S.* **61**, 768 (1968).

74. D. H. Shin and K. Moldave, Effect of ribosomes on the biosynthesis of ribonucleic acid *in vitro*. *J. Mol. Biol.* **21**, 231 (1966).

75. M. F. Singer, O. W. Jones, and M. W. Nirenberg, The effect of secondary structure on the template activity of polyribonucleotides. *Proc. Natl. Acad. Sci. U. S.* **49**, 392 (1963).

76. R. L. Sinsheimer, Is the nucleic acid message in a two-symbol code? *J. Mol. Biol.* **1**, 218 (1959).

77. D. Söll, J. D. Cherayil, and R. M. Bock, Studies on polynucleotides. LXXV. Specificity of tRNA for codon recognition as studied by the ribosomal binding technique. *J. Mol. Biol.* **29**, 97 (1967).

78. D. Söll, D. S. Jones, E. Ohtsuka, R. D. Faulkner, R. Lohrmann, H. Hayatsu, H. G. Khorana, J. D. Cherayil, A. Hampel, and R. M. Bock, Specificity of sRNA for recognition of codons as studied by the ribosomal binding technique. *J. Mol. Biol.* **19**, 556 (1966).

79. D. Söll, E. Ohtsuka, D. S. Jones, R. Lohrmann, H. Hayatsu, S. Nishimura, and H. G. Khorana. Studies on polynucleotides. XLIX. Stimulation of the binding of aminoacyl-sRNA's to ribosomes by ribotrinucleotides and a survey of codon assignments for 20 amino acids. *Proc. Natl. Acad. Sci. U. S.* **54**, 1378 (1965).

80. D. Söll and U. L. RajBhandary, Studies on polynucleotides. LXXVI. Specificity of transfer RNA for codon recognition as studied by amino acid incorporation. *J. Mol. Biol.* **29**, 113 (1967).

81. T. M. Sonneborn, Degeneracy of the genetic code: Extent, nature, and genetic implications. *In* "Evolving Genes and Proteins" (V. Bryson and H. J. Vogel, eds.), p. 377. Academic Press, New York, 1965.

82. G. S. Stent, The Operon: On its third anniversary. *Science* **144**, 816 (1964).

83. A. R. Subramanian, E. Z. Ron, and B. D. Davis, A factor required for ribosome dissociation in *Escherichia coli*. *Proc. Natl. Acad. Sci. U. S.* **61**, 761 (1968).

84. N. Sueoka and T. Kano-Sueoka, A specific modification of leucyl-sRNA of *Escherichia coli* after phage T2 infection. *Proc. Natl. Acad. Sci. U. S.* **52**, 1535 (1964).

85. W. Szer and S. Ochoa, Complexing ability and coding properties of synthetic polynucleotides. *J. Mol. Biol.* **8**, 823 (1964).

86. M. Takanami and Y. Yan, The release of polypeptide chains from ribosomes in cell-free amino acid-incorporating systems by specific combinations of bases in synthetic polyribonucleotides. *Proc. Natl. Acad. Sci. U. S.* **54**, 1450 (1965).

87. N. Tanaka, H. Masukawa, and H. Umezama, Structural basis of kanamycin for miscoding activity. *Biochem. Biophys. Res. Commun.* **26**, 544 (1967).

88. R. E. Thach and P. Doty, Synthesis of block oligonucleotides. *Science* **147**, 1310 (1965).

89. A. Tissieres, D. Schlessinger, and F. Gros, Amino acid incorporation into proteins by *Escherichia coli* ribosomes. *Proc. Natl. Acad. Sci. U. S.* **46**, 1450 (1960).

90. E. Volkin and L. Astrachan, Phosphorus incorporation in *Escherichia coli* ribonucleic acid after infection with bacteriophage T2. *Virology* **2**, 149 (1956).

91. G. von Ehrenstein and F. Lipmann, Experiments on hemoglobin biosynthesis. *Proc. Natl. Acad. Sci. U. S.* **47**, 941 (1961).

92. J. D. Watson and F. H. C. Crick, Molecular structure of nucleic acids. A structure for deoxyribose nucleic acid. *Nature* **171,** 737 (1953).

93. M. G. Weigert and A. Garen, Base composition of nonsense codons in *E. coli.* Evidence from amino acid substitutions at a tryptophan site in alkaline phosphatase. *Nature* **206,** 992 (1965).

94. M. G. Weigert, E. Lanka, and A. Garen, Base composition of nonsense codons in *Escherichia coli.* II. The N₂ codon UAA. *J. Mol. Biol.* **23,** 391 (1967).

95. I. B. Weinstein, S. M. Friedman, and M. Ochoa, Jr., Fidelity during translation of the genetic code. *Cold Spring Harbor Symp. Quant. Biol.* **31,** 671 (1966).

96. I. B. Weinstein, M. Ochoa, Jr., and S. M. Friedman, Fidelity in the translation of messenger ribonucleic acids in mammalian subcellular systems. *Biochemistry* **5,** 3332 (1966).

97. B. Weisblum, S. Benzer, and R. W. Holley, A physical basis for degeneracy in the amino acid code. *Proc. Natl. Acad. Sci. U. S.* **48,** 1449 (1962).

98. C. Woese, Nature of the biological code. *Nature* **194,** 1114 (1962).

99. C. R. Woese, Historical development. *In* "The Genetic Code" (H. O. Halvorson, H. L. Roman, and E. Bell, eds.), Chapter 2. Harper, New York, 1967.

100. C. Yanofsky and P. St. Lawrence, Gene action. *Ann. Rev. Microbiol.* **14,** 311 (1960).

V

Biosynthetic Reactions in the Cell Nucleus

Vincent G. Allfrey

I. INTRODUCTION

All cells of higher organisms, with few exceptions, have a well-defined and functional nucleus. The nucleus is our subject—to be examined in terms of its composition, enzymic activity, ultrastructure, and biological function. Our aim is to depict the structure and function of the nucleus in chemical terms—to relate its biosynthetic capacities to its biological role—to explain how this central commanding subcellular organelle directs and chooses the events to take place in the cytoplasm, and why it responds, in turn, to the changing demands of the cell and the environment.

The nucleus is, first and foremost, a repository for DNA and the major site of ribonucleic acid synthesis in the cell. These biochemical facts alone establish it as a control center for virtually all the other biosynthetic activities of the cell. But it should be pointed out that the importance of a nucleus to cell life and function was recognized by biologists many years ago. E. B. Wilson, in his classic work, *The Cell in Development and Heredity* (*97*), brilliantly summarized much of the early evidence that "the nucleus plays an important part in organic synthesis." He cited early experiments on the microdissection of unicellular organisms, particularly those of Brandt in 1877 and of M. Nussbaum in 1884, comparing the nucleated and nonnucleated halves of protozoans. "The nucleated fragments quickly heal the wound, regenerate the missing portions and thus produce a perfect animal, while enucleated fragments, consisting of cytoplasm only, quickly perish."

In 1896, F. R. Lillie observed that the large ciliate *Stentor polymorphus* could be shaken into fragments of all sizes and that nucleated fragments as small as 1/27 of the volume of the entire animal were still capable of complete regeneration; all nonnucleated fragments, regardless of size, died. Five years later, T. H. Morgan, experimenting along the same lines, reported that nucleated fragments not more than 1/64 of the normal cell size might still produce a complete individual. Similar experiments indicating the need for a nucleus were carried out on rhizopods, algae, and plant cells. In plants, some experiments made by C. O. Townsend in 1897 were of particular interest. He studied the effects of tearing the membrane in cells of the root hairs of *Cucurbita* and found that the influence of the nucleus—as evidenced by the capacity of these cells to repair the damaged membrane—could extend from a nucleated to a nonnucleated cell provided they were joined by intercellular bridges. It is no wonder that E. B. Wilson concluded . . . "The nucleus must, therefore, play an essential part in the operations of synthetic metabolism or chemical synthesis."

These impressive demonstrations of the role of the nucleus in the synthetic activities of interphase cells were matched by the simultaneous development of genetic theory. The view that hereditary characters are transmitted by the nucleus was advanced independently in 1884 by O. Hertwig and E. Strasburger. They based their argument on the equivalence of gametic nuclei in fertilization, noting that both parent germ cells make equivalent contributions to the hereditary endowments of their offspring, despite the great discrepancies in the amount of cytoplasm in oocyte and sperm cells. Arguments of this sort led to widespread acceptance of the nuclear theory of heredity before 1900. From 1900 on, the theory became a certainty as cytological studies of chromosome movements during mitosis and meiosis offered the physical basis underlying Mendel's laws of heredity. Once genetic mapping on chromosomes became a reality in the fruit fly *Drosophila*, in maize, and in other organisms, it became apparent that the nucleus, as a repository of the chromosomes, was the seat and source of genetic information. Moreover, there was good reason to believe that the nucleic acids (nucleins) discovered by Friedrich Miescher in 1869–71 were essential elements of nuclear structure and that they were localized in the chromatin. Long before the turn of the century it had been both surmised and stated that the nucleic acids were responsible for the transmission of hereditary characteristics. Thus the groundwork was laid relating the nucleus, the chromosomes, and the nucleic acids to the transmission and expression of hereditary factors.

The morphological basis of "the flow of genetic information" was clear.

The evolution of these concepts in chemical terms proceeded slowly at first, gaining in momentum with the simultaneous growth of biochemical knowledge and with the introduction of new and powerful experimental techniques. Consider the technical revolution represented by isotopic labeling procedures, cell fractionation, chromatography, ultracentrifugal analysis, electron microscopy, autoradiography, and X-ray diffraction techniques. These methods ushered in the series of important discoveries relating the nucleic acids to the synthesis of protein, and, eventually, leading to the explanation of biological specificity in terms of nucleotide sequences in DNA and RNA.

This latter history, the advent of molecular biology, owes much of its origins to pioneering studies of cells of higher organisms. For example, the requirement for ribonucleic acids in protein synthesis was first indicated by the observations of Caspersson in 1941 and of Brachet in 1942 who were the first to point out the striking correlations between the RNA contents of cells in different tissues and their capacities for protein synthesis (15, 16). From 1950 onward, important discoveries came in rapid succession: the localization of newly synthesized protein on microsomes; the development of *in vitro* systems for the study of the mechanism of protein synthesis; the demonstration that ribonuclease would inhibit amino acid incorporation by ribosomes; the discovery and characterization of ribosomes and their visualization under the electron microscope; the discovery of the amino acid "activating" enzymes, of transfer RNA's, and of the ATP and GTP requirements for protein synthesis; the finding that protein synthesis begins at the amino terminal end and proceeds toward the carboxyl terminal end; the isolation and visualization of polysomes; the discovery of RNA polymerase and its DNA dependence— all were first achieved in studies of cell function in higher organisms.

In addition, the demonstration that a single gene mutation in humans with sickle-cell anemia gave rise to an alteration in the amino acid sequence of the hemoglobin of their red blood cells (45) was of the greatest importance because it linked nucleotide sequences in DNA to amino acid sequences in a particular polypeptide chain. It also suggested that the role of DNA in controlling protein synthesis must be an indirect one, probably involving RNA as an intermediate, because the observation was made that mammalian erythrocytes continue to synthesize hemoglobin after they have discarded the nucleus, but only as long as functioning ribonucleoprotein particles are present in their cytoplasms. When this was considered together with the observations that chemical alteration (by nitrous acid) of a single base in the RNA of tobacco mosaic virus could cause a single amino acid substitution in the viral coat protein, the role of nucleic acids in specifying amino acid sequences became clearer. The culmination of such studies—the assignment of nucleotide triplets to each amino acid—came, however, from quite a different source, the study of the coding properties of synthetic polynucleotides and oligonucleotides in cellfree systems derived from bacteria. (See Chapter IV.)

The many discoveries made in the course of study of animal and plant cells have been stressed here because of a growing lack of awareness of the contributions made by research on higher organisms to the current state of molecular biology. However, the point to be made is not that higher cells are better than simpler forms, but rather that all living systems have their uses and intrinsic values in research. Many contributions which have great relevance to the study of nuclear function have come with increasing frequency from the work on the genetics and biochemistry of microorganisms and viruses, among them the demonstration that hereditary transformation of bacterial cells can be effected by DNA and nothing else; the discovery of the DNA-dependent enzymic mechanism for DNA synthesis; the detection of specific DNA-like RNA's produced after infection by DNA phages; the experimental and theoretical analysis of bacterial enzyme induction and repression; the discovery of specific sites and mechanisms of feedback control; the discovery of suppressor mutations affecting transfer RNA structure and coding characteristics; the proof of the colinearity of DNA nucleotide sequences and the order of amino acids in proteins; and the isolation of specific gene repressor proteins. All have helped set the stage for the further investigation of the mechanisms by which DNA directs the synthetic reactions in the nuclei of higher cells and is, itself, subject to control.

The simultaneous growth of genetic and biochemical insights has revolutionized experimental approaches to the major problems in cell biology. It has introduced an incisiveness of purpose in experimental design coupled to techniques of unparalleled sensitivity and resolution, such as the combination of isotopic labeling procedures and the hybridization of complementary DNA and RNA strands. Techniques of density-gradient centrifugation and newer chromatographic and electrophoretic procedures have made possible separations of DNA's and RNA's; of polysomes, ribosomes, and ribosomal subunits; of viruses; and of nucleoli and chromatin fractions differing in their RNA synthetic capacities. At the cytological level, autoradiography under the electron microscope, new methods of quantitative histochemistry, and new approaches made possible by cell culture techniques have introduced high precision to the study of the localization and timing of intracellular processes. The results have, if anything, reinforced the earlier conviction that the cell nucleus is a unique center of biosynthetic activity, and, given the fact that the genetic information encoded in deoxyribonucleic acid is localized almost exclusively in the cell nucleus, nuclear biochemistry has extended its relevance to nearly all areas of cell biology.

As will be shown below, the chemical responsibilities of the nucleus are enormous. Utilizing the information encoded in the nucleotide sequences of its DNA, and drawing upon the cytoplasm for substrates and other essentials, we see that the nucleus must turn out most of the synthetic apparatus necessary for protein synthesis in the cell. This includes the ribosomes, the messenger RNA's, the amino acid transfer RNA's, and a host of regulatory factors which determine which genes will be functional and when; which patterns of growth and metabolism to follow; which hormones to respond to; which substrates, enzymes, and coenzymes are required at a particular time; and which products must be synthesized and exported to meet the demands of other cells in the organism. In brief, the nucleus must direct the present and future per- formance of the cell. It is the nucleus that tells the pancreatic islet cells to make insulin, while other nuclei instruct pancreatic acinar cells to synthesize digestive enzymes. It is the nucleus that controls antibody synthesis in the plasma cell, albumin synthesis in liver, and myosin synthesis in muscle. It is the nucleus that dominates the patterns of growth and development after the blastula stage in morphogenesis. Of course, nuclei do not do this alone, nor are they immune from environ- mental influences, despite the simplistic notion that DNA makes RNA which makes protein. The biosynthetic activities of the nucleus are continually dependent upon the cytoplasm and ever responsive to the changing demands of the organism and its environment. These demands affect both the form and the function of the genetic material and of the accessory nuclear structures involved in its operation.

II. ELEMENTS OF NUCLEAR STRUCTURE

In considering synthetic reactions which take place within nuclei, emphasis will be placed on the close relationship between ultrastructure and biochemical function. Enzymology alone will not suffice to explain the more complex aspects of nuclear activity such as, for example, the selective patterns of gene activation and repression in different cells; nor will a catalog of chemical reactions known to occur in nuclei com- pletely illuminate the way in which so many diverse activities become organized into a precisely regulated biological whole. In the nucleus, as in all biological systems, chemical reactivity is intimately geared to the physical state, and it is restricted by conditions which are often quite peripheral.

This is illustrated most graphically by the importance of intracellular membrane systems—barriers to free flow which have evolved to maintain highly specialized internal environments, which limit and direct the transport of metabolites, and which serve to bar the passage of unwanted molecules and ions. The types of metabolic control exerted by the nuclear envelope will be mentioned below, but other considerations are equally important.

The molecules which take part in nuclear reactions do not always occur in free solution, nor are they uniformly dispersed throughout a homogeneous matrix. On the contrary, the molecules of greatest interest, DNA, histones, phosphoproteins, RNA's, and ribosomal subunits exist in complex and inhomogeneous distributions. They may be sequestered into pools, taken out of phase, packed in fibrous, paracrystalline, or supercoiled arrays, assembled into complex interacting particles, and affected in form and function by local fluctuations in pH or surface charge density. To add to this complexity, most of the components of the nucleus (with the exception of DNA) are subject to continuous breakdown, replacement, and decay. Many components never leave the nucleus; others are destined for the cytoplasm; and still others leave the cytoplasm and enter the nucleus.

The molecular biology of the nucleus must attempt to take these structural factors into account, especially in dealing with the nucleic acids, whose molecular weights range into the millions and whose extended linear dimensions would be measurable in hundreds of micra—too long to fit into a nucleus without extensive coiling and folding. The organization of these macromolecules into chromosomes or chromatin fibrils, nucleoli, ribosomes, etc., constitutes the morphological basis of ordered nuclear activity and is an important key to the understanding of nuclear RNA synthesis and its control. Much of what follows is an attempt to relate these details of nuclear morphology to biochemical activity, and vice versa, stressing the fact that in a dynamic living system neither structure nor composition are inflexible, permanent, or invariable aspects of cell design.

A. Morphology of the Nucleus

The interphase nucleus, as seen in the light microscope, shows 3 characteristic components, referred to by classical cytologists as the chromatin, the nucleolus, and the nuclear sap or ground substance.

1. *Chromatin*

When cells are treated with basic dyes, such as Crystal Violet, the nuclei take on a characteristically heavy strain, revealing clumps, fibrils, and granular material described in the early literature as chromatin. With the introduction of the Feulgen reaction, which is specific for deoxyribonucleic acid, the presence of DNA in chromatin was revealed, and the term is now used to refer to DNA-containing regions of the nucleus in which the chromosomes exist in their interphase state. The physical state of the chromatin is probably best described as a mass of intertwined fibrils, some extended, some tightly coiled, that differ in thickness and are often condensed into the tight bundles typical of the chromosomes themselves. Though it is not usual to find in the nucleus the discrete chromosomal structures seen in dividing cells at metaphase, portions of the chromosomes may remain condensed during the interphase period—these are often called the heterochromatin. As will be shown below, this form of the chromatin is probably inactive in RNA synthesis.

Some DNA, together with associated proteins and RNA, is dispersed throughout other regions of the nucleus as a more diffuse, lightly staining form of chromatin, sometimes called euchromatin. DNA in this form has been shown to be active as a template for RNA synthesis (*52*) and also for DNA replication (*35*).

The morphological distinction between the two forms of chromatin is not a permanent one, since all the chromatin becomes condensed at cell division, when the chromosomes become visible, and because DNA replication requires at least a temporary uncoiling of the DNA double helix at the region of the replication fork. Indeed, it is likely that an active mechanism exists for converting the chromatin from one state to the other.

Visualization of the chromatin under the electron microscope depends a good deal on the techniques of fixation and staining employed. However, studies of thin nuclear sections under the electron microscope and adjacent thick sections under the light microscope have shown that the electron-dense regions of the chromatin correspond to DNA-rich regions of the nucleus as judged by the intensity of Feulgen staining or ultraviolet absorption. Differences in texture and density in different regions of the nucleus, corresponding to the condensed and diffuse states of chromatin, are readily recognized in electron micrographs when nuclei are fixed in glutaraldehyde and stained with uranyl ions. This is shown in Fig. 1(A). A high-power view showing the continuity of the nucleoprotein fibrils leaving the condensed chromatin clumps and entering the diffuse chromatin region is shown in Fig. 1(B). Thus, deoxyribonucleo-

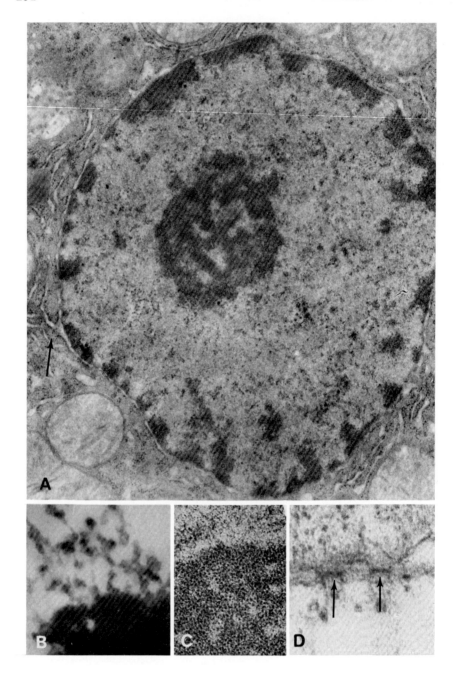

protein has a filamentous form wherever it occurs in the nucleus. The finest filaments are about 50 Å in diameter, a dimension not incompatible with the diameter of the DNA double helix (about 20 Å) when it is surrounded by associated proteins.

The appearance of the chromatin, as well as the size and shape of the nucleus, may undergo extensive changes during the differentiation of a given cell type. The change reflects a changing capacity for nuclear RNA synthesis as well. This is well illustrated in the course of maturation of erythroblasts in the bone marrow. Much of the chromatin in the metabolically active, primitive erythroblast exists in the diffuse extended state. This proportion diminishes as the cell ages, and, at later stages, the nuclear chromatin is nearly fully condensed (25). In mammals, the nucleus is then extruded from the cell, leaving an erythrocyte. In birds, fish, and amphibians, the nucleus is retained by the red blood cell. Both mammalian erythrocytes, which have lost their nuclei, and bird erythrocytes, which are nucleated but in which the chromatin is highly condensed, have lost the capacity for RNA synthesis. In the former case the loss is irreversible, as would be expected for an enucleated cell. However, it is an interesting fact that the bird nucleus may resume RNA synthesis when it is transferred by hybridization experiments to the cytoplasm of macrophages or HeLa cells (34).

2. Nucleolus

The nucleolus, as seen in the light microscope, is a refractile rounded body usually placed eccentrically in the nucleus and sometimes closely associated with the nuclear envelope. Most cell types have 1 to 3 such structures in their nuclei, but a few cells, oocytes in particular, may have hundreds of highly specialized and independent nucleoli.

Characteristically, nucleoli stain intensely in histochemical tests for

---

FIG. 1. Nuclear morphology as seen in the electron microscope. (A) Section of a rat liver nucleus *in situ*, showing the nucleolus, the condensed and diffuse states of the chromatin, the ribosomes on the outer nuclear membrane, and the continuity of the outer nuclear membrane with the membrane system of the cytoplasm (arrow). Micrograph by Dr. V. C. Littau. Nuclei fixed in glutaraldehyde followed by OsO₄ (22,000 ✕). (B) High-power magnification of the chromatin in isolated calf thymus nuclei. Note the continuity of the fibrils leaving the condensed chromatin clump and entering the diffuse chromatin region. Micrograph by Dr. V. C. Littau. Nuclei fixed in OsO₄ buffered in neutral phosphate (100,000 ✕). (C) Ribosomal particles (or subunits) in the nucleolus of the leech ganglion cell. Micrograph by Dr. D. W. Fawcett (25). (D) Cross section through pores of the nuclear membrane of the frog oocyte. Note the septum bridging the discontinuity in the nuclear envelope (arrows). Micrograph by Dr. V. C. Littau. Nuclei fixed in OsO₄ buffered in veronal acetate (88,000 ✕).

RNA but fail to give a strong Feulgen reaction for DNA. The staining intensity is largely a matter of relative concentrations, for electron microscopy reveals that the finely dispersed chromatin does extend into the nucleolar regions, and it is known from genetic and cytological studies that the DNA in a particular part of the chromosome, the so-called nucleolar-organizing region, has control over the appearance and function of the nucleolus. In oocytes, each of the many independent nucleoli has its own fragment of DNA which presumably directs the synthesis of ribosomal RNA's needed in great quantity for the developing egg (62).

Electron micrographs of thin nuclear sections often show the nucleolus as consisting of a threadlike element (the nucleolonema) loosely coiled about a virtually structureless zone called the pars amorpha. Higher magnifications reveal the nucleolonema to be a fine-textured matrix in which are embedded many dense granules about 150 Å in diameter [see Fig. 1(C)]. These granules resemble ribosomal particles in the cytoplasm, and there is good reason to believe that they are, in fact, ribosomal subunits or precursor particles. Much of the mass of the nucleolus is composed of protein, some of which may represent an accumulation of ribosomal structural proteins. The nucleolus is the main site of ribosomal RNA synthesis, as will be discussed below in detail.

The nucleolus is not membrane enclosed. Its size and shape vary with the metabolic state of the cell. Indeed, the volume of the nucleolus in plant cells has been observed to fluctuate depending upon exposure of the plant to light or darkness. When RNA synthesis in the nucleus is selectively inhibited, the normal nucleolar structure breaks down. Thus, the functional activity of this organelle and its structure appear to be closely interdependent.

When cells divide, the nuclear membrane breaks down to make way for chromosome movements. At this time, the contents of the nucleolus are discharged into the cytoplasm. A nucleolar remnant may persist, but, as a rule a new nucleolus makes its appearance in each daughter cell once mitosis is completed. Though it is clear that many ribosomes are released into the cytoplasm of dividing cells in this way, the precise mechanism of ribosome transfer from the nucleus to the cytoplasm of nondividing cells remains to be established, though there is good reason to believe that the ribosomal subunits are assembled and released in different ways (see below).

3. Nuclear Envelope

Cell nuclei are enclosed by a complex double-membrane system which serves to regulate the exchange of materials between nucleus and cytoplasm. Each of the nuclear membranes is about 75 Å thick and is separated from its partner by a space (called the perinuclear cisterna) that ranges in width from 400–700 Å.

The outer nuclear membrane can often be seen to be continuous with the membrane system of the cytoplasm (the so-called endoplasmic reticulum). [See Fig. 1(A), arrow.] Like many of the cytoplasmic membranes, the outer membrane of the nucleus is studded with ribonucleoprotein particles. These perinuclear ribosomes are evident in electron micrographs of many types of nuclei and can be seen clearly in Fig. 1(A). Their presence sometimes complicates attempts to study the formation and biosynthetic properties of ribosomes within the nucleus. For this reason, the most convincing studies of intranuclear ribosomes have employed techniques which remove the outer nuclear membrane (82), or, alternatively, the perinuclear ribosomes have been detached and selectively degraded by exposing isolated nuclei to chelating agents and ribonuclease (60).

The morphology of the nuclear envelope includes a number of other distinctive features, perhaps the most striking of which is the presence of large numbers of apparent pores, each about 660–780 Å in diameter (30). These are clearly seen in tangential sections of the nuclear membrane and are dramatically highlighted by shadow-casting of isolated nuclear membranes. (See Fig. 2.) Nuclear pore complexes were first seen

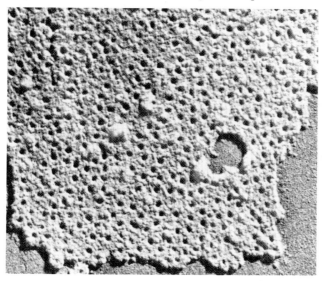

FIG. 2. Electron micrograph of nuclear envelopes isolated from *Allium* root tip nuclei. Air-dried and shadowed with gold and palladium. Micrograph by Dr. W. W. Franke (27a). (26,000 ×.)

in oocyte nuclei. Recent studies of oocytes of the newt, starfish, and frog indicate that each pore is actually octagonal in outline and has an associated annulus of about 1200 Å outer diameter (*30*).

The term nuclear pore is sometimes misleading because it implies an open channel of communication between the nucleus and the cytoplasm. Actually, the pore appears to be closed by a septum or diaphragm which joins the outer and inner nuclear membranes. This is seen in cross sections of the nuclear membrane [see Fig. 1(D), in which the septum is indicated by the arrow]. This diaphragm is sometimes seen with a prominent flange extending into the perinuclear cisterna.

Nuclear membranes exhibit a surprising selectivity in regulating the entry of proteins, small molecules, and ions into the nucleus. The electrical properties of the nuclear envelope in large cells can be studied with the aid of microelectrodes (*54*). Nuclei of *Drosophila* salivary gland cells, for example, have been found to maintain a resting potential more negative than that of the cytoplasm, and their resistance to the free diffusion of ions indicates that their pores are not simply open channels to the cytoplasm.

The permeability properties of the nuclear membrane will be discussed below in connection with amino acid transport reactions and the impermeability of nuclei to certain metabolites and inhibitors, but it should be stressed here that the complex morphology of the nuclear envelope is itself indicative of a highly selective mechanism for regulating nucleocytoplasmic interactions. The pore is, presumably, one of the more active avenues of communication between the nucleus and the cytoplasm it serves.

4. *Fibrous Lamina*

In many cell types the inner membrane of the nucleus is reinforced by a thick filamentous layer called the fibrous lamina (see Fawcett, *25*). This structure is sometimes very highly developed, as in the nucleus of *Amoeba proteus,* and in the ganglial cells of invertebrates, in which a honeycomb-like bed supports the nuclear membrane system. A cross-sectional view of this structure in the leech ganglion cell is shown in Fig. 3. The fibrous lamina has been found to exist in a variety of forms in the cells of both vertebrates and invertebrates. Though its role is not yet understood, it is presumed to be largely structural.

5. *Nuclear Ribosomes*

As mentioned above, ribosomal subunits or precursor particles of about 150 Å diameter are usually visible in electron micrographs of the nucleolus. Ribonucleoprotein (RNP) particles are also seen elsewhere in

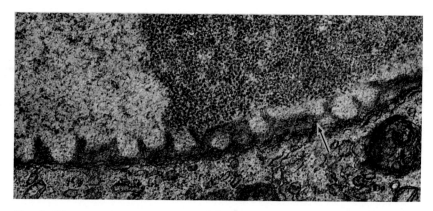

Fig. 3. Electron micrograph showing the fibrous lamina underlying the nuclear envelope in a ganglionic cell of the leech. Micrograph by Dr. D. W. Fawcett (*25*).

the nucleus, but it is usually difficult to identify the latter as functional ribosomes. Moreover, many RNP particles seen in nuclei are larger than free ribosomes (*65a*). The inner nuclear membrane often appears to contain an occasional cluster of ribosomal particles, indistinguishable in appearance from the ribosomes on the outer nuclear membrane. The morphological evidence can be misleading because cross sections through coils of chromatin fibrils may reveal an apparently granular structure resembling a row of ribosomes.

There is good biochemical evidence that intranuclear ribosomes do exist, and some of this will be presented below in connection with the mechanisms of protein synthesis in the nucleus. One of the most convincing electron micrographs of nuclear ribosomes *in situ* is that of Cronshaw, Hoefert, and Esau (*19*) showing high concentrations of ribosomes surrounding a viral inclusion within the nucleus of a leaf cell (Fig. 4).

III. ISOLATION OF CELL NUCLEI

Methods now exist for the bulk isolation of highly purified nuclei from a wide variety of animal and plant tissues; indeed, the investigator may often be faced with a choice of procedures for the isolation of nuclei from a particular tissue and must then consider questions of yield, retention of function, loss of contents, and reliability of the method. It should be emphasized at the outset that bulk isolations of nuclei and other subcellular components can be carried out in quite different ways, often

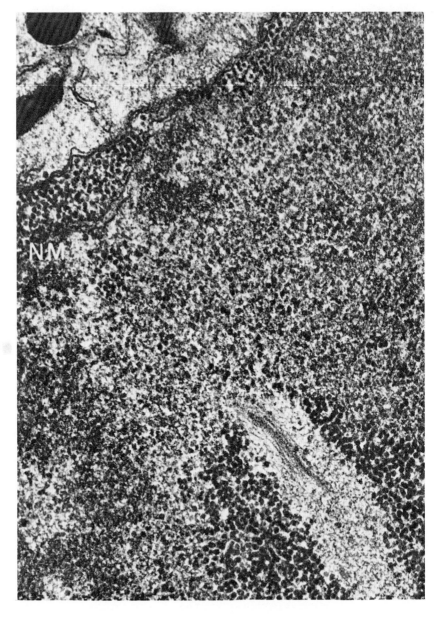

Fig. 4. Nuclear ribosomes in a section of the nucleus in a leaf cell of *Beta vulgaris* infected by beet yellows virus. The nucleus has an inclusion of fine fibrils surrounded by a clear region which, in turn, is surrounded by a region containing ribosome-like particles. NM = nuclear membrane (50,000 ×) [from Cronshaw *et al.* (*19*)].

with different results. For this reason, a brief outline of some of the major considerations affecting the choice of an isolation method will be given below. [More detailed treatments of the problems and rationale of cell fractionation technique are presented in recent reviews (e.g., *1*, *88*).]

All the major nuclear isolation procedures can be grouped into 2 categories—the aqueous and nonaqueous isolation methods—and each has its advantages and disadvantages. A knowledge of the limitations inherent in the methods is essential for critical work on the intracellular distribution of enzymes and is especially important in studies of nuclear function *in vitro*.

When cells are broken in an aqueous medium, e.g., in sucrose, citric acid, or salt solutions, 2 problems are introduced. The first is the extraction and loss of water-soluble components from the nucleus. The second problem is the transfer or exchange of soluble enzymes or other substances between the nucleus and the soluble phase of the homogenate. The ion-exchange properties of the basic proteins and the nucleic acids within the nucleus greatly complicate studies of the distribution of enzymes, substrates, and other adsorbable substances in tissue homogenates, The possibilities of loss and exchange of proteins, nucleic acids, small molecules, and ions are legion, depending upon the pH and composition of the medium and on the nature and concentration of the homogenate. A few examples of the pitfalls encountered will be cited below in connection with the comparison of aqueous and nonaqueous isolation procedures.

A. Isolation of Nuclei in Nonaqueous Media

The exchange of water-soluble components between the nucleus and the cytoplasm is precluded if the water is first removed from the tissue, and if all subsequent steps in the isolation are carried out in organic solvents. This is the rationale of the nonaqueous isolation methods. Such procedures were actually the first to be used for the large-scale isolation of subcellular components. After their introduction by Martin Behrens in 1932 (see Allfrey, *1*), they were employed to demonstrate the fact that virtually all the DNA of the cell can be recovered in the isolated nucleus. Nonaqueous isolation methods, as modified by later workers (see Allfrey, *1*), still constitute the basis for the most reliable estimates of nuclear composition and enzymatic activity.

Typical steps in the isolation are as follows: (1) Thin slices of tissue or cells in suspension are rapidly frozen. Freezing is best achieved by immersion in liquid nitrogen, but it is sometimes carried out by pulverizing the tissue with CO_2 powder, or by freezing thin films of suspended cells on the walls of a lyophilizing flask immersed in CO_2–acetone. (2) The frozen tissue is then lyophilized. High vacuum is employed, and precautions are taken to avoid any thawing of the tissue during the desiccation process. (3) The dry powder so obtained is ground to fragment the cells and liberate the nuclei. In most of the nonaqueous isolation procedures, this involves a milling or grinding procedure which breaks cells as a result of impact, or a tearing and maceration between hard grinding surfaces, an effect which is conveniently achieved with a ball mill (see Allfrey, 1). (4) The nuclei are then separated from the lighter and heavier components of the ground tissue suspension by alternate sedimentations and flotations in mixtures of organic solvents. Suspending media for these centrifugations are made by blending a light organic solvent (such as cyclohexane) with a heavier one (such as carbon tetrachloride), and the density of the mixture depends on the relative amounts of the 2 liquids. Since these solvents do not extract proteins, peptides, or amino acids; nucleic acids or nucleotides; water-soluble coenzymes; or mineral salts, nuclei obtained in this way retain their original complement of water-soluble components. They can then be used as standards to compare the composition and enzymic activities of nuclei isolated in sucrose solutions or in other aqueous systems.

Because many enzymes can withstand lyophilization and exposure to the nonpolar organic solvents used, the method has also become an important check on the localization of enzymes in the nucleus. The results can often be unequivocal. For example, several studies of mammalian nuclei isolated in aqueous media had failed to detect appreciable amounts of DNA polymerase (DNA-nucleotidyl transferase) in the nuclear fraction. A reexamination of the enzyme distribution using nonaqueous procedures showed high concentrations in the nuclei of regenerating liver, as would be expected for a dividing cell population which is increasing its DNA content (47).

Many other examples could be cited of the loss of nuclear enzymes during nuclear isolations in aqueous media, and their retention during nonaqueous isolations. Since much of this information is reviewed elsewhere (1), other cases of enzyme redistribution during the course of nuclear isolations in aqueous media need not be given here.

The value of the nonaqueous isolation procedures can hardly be exaggerated when one is faced with the problem of the intracellular distribution of small molecules (such as ATP) and inorganic ions (such as Na^+). Nonaqueous nuclei are the material of choice for measurements of nucleotide and amino acid pools, for determining which substrates of glycolysis and oxidative metabolism really occur inside the nucleus, and for studies of the intranuclear coenzymes, such as NAD, FAD, and co-enzyme A.

Unfortunately, the nonaqueous isolation procedures have their dis-advantages as well. They are relatively time consuming, since the operations require several days of lyophilization, grinding, differential centrifugation, and drying to remove the organic solvents. However, considering the unmatched utility of the method in many studies of nuclear composition, this is hardly a serious obstacle. A major dis-advantage is that many enzymes cannot withstand lyophilization or treatment with organic solvents(e.g., more than 99% of the aspartate carbamoyl transferase activity of liver tissue is lost on freeze-drying, and 60–80% of the uridine kinase activity is destroyed by treatment with the organic solvents). A further disadvantage of the nonaqueous methods is damage to membranes and to functions which depend on the contri-bution of essential lipids (which may be removed by organic solvents). Structural damage, enzyme inactivation, and protein denaturation may take a high toll in integrated function. For example, calf thymus nuclei prepared in nonaqueous media do not synthesize ATP or incorporate amino acids into protein, while those prepared in 0.25 M sucrose solutions do. However, nonaqueous thymus nuclei do contain the ATP-dependent amino acid-activating enzymes, tRNA's, amino acids, and ATP; presum-ably, the loss of function reflects damage to ribosomal proteins or en-zymes required for chain initiation, tRNA binding, or peptide bond formation.

Obviously, the organic solvent techniques preclude many studies of nuclear lipid composition. It is also unlikely that the lipid-soluble vita-mins (e.g., vitamin A) would be retained in the course of treatment with cyclohexane–CCl$_4$ mixtures. Yet, despite these objections, nonaqueous nuclei retain a surprisingly large number of enzymic activities, and these sometimes involve lipid or steroid metabolism. For example, it has been shown recently that liver nuclei isolated in this way retain a capacity for NAD$^+$-dependent steroid oxidation (87).

B. Nuclear Isolations in Citric Acid

Early methods for the preparation of cell nuclei, beginning with the work of Miescher, employed acid media. The use of citric acid by Stone-burg in 1939 marked the beginning of a series of citric acid isolations which had far-reaching results. For example, the work of Mirsky and Ris on the constancy in DNA content of somatic cell nuclei was based on the analysis and counting of nuclei isolated from different tissues in dilute citric acid (65). Citric acid nuclei were also employed in pioneering studies of RNA synthesis which revealed the rapid incorporation of RNA precursors by the cell nucleus (see below). They have also been used for studies of nuclear lipids.

Citric acid isolations have many advantages. They are usually fast, and they supply clean nuclei in high yields from a wide variety of tissues. There are many isolation procedures available (see Allfrey, 1), but the essential operations are: (1) homogenization of the tissue in dilute citric acid with the aid of a blender or coaxial homogenizer; (2) filtrations to remove connective tissue and fibrous contaminants; and (3) differential centrifugation at low speed to sediment the nuclei and remove the cytoplasm.

The method has serious disadvantages which limit its widespread application. Apart from the loss of enzyme activity and the damage to ultrastructure due to the low pH values usually employed or to the chelation properties of citric acid solutions, there is a serious extraction problem. As much as 55% of the protein of liver nuclei is lost during citric acid isolations and some nuclear ribonucleic acid is lost as well (see Allfrey, 1). However, some nuclear proteins, histones in particular, are retained provided the acid concentration is kept below 0.01 M. In practice, this has become a useful procedure for removing much of the nonhistone protein from the nucleus before attempting an isolation or fractionation of the histones themselves.

C. Nuclear Isolations in Sucrose

Sucrose solutions were first introduced for the study of cytoplasmic particulates by Hogeboom and co-workers (37). Since that time many methods have been devised for the isolation of nuclei in sucrose-containing media. The methods differ in many ways, the most important variables being the concentration of sucrose employed, the presence of added salts (particularly divalent cations), the pH of the suspension, and the use of detergents or chelating agents.

The essential operations are: (1) homogenization of the tissue with the aid of a blender or coaxial homogenizer (see Allfrey, 1); (2) filtration to remove fibrous contamination; and (3) differential centrifugation to sediment the nuclei. The latter may be further purified by density-gradient centrifugations or by treatment with chelating agents, detergents, or enzymes.

There is good reason to believe that the most desirable isolation conditions are those which maintain the sucrose concentration at the isotonic level (i.e., at about 0.25 M for mammalian tissues). This conclusion is based on studies of RNA and protein synthesis in isolated calf thymus nuclei that show a definite functional optimum in the isotonic range and an irreversible loss of activity when transferred to hypotonic or hypertonic media. The harmful effects of hypotonic shock are worth emphasizing because the procedure is being used with increasing frequency to break cells (see below). Although thymus nuclei are readily obtainable in isotonic sucrose, attempts to prepare nuclei of comparable purity from other cell types, such as liver, usually results in extensive contamination of the nuclear fraction by whole cells or cytoplasmic particulates. For this reason, isolation procedures employing concentrated sucrose have taken precedence in many studies of nuclear composition and function.

These methods often involve layering procedures in which the tissue homogenate is layered over a denser medium in the centrifuge tube (1). The denser solution acts as a barrier to prevent the sedimentation of whole cells and other cell components which are lighter than the nuclei. Several such density barriers may be used in a single centrifuge tube. An interesting application of the method has recently been described in which liver nuclei were sedimented through 2 layers of dense sucrose, the upper layer containing EDTA and ribonuclease which destroyed the ribosomes on the outer nuclear membrane. Nuclei passing this barrier and sedimenting through a more dense, EDTA- and nuclease-free solution were then used for the preparation of a nuclear ribosome fraction (60).

Alternatively, the tissue may be homogenized directly in dense sucrose solutions (e.g., 2.2 M) and the nuclei collected by centrifugation at high speeds. This is the procedure introduced by Chauveau and co-workers (17), which is now widely applied. It yields nuclei of high purity from many types of tissue and has proven to be fast and generally reproducible. Yields are sometimes too low, but are usually improved by the addition of divalent cations (e.g., 1–3 mM $CaCl_2$ or $MgCl_2$) to the dense sucrose medium. Nuclei isolated by modifications of the Chauveau procedure have been useful in studies of the RNA-polymerase activity and DNA-template activity in cells during periods of gene activation. (These experiments will be discussed in more detail below.) Although nuclei isolated in dense sucrose usually lose their nucleotide pools and other low molecular weight constituents, they usually retain more enzymic activity and have a higher protein content than nuclei isolated from the same tissue using dilute sucrose or salt solution.

Nuclei prepared from a mixed cell population can differ in density, depending upon the cell type and the physiological state of the cell. In any given cell type the density may vary depending on the ratio of DNA to protein and other components of the nucleus. When cells divide this proportion changes. A fractionation of nuclei according to DNA content becomes possible when density gradients are used, and the nuclei sediment into zones of different specific gravities. Such techniques have been employed to fractionate nuclei at different times during regeneration of the liver (27). Density gradients have also been used to separate calf thymus nuclei depending upon their residual DNA contents after exposure to DNAase in experiments relating the DNA content of the nucleus to its capacity for RNA synthesis (8).

The dense sucrose isolation procedures have many advantages and some disadvantages. Liver nuclei isolated in this way are good material for the study of RNA polymerase activity, but they are virtually inactive in protein synthesis. Fortunately, a very good method does exist for the preparation of liver nuclei which can incorporate amino acids into their proteins. This is a recent modification (90b) of the procedure of Hubert et al. (42). The method employs a nonionic detergent (Cemulsol) in 0.25 M sucrose containing divalent cations and phosphate buffer. Nuclei of very high purity are obtained by differential centrifugation. They have been found to be active in the incorporation of several amino acids into proteins and to carry out RNA synthesis.

D. Other Isolation Procedures

Although nuclei can be prepared in many ways from a wide variety of cell types, the isolation is sometimes complicated at the outset by the resistance of cells to breakage. Hypotonic shock has offered a partial solution to this difficulty, because swollen cells are more susceptible to the shearing forces produced in a blender or coaxial homogenizer. The procedure has been especially valuable in dealing with tumor cells, such as Ehrlich ascites tumor and HeLa cells grown in tissue culture. This was the method used to break Ehrlich ascites cells in the first direct demonstration that amino acids are transferred from sRNA to the microsome fraction (36). It is now being used for studies of ribosomal RNA synthesis in HeLa cells, which are very difficult to break by shearing forces alone; even hypotonic shock in very dilute salt solutions is inefficient unless supplemented by the use of both ionic detergents (e.g., sodium deoxycholate) and nonionic detergents (Tween 40). HeLa cell nuclei isolated after this rigorous treatment have few intact ribosomes left, an observation which has led some workers to conclude that ribosomes (i.e., 78 S ribosomes) do not occur in the nucleus of the HeLa cell (e.g., 71). It should be pointed out, however, that if one tests the effects of hypotonic shock on other nuclear types, on calf thymus nuclei in particular, hypotonic salt solutions prove to be an effective way to remove nuclear ribosomes plus most of the soluble proteins of the nucleus (75). Indeed, it is quite a common observation that nuclei isolated in sucrose and then exposed to salt solutions lose a large part of their contents (see Allfrey, *1*). Moreover, the convincing nature of the biochemical and autoradiographic evidence for protein synthesis in the nuclei of other cell types (see below) and in isolated HeLa cell nuclei (*31a, 99a*), plus the fact that ribosomes have been prepared from liver (*60*) and thymus nuclei (*75*) make it doubtful that HeLa nuclei *in vivo* are entirely without functional ribosomes or polysomes. Indeed, recent reports indicate that functional ribosomes are retained by the isolated HeLa nucleus even after removal of the outer nuclear membrane. Large amounts of ribosomal RNA have also been clearly identified on HeLa metaphase chromosomes (*56*). In any case, there is no question about the presence in nuclei of ribosomal subunits which are synthesized there. These remain in the nucleoli despite exposure to hypotonic media and detergents (*72*) (although low recoveries of 18 S RNA suggest losses of the small subunit). HeLa cell nuclei have been an important source material for some of the most interesting kinetic studies of the synthesis and maturation of ribosomal RNA's. This work will be considered below in connection with nuclear RNA synthesis.

Apart from isolations in salt solutions—with or without detergents—nuclei have been isolated in glycerol, ethylene glycol, sugars other than sucrose, dextrans, Ficoll (a high molecular weight, branched-chain carbohydrate), and concentrated albumin solutions. These procedures will not be considered here except to note an interesting observation: Exposure of thymus nuclei to isotonic glycerol or ethylene glycol solutions leads to a loss of function, while equivalent concentrations of many sugars permit the isolation of nuclei that are capable of protein and RNA synthesis.

E. CRITERIA OF PURITY

In any nuclear isolation it is important to have some estimate of the degree to which the final nuclear sediment is contaminated by other cell components. Electron microscopy of sections through the nuclear pellet is one of the most sensitive indicators of nuclear purity; it will reveal contaminants which easily escape detection in the light microscope, such as fragments of the endoplasmic reticulum or bits of cytoplasm or mitochondria adhering to the outer nuclear membrane. Chemical and enzymatic tests for characteristic cytoplasmic components, such as succinoxidase, β-glucuronidase, or glucose-6-phosphatase are also used to test for mitochondrial, lysosomal, or membranous contamination, respectively (see Allfrey, *1*).

It is sometimes possible to verify the results obtained in bulk isolations of nuclei by microtechniques which test for function in single nuclei obtained by microdissection. Two such applications will be discussed below: one, in connection with the proof that glycolytic and oxidative enzymes do exist in the nuclei of nerve cells (*51*) and two, with direct evidence that the nucleus of the oocyte concentrates sodium ions (*69*).

Other criteria and tests for function in isolated nuclei have been discussed in detail in recent reviews (*1, 88*).

IV. ISOLATION OF SUBNUCLEAR COMPONENTS

A. NUCLEOLI

Several methods exist for the isolation of nucleoli or nucleolar fragments from mammalian tissues and from cells in culture. These procedures have been applied with increasing success to studies of ribosomal RNA and protein synthesis. A few of the methods will be briefly described as a background to the later consideration of the nucleolus as a source of the cell's ribosomes.

One isolation of nucleoli from liver (*66*) begins with a preparation of the nuclei by centrifugation through dense (2.4 M) sucrose solutions. The nuclei are then disrupted by sonication and the sonicate is layered over 0.88 M sucrose. Centrifugation at low speeds yields a pellet of nucleoli and nucleolar fragments. These contain ribosomal subunits and newly synthesized ribosomal RNA precursors.

HeLa cell nucleoli have been prepared from nuclei isolated in hypotonic salt solutions plus detergents. The nuclei are lysed by exposing them to 0.5 M salt at pH 7.4. The resulting nucleohistone gel is broken by treatment with deoxyribonuclease. High speed centrifugation then yields a sediment containing the nucleoli together with membranes and bits of chromatin (*72*).

Nucleoli have also been isolated from tumor cells, from pea-seedling nuclei, and from starfish oocytes (see Vincent and Miller, *94*).

B. METAPHASE CHROMOSOMES

A number of useful methods exist for the isolation of metaphase chromosomes from mammalian cells grown in culture. Milligram quantities of intact chromosomes have been prepared from HeLa cells in which mitosis was arrested by the use of colchicine (*41*). The method involves: (1) hypotonic swelling of the metaphase cells in dilute sucrose–salt solutions at pH 3; (2) homogenization of the swollen cells in a glass homogenizer; (3) centrifugation at low speeds to sediment chromosomes and some unbroken nuclei; (4) separation of chromosomes from nuclei in a sucrose density gradient at pH 3; and (5) layering of the chromosome suspension over 2.2 *M* sucrose and high speed centrifugation to sediment the chromosomes, leaving debris in suspension.

Chromosomes isolated by this procedure retain DNA, histones, and other proteins, though some proteins may be lost in the course of isolation. They also contain a surprisingly high content of RNA (RNA/DNA ratio = 0.66). This finding of RNA on chromosomes is in accord with many previous cytological observations. The demonstration that most of this RNA is ribosomal in nature (*41*) is of some interest because of its relevance to the question of the localization and function of intranuclear ribosomes.

An isolation procedure permitting the preparation of metaphase chromosomes at neutral pH values has recently been described (*56*). Cells arrested at metaphase were swollen in a hypotonic medium containing saponin and then disrupted in a small homogenizer. Unbroken cells and nuclei were removed by low speed centrifugation, leaving the chromosomes in suspension. The chromosomes were sedimented at higher speeds, resuspended, and then layered over 2.2 *M* sucrose buffered at pH 7. Centrifugation at very high speed sedimented the chromosomes, leaving debris in the supernatant phase.

The method has been successfully applied to 4 types of mammalian cells; HeLa, mouse L-cells, and Chinese and Syrian hamster cell cultures. The isolated chromosomes are an impressive sight (see Fig. 5). They contain about 16% DNA, 12% RNA, and 72% protein, half of which is acid soluble and includes the histones. More than 80% of the RNA associated with the chromatin is ribosomal RNA. Its origins and functions on the chromosomes remain to be established.

C. CHROMATIN AND CHROMATIN SUBFRACTIONS

In Miescher's original work the nucleins were prepared under acid conditions. Lilienfeld in 1894 and Huiskamp in 1901 introduced more gentle procedures for the isolation of calf thymus nucleohistones based on their extractability in water. Since the turn of the century, and especially since the pioneering investigations of Mirsky and Pollister (*63*) and Mirsky and Ris (*64*), knowledge of the chemistry and metabolism of the

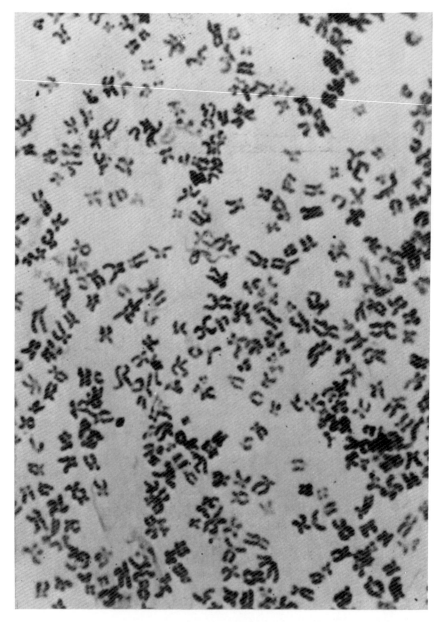

Fig. 5. Isolated metaphase chromosomes of HeLa cells, as seen in the light microscope. Chromosomes fixed in methanol–acetic acid (3:1) and stained with methylene blue (Unna) [from Maio and Schildkraut (56)].

genetic material in interphase cells has become increasingly dependent on the analysis of isolated chromatin fractions and nucleohistone complexes prepared by gentle procedures under neutral pH conditions.

The solubility characteristics of nucleohistone complexes are an effective guide to the rationale of chromatin isolations. The deoxyribonucleoproteins of higher organisms are least soluble in isotonic salt solutions, but they will dissolve or disperse to form viscous gels in distilled water. Such gels are readily precipitated by raising the salt concentration slightly; even 0.01 M salt will precipitate nucleoprotein dispersed in the gel phase.

Chromatin, being largely insoluble in isotonic salt solutions, can be prepared by exposing isolated nuclei in 0.14 M NaCl to the high shearing forces of a colloid mill (64). The resulting chromatin fibrils are then collected by differential centrifugation and washed extensively with isotonic saline to remove contaminants. Classic studies of chromatin fractions isolated in this way (64) indicated the presence of DNA, RNA, and associated proteins. The histones (highly basic proteins lacking in tryptophan) were shown to be present in close association with the DNA, held by saltlike linkages that could be dissociated in media of high ionic strengths (e.g., 1 M NaCl). Also present were less basic, tryptophan-containing proteins, the proportion of which varied in the chromatin of different cell types. The amount of this residual protein (so called because of its insolubility at the time nucleohistone was removed in 1 M NaCl) was greater in metabolically active cells such as liver, than in relatively inactive cells such as small lymphocytes and nucleated erythrocytes (64). A similar correlation between the nonhistone protein content of chromatin fractions and nuclear activity in the cells of origin has since been observed in cells at different stages of development (22). Among the proteins now known to be present in chromatin are enzymes, such as RNA polymerase, the NAD-synthesizing enzyme, and histone acetylase, as well as the histones and nuclear phosphoproteins. These components of chromatin will be discussed in detail below in connection with chromatin structure and the control of RNA synthesis.

A widely used method for the preparation of nucleohistone complexes was introduced by Zubay and Doty (100) who employed NaCl–versene (EDTA) mixtures at pH 8 for preliminary washing of the chromatin and then dispersed the nucleohistone by homogenization in distilled water. Versene was introduced for good reasons: It was felt that a chelating agent would inhibit the activity of nucleases, such as deoxyribonuclease I, which require divalent cations for their function. It should be pointed out, however, that the use of chelating agents raises new problems, because it is known that the structure and activity of chromatin depend a great deal on the nature and concentration of the divalent cations present. Nuclei concentrate divalent cations. It is also known that the extractability of certain histones can be changed by chelating agents. For example, thymus and liver nuclei isolated in sucrose solutions will normally retain their lysine-rich histones until the pH drops to well below pH 3, but if such nuclei are pretreated with EDTA, much of the lysine-rich histone fraction is extractable at pH 5. Moreover, there is good evidence that this particular histone fraction plays a special role in the organization of the condensed state of chromatin as it occurs *in vivo* (*53*). It follows that histone-DNA interactions in the absence of the divalent cations may not be those which would have been present in native chromatin. Similar considerations apply to the role of chromatin as a template for RNA synthesis, for it will be shown below that the types of RNA synthesized in isolated nuclei, and their sites of synthesis within the nucleus, can be varied depending on the presence of magnesium or manganous ions. Removal of such ions by chelation would thus be expected to lead to changes in the functional organization of chromatin. Although the use of chelating agents in the preparation of nucleohistones is not without its disadvantages, it does provide one of the very few ways to bring nucleohistones into the soluble phase for physical studies, and the method has produced interesting and important results (*100*).

Another major isolation procedure, introduced by Mirsky and Pollister (*63*), uses high salt concentrations to dissociate nucleoprotein complexes. Isolated nuclei are suspended in neutral $1 M$ NaCl and the resulting gel is broken by shearing forces in a blender or high-speed stirrer. Nucleoli and membranes are then removed by centrifugation. When the salt concentration is reduced to isotonic levels, the nucleohistone reprecipitates; this step is conveniently carried out by pouring the $1 M$ solution into 6 volumes of water. The nucleohistone threads which form rapidly are wound around a rod and easily transferred for further operations. Deproteinization of such nucleohistone complexes is one of the most effective procedures for the isolation of DNA's.

The methods described have dealt exclusively with the preparation of chromatin from animal tissues. In recent years, due largely to work in James Bonner's laboratory, methods have been devised for the study of plant chromatin. In a procedure for pea-seedling chromatin, embryos are subjected to high shearing forces in $0.25 M$ sucrose solution, buffered at pH 8 and containing mercaptoethanol. Low speed centrifugation yields a pellet of chromatin overlying a layer of starch granules. The chromatin is scraped off, resuspended, and layered over $2 M$ sucrose. High speed centrifugation then yields a pellet of purified chromatin (*40*). This material has been used for the isolation of the plant histones (*24*) and for many studies of chromatin and DNA template activity in RNA synthesis (*40*).

D. ACTIVE AND INACTIVE CHROMATIN

In some cell types, lymphocytes in particular, much of the chromatin occurs in a highly condensed state. Tightly coiled bundles of nucleoprotein, containing most of the DNA of the nucleus, are visible under both the ultraviolet and electron microscopes. As mentioned earlier, nucleoprotein fibrils originating in the condensed chromatin may extend out of the clumps into the diffuse chromatin region [Fig. 1(B)]. High resolution autoradiography of thymus nuclei after brief incubations in the presence of tritium-labeled RNA precursors has shown that nearly all the newly synthesized RNA is made in the diffuse regions of the chromatin and that few, if any, grains are localized over the clumps. It follows that the DNA in the condensed chromatin is not being used as a template for RNA synthesis, while the DNA in the diffuse chromatin is active.

A method has been devised which permits a partial separation of the 2 states of the chromatin (*29*). Calf thymus nuclei are isolated in isotonic sucrose containing divalent cations and then washed with dilute buffers. The nuclei are resuspended in cation-free sucrose for about 10 minutes; in that time they expand to about twice their normal size. Sonication at 20,000 cycles for a few seconds disrupts the swollen chromatin, breaking many of the fibrils where they emerge from the clumps. Low speed centrifugation is then used to sediment the large clumps of chromatin, while the more diffuse chromatin remains in suspension. It is then collected by centrifugation at high speeds.

When such fractionations are carried out after labeling experiments with RNA precursors it is found that the RNA of the diffuse chromatin fraction is much more radioactive than the RNA in the clumped chromatin fraction. RNA polymerase assays of the 2 fractions have also indicated differences in their template activity. Other studies have shown that hormone induction of RNA synthesis can be followed by such a fractionation procedure. This will be discussed in more detail below.

V. ENERGY METABOLISM IN THE NUCLEUS

The major biosynthetic reactions which take place in the nucleus—RNA, DNA, and protein synthesis—all involve the joining together of small units to form polymers of high molecular weight. These are enzymic processes and they depend on a supply of appropriate low molecular weight precursors—deoxyribonucleotides, ribonucleotides, and amino acids. Yet, it is an important fact that the condensation of monomeric units or small molecules to form polymers is usually thermodynamically unfavorable. The synthetic reaction is not a spontaneous one despite the presence of enzymes or their apparent substrates; on the contrary, it is the hydrolysis of biological macromolecules which proceeds spontaneously with a net release of energy. The synthesis of complex molecules requires a corresponding input of energy to drive the reaction forward. This requirement has obliged the cell to couple energy-yielding reactions to the energy-requiring reactions of biosynthesis. In the nucleus, as in the cytoplasm, the energy coupling involves the participation of adenosine triphosphate (ATP).

ATP is the main energy intermediate of the nucleus. It has been shown to occur in many types of nuclei in high concentration and to supply the driving force needed for the synthesis of nuclear proteins and nucleic acids. Thus, the first step in nuclear protein synthesis, as in the cytoplasm, is an ATP-dependent activation of the amino acids. [At least 17 amino acid "activating enzymes" have been detected in cell nuclei isolated in either aqueous or nonaqueous media (see below).] Similarly, ATP is the phosphoryl group donor in the kinase reactions essential to the formation of the nucleotides and nucleoside triphosphates needed for RNA and DNA synthesis. ATP itself is one of the 4 direct precursors of RNA in the RNA polymerase reaction.

What is the source of nuclear ATP and how is it formed?

A. NUCLEOTIDE POOLS IN THE NUCLEUS

The presence of free nucleotides in the cell nucleus was first suggested by microspectrophotometry of single cells. Measurements of the ultraviolet absorption of the nuclei in cultured cells indicated that there were ultraviolet-absorbing substances present which were readily soluble in acid–alcohol fixatives; these were believed to be deoxynucleotides needed for DNA synthesis. In 1954, Naora and Takeda recognized the occurrence of labile phosphate compounds, presumed to be ATP, in rat liver nuclei isolated in nonaqueous media.

In 1955, it was observed that nuclei isolated from calf thymus tissue by nonaqueous isolation procedures contained mononucleotides of adenine, cytosine, uracil, and guanine, together with a number of uridine diphosphate derivatives. The nucleotides of adenine predominated in amount, with nuclear levels of AMP, ADP, and ATP quite comparable to those measured in the cytoplasmic fraction (70). Similar observations were made on calf liver nuclei and other types of nuclei as well (46).

An analysis of the nucleotide content of calf thymus nuclei isolated in isotonic sucrose solutions indicated that they had retained 60–80% of their original soluble nucleotides. The nucleotide levels in the isolated nuclei were far above the possible contamination due to cytoplasmic fragments or mitochondria adhering to the outer nuclear membrane. The retention of nucleotides by thymus nuclei in the course of sucrose isolations is exceptional, but it is not unique. Liver nuclei, for example, lose their nucleotides during isolations in isotonic sucrose, although they will retain them in nonaqueous media. Other nuclear types (e.g., AKR lymphoma cell nuclei, rabbit appendix nuclei, and Novikoff hepatoma nuclei) will carry out ATP-dependent synthetic reactions after isolation in sucrose, and it can be concluded that they retain a functional ATP pool. A deoxyribonucleotide pool has been observed and characterized in the nuclei of Flexner-Jobling carcinoma cells.

B. ATP Synthesis in the Nucleus

The question arises as to whether the nucleotides found in the nucleus, ATP in particular, originate there, or whether they are transported into the nucleus from the cytoplasm. In some types of nuclei, there is very good evidence for intranuclear ATP synthesis.

When calf thymus nuclei are isolated rapidly, and their nucleotides extracted and fractionated on Dowex-1 columns, it is found that most of the nucleotides are present in the monophosphate form. This is illustrated in the upper curve of Fig. 6. If the nuclei are then incubated in air, there is an increase in the amount of ATP and a corresponding decrease in the concentrations of AMP and ADP (70) (Fig. 6, lower curve). It should be pointed out that this increase in ATP content takes place under aerobic conditions, and not under nitrogen, and that neither the cytoplasmic fraction nor the isolated mitochondria can carry out ATP synthesis under these conditions of high Ca^{2+} concentrations in the medium. Independent studies of the rate of ATP synthesis in nuclei from calf and rat thymus show similar figures; 0.9 mμmoles of ATP is formed per minute per milligram of nuclear protein. An uptake of ^{32}P-orthophosphate into acid-soluble nucleotides by isolated rat liver nuclei has also been reported, but its oxygen dependence remains to be established.

The phosphorylation reactions detected in isolated thymus nuclei appear to be limited to nucleotides already present in the intranuclear

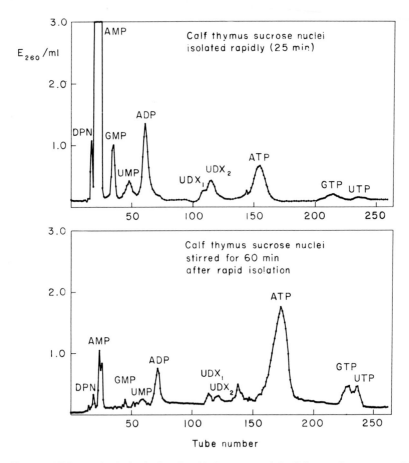

Fig. 6. ATP synthesis in isolated calf thymus nuclei. Column chromatographic separation of the nucleotides in freshly prepared nuclei shows a predominance of AMP and relatively little ATP (upper curve). Nucleotide analysis of the same nuclear preparation after being stirred aerobically in the cold for an additional 60 minutes shows an increase in the amount of ATP and a decrease in AMP content (from Osawa *et al., 70*).

pool. Such nuclei will not phosphorylate AMP added to the incubation medium, and there is good evidence that added nucleotides do not penetrate the thymus nucleus readily (*61*). This is not true for isolated liver nuclei, which lose their nucleotides in the course of sucrose isolations, but which can also admit and utilize added nucleoside triphosphates for the RNA polymerase reaction (e.g., *74*). The selective permeability of the nuclear envelope in thymus lymphocyte nuclei is also evident in studies of the effects of different substrates and inhibitors on nuclear carbohydrate catabolism and in amino acid and nucleoside transport reactions. (These properties will be discussed below.)

If thymus nuclei are allowed to synthesize ATP aerobically and are then placed in a nitrogen atmosphere, the ATP levels decrease. The rate of loss of ATP under such conditions can then be taken as a rough estimate of the rate of its utilization. When this figure is added to the rate of ATP formation under aerobic conditions and compared to the measured rates of O_2 consumption by nuclei (1.6–3.4 mμmoles/minute/mg protein), the P/O ratio can be estimated. In the thymus nuclear preparations, this ratio approaches 1. However, such estimates of P/O ratios in isolated nuclei are made difficult by the presence of ATPases, which have been detected and characterized in nuclei isolated by nonaqueous techniques from a variety of cell types (26). It has also been shown that when radioactive ATP is added to suspensions of thymus nuclei, more than 80% of the added ^{14}C-ATP is hydrolyzed to ^{14}C-ADP and ^{14}C-AMP within 5 minutes, and in 30 minutes 40% of the original material is degraded to ^{14}C-adenosine and ^{14}C-adenine. The latter 2 compounds do enter the nuclear pool and are used again as precursors of intranuclear ATP and nucleic acids. These findings are of some interest because they make it very improbable that the increase in nuclear ATP content occurring when thymus nuclei are incubated in air could represent a transfer of ATP into the nucleus from contaminating mitochondria or some other extranuclear phosphorylating system.

1. *Differences between Nuclear ATP Synthesis and Mitochondrial Oxidative Phosphorylation*

A comparative study has been made of ATP synthesis in thymus nuclei and in mitochondria isolated from the same tissue (61). The results show many points of similarity but also reveal striking differences in the 2 systems. Particularly informative are the effects of a number of inhibitors of electron transport systems, glycolysis, and the citric acid cycle.

Some effects of uncoupling agents and inhibitors of mitochondrial electron transport on ATP synthesis in the nucleus and on amino acid incorporation into nuclear proteins are summarized in Table I. It is clear that thymus nuclei resemble mitochondria in their susceptibility to a large number of inhibitors, including azide, 2,4-dinitrophenol, cyanide, Antimycin A, dicumarol, and amytal. It is also obvious that agents which block nuclear ATP synthesis also depress amino acid uptake, thus indicating the ATP dependence of nuclear protein synthesis. The significance of some of these inhibitor effects will be discussed in more detail below; what is important here is to stress some of the negative results—

TABLE I

COMPARATIVE EFFECTS OF DIFFERENT METABOLIC INHIBITORS ON NUCLEAR AND
MITOCHONDRIAL ATP SYNTHESIS AND BIOSYNTHETIC REACTIONS
IN ISOLATED THYMUS NUCLEI

Inhibitor	Concentration	Change in nuclear function (%)			Uncoupling of mitochondrial oxidative phosphorylation (%)
		ATP level	Protein synthesis[a]	RNA synthesis[b]	
NaCN	$1 \times 10^{-3} M$	−100	−76	—	—
2,4-Dinitrophenol	$2 \times 10^{-4} M$	−100	−84	−87	—
NaN$_3$	$1 \times 10^{-3} M$	−100	−91	—	—
Antimycin A	1 μg/ml	−61	−89	−73	—
Dicumarol	$3 \times 10^{-5} M$	−47	−96	−77	—
Amytal (amobarbital)	$1 \times 10^{-3} M$	—	−86	—	—
Methylene blue	$2 \times 10^{-5} M$	+31	+3	—	−71
Ca^{2+} ions	$2 \times 10^{-3} M$	0	0	0	−72
CO	95% CO–5% O$_2$	0	0	0	−100
DNase	0.5 mg/ml	−80	—	—	0
Thymus histone	0.5 mg/ml	−3	—	—	−100
Iodoacetic acid	$1 \times 10^{-3} M$	−89	—	−87	—
NaF	$1 \times 10^{-2} M$	−76	—	−85	—
Fluoroacetic acid	$1 \times 10^{-2} M$	−43	—	−80	—
Dehydroacetic acid	$5 \times 10^{-2} M$	−27	—	−77	—

[a] Percent inhibition of 1-^{14}C-alanine incorporation in a 60-minute incubation period.
[b] Percent inhibiton of 8-^{14}C-adenosine incorporation in a 30-minute incubation period.

particularly the failure of carbon monoxide (90–95%), methylene blue, and Ca^{2+} ions to interfere with nuclear phosphorylation, while parallel experiments on isolated thymus mitochondria gave complete uncoupling or inhibition (61).

Nuclear phosphorylation also differs from mitochondrial phosphorylation in its susceptibility to attack by deoxyribonuclease. Mitochondrial respiration and phosphate uptake are not affected by this enzyme, while nuclear ATP synthesis is strongly inhibited. The inhibition can be reversed by adding back DNA or other polyanionic molecules, such as polyethylene sulfonate, polyacrylic acid, or RNA. This makes it likely that the DNase effect on nuclear phosphorylation is an indirect one. It probably represents a sudden inhibition of nuclear function due to the release of histones which are no longer held in salt linkage to DNA. The addition of polyanions, by binding the histones, would permit nuclear phosphorylation to resume. This inhibitory effect of histones will be discussed below in connection with the problem of histone function; it need only be noted here that added histones do inhibit ATP synthesis in nuclear suspensions.

2. *Selective Inhibition of Protein Synthesis in the Cytoplasm*

In view of the ability of CO to block ATP synthesis selectively in mitochondria without preventing ATP synthesis in thymus nuclei, experiments were undertaken to establish whether or not the major energy-requiring reactions in the nucleus depend upon a supply of ATP from the cytoplasm. Protein synthesis takes place in both the nucleus and the cytoplasm of thymus cells, and that portion occurring in the cytoplasm would be expected to depend for energy on mitochondrial ATP synthesis. Carbon monoxide, by selectively inhibiting mitochondrial phosphorylation, should stop protein synthesis in the cytoplasm. This was tested by comparing the effects of CO on a preparation of purified nuclei and on cell suspensions. (Nuclei and cells were separated by density-barrier centrifugation in Ficoll–sucrose solutions.) CO had no effect on ^{14}C-alanine uptake into the proteins of isolated nuclei, but, in intact cells, protein synthesis fell to levels characteristic of nuclear synthesis alone, indicating a complete inhibition of amino acid uptake in the cytoplasm. Under completely anaerobic conditions (i.e., 100% N_2) neither nuclei nor intact cells were capable of protein synthesis. (See Fig. 7.)

3. *Evidence for Nuclear Localization of RNA Synthesis*

Similar experiments comparing the effects of CO on ^{14}C-adenosine uptake into RNA were carried out at the same time. As expected, CO had no effect on RNA synthesis in free nuclei, nor did it cause any major depression of RNA synthesis in intact cells (Fig. 7). This finding is consistent with the view that nearly all of the RNA synthesis occurring in the cell is confined to its nucleus. The results also establish the fact that the ATP needed for RNA synthesis in the nucleus is made there. It remains to be seen whether this is generally the case or whether lymphocyte nuclei have unusual capacities for ATP synthesis.

4. *Metabolism of Carbohydrates in Nuclei: Glycolysis*

Earlier experiments by Stern and Mirsky (see Allfrey, *1*) had stressed the importance of the glycolytic enzymes in nuclear metabolism. They observed that nuclei prepared in nonaqueous media from wheat germ, calf liver, and kidney had high concentrations of aldolase, phosphoglyceraldehyde dehydrogenase, enolase, and pyruvate kinase. More recent studies by Gunther Siebert and co-workers have established thoroughly and beyond doubt the presence of all the glycolytic enzymes and most of their substrates in the nucleus. The activities of 9 enzymes of

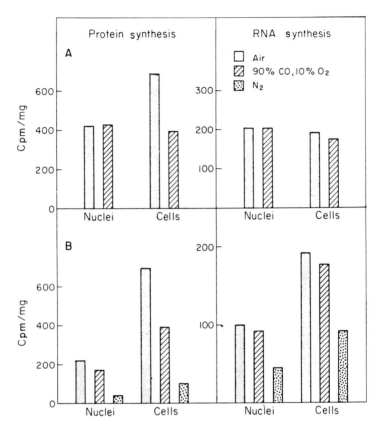

Fig. 7. Effects of CO and exposure to a nitrogen atmosphere on protein and RNA synthesis in isolated thymus nuclei and intact cells. Nuclei for (A) were isolated in 0.25 M sucrose–3 mM CaCl$_2$ and the incorporation of 1-^{14}C-alanine was measured over a 90-minute incorporation period. RNA synthesis was followed by the uptake of 6-^{14}C-orotic acid in a 30-minute incubation. The cell fraction was isolated by a Ficoll barrier density-gradient centrifugation technique and contained 60–70% intact cells. Protein synthesis was followed by 1-^{14}C-alanine uptake (60 minutes) and RNA synthesis by the incorporation of 8-^{14}C-adenosine (30 minutes). Note that CO has no effect on nuclear protein synthesis, while it depresses amino acid uptake in intact cells to that expected for nuclei alone. CO has little effect on RNA synthesis in isolated nuclei or in intact cells, indicating the nuclear localization of the energy supply in both cases (from McEwen et al., 61a).

the glycolytic pathway were measured in tissue homogenates and in nuclei isolated from rat liver, pig kidney, and beef brain. In some cases (e.g., aldolase) the nuclear enzyme concentration (expressed in units of activity per milligram of total protein) exceeded that in the cytoplasm. The activities of other enzymes, such as hexokinase and lactic dehydro-

genase, were lower in the nucleus than in the cytoplasm. Some, like phos-
phofructokinase, were equally concentrated in nucleus and cytoplasm
(86).

Similar differences were observed when metabolite concentrations were
compared. Liver nuclei had higher concentrations of lactate and pyruvate
than did the homogenate, but they had less dihydroxyacetone phosphate
and phosphoenol pyruvate. These differences are stressed here because
they form the basis of a strong argument against a free and rapid diffusion of the glycolytic enzymes or their substrates between the nucleus
and the cytoplasm.

The intracellular distribution of lactic dehydrogenase (LDH) is of
special interest because recent tracer studies (cited below) indicate that
this enzyme can be synthesized in the nucleus. The tracer evidence for
nuclear LDH synthesis is good because it is based on isolations in nonaqueous media, but it is strengthened considerably by supporting evidence from enzyme localization studies. First, it has been shown that the
properties of nuclear LDH in liver differ somewhat from those of the
cytoplasmic enzyme (e.g., with respect to pyruvate binding at different
pH values, and in the reduction of NAD analogs). Second, it has been
found that in the course of liver regeneration, nuclear LDH activity
doubles, while that of the cytoplasm decreases. Similar changes in the
relative activities of other glycolytic enzymes have also been detected.
The results make it most unlikely that the rapid appearance of radioactivity in nuclear LDH *in vivo* (50) represents a transfer of newly
synthesized enzyme into the nucleus from the cytoplasm.

Many types of evidence support the view that the nucleus has its own
glycolytic capacity. Some nuclei retain the glycolytic enzymes during
isolations in sucrose solutions. Calf thymus nuclei, for example, not only
retain the enzymes, but many of their substrates as well (61). They also
show a capacity for complete glycolysis; i.e., they will convert glucose-
6-^{14}C into ^{14}C-lactate and ^{14}C-pyruvate, the end products of glycolysis.
Moreover, the formation of lactate and pyruvate can be blocked by the
addition of inhibitors of specific glycolytic enzymes. Fluoride (10^{-2} M),
for example, inhibits enolase activity and reduces nuclear ATP levels by
76%. Iodoacetate (10^{-3} M) inhibits glyceraldehyde-3-phosphate dehydrogenase and lowers nuclear ATP concentrations by 89% (61).

The extensive loss of nuclear ATP in the presence of these inhibitors
is a revealing indication of the importance of glycolysis in nuclear energy
metabolism. It has also been shown that the glycolytic pathway of the
nucleus increases dramatically at times of increased nuclear activity; in
liver regeneration, it goes up by 400%.

Finally, the presence of glycolytic enzymes in the nucleus has been

verified by the microdissection experiments of Lehrer and Mathewson
(51). They examined nuclei removed from the neurons of puffer fish and
demonstrated the presence of hexokinase and lactic dehydrogenase
activity, as well as the glucose-shunt enzyme glucose-6-phosphate dehydrogenase and the tricarboxylic acid cycle enzyme malic dehydrogenase.

5. Evidence for a Nuclear Citric Acid Cycle

As mentioned above, nuclei prepared by microdissection techniques
contain at least 1 enzyme of the citric acid cycle. Malic dehydrogenase
and isocitric dehydrogenase (NADP dependent) have also been detected
in nuclei isolated from calf thymus, rat liver, pig kidney, and beef brain.
Succinic dehydrogenase activity has been observed and characterized in
nuclei isolated from the thymus and shown to exhibit pH optima distinct
from those of the mitochondrial enzyme (61). Moreover, purification of
the nuclei by density barrier centrifugation removed succinoxidase activity (an indicator of mitochondrial contamination) but did not eliminate
succinic dehydrogenase activity. This activity requires a flavin prosthetic
group; analysis for peptide-bound flavin gave positive results in thymus
nuclei isolated by both aqueous and nonaqueous procedures, and similar
levels of peptide-bound flavin were detected in rat liver nuclei. Succinic
dehydrogenase can also be determined by histochemical techniques, e.g.,
in the presence of succinate, tetrazolium dyes are converted to insoluble
formazan deposits. Such deposits have been described in bird erythrocyte
nuclei, tumor cell nuclei, and rat liver nuclei.

Malic dehydrogenase has been detected in neuronal nuclei after isolation by microdissection. Thus, the presence of at least 3 enzymes of
the citric acid cycle can be ascribed to the nucleus. This suggests that
the enzymic mechanism should also exist for taking the end products of
glycolysis (pyruvate and lactate) and bringing them into the citric acid
cycle. The entry of pyruvate into the citric acid cycle requires its oxidative decarboxylation and linkage to coenzyme A to form acetyl-CoA.
It has been established that thymus nuclei do synthesize ^{14}C-acetyl
CoA from added ^{14}C-pyruvate and ^{14}C-acetate. This is an important
point, not only with respect to its implications for a functional TCA
cycle in the cell nucleus, but also because acetyl-CoA is the acetyl group
donor in an interesting series of reactions involving acetyl transfer to
histones (see below).

Other studies testing the effects of inhibitors lend support to the view
that nuclear ATP synthesis is partly dependent upon oxidative reactions
in the citric acid cycle. McEwen et al. (61) employed dehydroacetic acid
as a specific inhibitor of nuclear succinic dehydrogenase; in its presence,
nuclear ATP synthesis fell by about one-third. (The inhibitor usually

used to block succinic dehydrogenase is malonate, but malonate failed to enter the isolated thymocyte nucleus.)

The tracer evidence for the operation of a complete citric acid cycle in isolated thymus nuclei is quite convincing and can be summarized briefly: (1) Added ^{14}C-acetate appears as ^{14}C-citrate in the nuclear pool, (2) lactate-3-^{14}C and pyruvate-3-^{14}C are both oxidized to $^{14}CO_2$, and (3) glucose-6-^{14}C appears as $^{14}CO_2$ under aerobic conditions. All the above reactions are inhibited when the nuclei are pretreated with deoxyribonuclease, suggesting that the release of histones from the degraded DNA has interfered with these metabolic activities within the nuclei. The latter conclusion is supported by the restoration of function to DNase-treated nuclei by the addition of polyanions; polyethylene sulfonate, for example, combines with the histones and restores both respiration and the nuclear capacity for ATP synthesis (7). (By way of contrast, deoxyribonuclease has no effect on oxidative pathways or ATP synthesis in isolated mitochondria.)

Finally, the analysis of nuclei isolated in nonaqueous media has revealed the presence of several substrates of the TCA cycle. The concentration of malate in rat liver nuclei approaches that in the whole homogenate. In calf thymus nuclei, the concentrations of malate, succinate, and α-ketoglutarate equal or exceed those in the cytoplasm. The conversion of acetate to citrate is also convincing evidence for a functional citric acid cycle in the thymus nucleus.

The apparent presence of a complete citric acid cycle in thymus nuclei raises questions about its contribution to nuclear ATP synthesis (and to other energy-dependent reactions). It has already been pointed out that the inhibition of nuclear glycolysis (by iodoacetate) reduces ATP levels by 93%. Similarly, fluoroacetate and dehydroacetic acid have been used to block the citric acid cycle. The evolution of $^{14}CO_2$ from glucose-6-^{14}C was inhibited by 87%, while nuclear ATP concentrations were 43% lower than in corresponding controls. Fluoroacetate and dehydroacetic acid, by reducing nuclear ATP concentrations, also reduce the incorporation of adenosine-8-^{14}C into nuclear RNA.

6. Electron Transport Systems in Nuclei

The reactions of glycolysis and the citric acid cycle give rise to a pool of reduced nucleotide coenzymes (NADH and NADPH). How are these nucleotides reoxidized, and what is the route of electron transfer? Is electron transport in the nucleus coupled to the phosphorylation of ADP to form ATP, as in the oxidative phosphorylations in mitochondria?

It has already been noted that the production of ATP in the thymus nucleus is not inhibited by some agents that effectively block oxidative

phosphorylation in mitochondria, such as Ca^{2+} ions, methylene blue, and
90% carbon monoxide, so the systems are clearly not identical. Yet there
is good reason to believe that some steps in the transport of electrons are
similar in the nucleus and mitochondria. This conclusion is based largely
on the effects of different inhibitors on ATP synthesis or ATP-dependent
reactions in isolated thymus nuclei. (See Table I.)

Nuclei, like mitochondria, are inhibited by uncouplers of oxidative
phosphorylation such as 2,4-dinitrophenol and dicumarol. They are also
inhibited by the 2 terminal inhibitors of the mitochondrial respiratory
chain, cyanide and azide.

One of most revealing inhibitor effects is that of amobarbital, a
barbiturate which prevents the transfer of electrons from NADH to
flavins. Amobarbital effectively inhibits ATP-dependent synthetic reactions in thymus nuclei, and it may be presumed that it prevents the
reoxidation of NADH and thus blocks electron transport and any
coupled phosphorylations. Fortunately, this assumption is supported by
the most direct type of evidence. Amobarbital has been found to block
the reoxidation of NADH in the nuclei of intact EL2 ascites tumor cells.
Its effect was observed in studies employing a microfluorimeter to follow
the changes in reduced pyridine nucleotides in the nucleus and the cytoplasm of single cells in culture (49). These experiments conclusively
established the correctness of the view that pyridine nucleotides participate in the transport of electrons released in the course of nuclear carbohydrate catabolism. They also revealed a surprisingly rapid penetration
and metabolism of glucose by the nucleus. More importantly, they demonstrated that nuclear metabolism is exceedingly dependent on the glucose
concentration of the medium surrounding the cell.

In mitochondria, the transfer of electrons from NADH to flavin is one
of the 3 sites coupled to phosphorylation. Phosphorylations which are
tightly coupled to electron transport can be blocked by certain types of
inhibitors, such as oligomycin. It has recently been shown by Betel and
Klouwen that oligomycin inhibits both nuclear phosphorylation and
nuclear respiration.

It was pointed out above that at least 1 flavoprotein (succinic dehydrogenase) exists in cell nuclei. Further evidence for the participation
of flavoproteins in electron transport is supplied by the use of flavin
antagonists, such as quinacrine and chlorpromazine. Both these substances
inhibit the ATP-dependent uptake of adenosine-8-^{14}C into RNA by
isolated thymus nuclei.

Granted that electrons produced in the course of nuclear carbohydrate
catabolism are passed on to NADH (or NADPH in the case of isocitric
dehydrogenase) and from pyridine coenzymes to flavin coenzymes, what

are the next steps in electron transport? In the mitochondria, ubiquinone and the cytochrome pigments transfer the electrons to molecular oxygen. What is the evidence for nuclear cytochromes? As yet, it is far from conclusive, and some of it is controversial, but the importance of the subject of ATP production and its relevance to other biosynthetic processes in the nucleus warrant at least a brief consideration of the nuclear cytochrome problem.

Inhibitor studies have been helpful; Antimycin A, which blocks electron transport from cytochrome b (or ubiquinone) to the higher cytochromes in mitochondria, also lowers nuclear ATP levels by 61%, a result which suggests that a block in electron transport at the cytochrome level can influence nuclear phosphorylations.

A more direct approach to the identification of respiratory pigments in nuclei is based on low temperature spectroscopy, searching out the characteristic absorption maxima of the cytochrome pigments and observing differences in the spectra before and after reduction. The method has been applied by Conover and Siebert to rat liver nuclei isolated in nonaqueous media and in dense sucrose solutions. The spectra of both types of nuclei at $77°\text{K}$ showed the presence of cytochrome c and cytochrome b_5. The presence of cytochrome b_5 in isolated nuclei is not surprising, since it would be expected to be a component of the outer nuclear membrane (which is continuous with the cell membrane system upon which cytochrome b_5 has been localized). In fact, it was found that the cytochrome b_5 vanished when the outer nuclear membrane was removed with the aid of detergents; but the cytochrome c remained (18). The spectra showed a total absence of the absorption maxima characteristic of cytochromes a, a_3, and b. It must be concluded that isolated rat liver nuclei do not contain a complete respiratory chain of the type found in mitochondria. Similar conclusions about the absence of cytochromes a, a_3, and b, and the presence of cytochrome c, in the nucleus were reached in several laboratories. It should be noted that the cytochromes c and b_5 which were found in the nuclear preparations were functional, as judged by the fact that the addition of NADH under anaerobic conditions led to complete reduction of both cytochromes. Rat liver nuclei and isolated nucleoli have been found to catalyze the reduction of oxygen with NADH as a substrate. This reaction requires the presence of cytochrome c but not of cytochrome b_5 (20). The possibility is thus raised of a relatively simple electron transport sequence: NADH → flavins → cytochrome c → oxygen.

Conover has recently carried out a low temperature spectral analysis of calf thymus nuclei, both before and after purification by centrifugation through dense sucrose solutions (18a). The results are of particular interest because of the evidence favoring a more complete electron transport system in the thymus nucleus. Difference spectra showed the presence of cytochromes c and b_5, as in the case of liver nuclei. However, thymus nuclei, in contrast to liver nuclei, also showed the presence of other cytochromes, including cytochrome a and cytochrome a_3. Moreover, once cytochrome b_5 had been removed by stripping the outer nuclear membrane, the difference spectra revealed the presence of cytochrome b, and, possibly, c_1. (18a). Some indication of the intranuclear localization of the cytochrome pigments is provided by recent studies of thymus nuclei by Ueda et al. (92g). They concluded that the cytochromes associated with isolated nuclei are probably localized on the inner nuclear membrane, and that the nuclear cytochrome system of the thymus lymphocyte is composed at least of cytochromes c, a + a_3, and another b type cytochrome. Thus, the data strongly suggest the presence of a complete respiratory chain in the calf thymus nucleus. This may be an unusual state of affairs; the thymus lymphocyte is a small cell in which the nucleus comprises about 60% of the cell mass, and such a nucleus may have properties not commonly encountered in cells with more abundant and more active cytoplasmic compartments. In any case, the cytochromes which are present in the thymus nucleus are active, and they are reduced by the addition of NADH.

In mitochondria, electrons are eventually transported down the cytochrome chain to molecular oxygen. Whether this takes place in the generality of cell nuclei is not known, but isolated calf and rat thymus nuclei do respire, and their capacity to produce ATP is impaired in the presence of inhibitors such as cyanide and azide, which are known to block the function of the terminal cytochromes of mitochondria. Considering the spectral evidence for the presence of cytochrome a_3 in thymus nuclei, and the inhibitory effects of cyanide and azide on nuclear ATP synthesis, it is tempting to propose that the electron transport system is the same in the thymus nucleus as in the cytoplasm. However, the insensitivity of nuclear phosphorylation to 95% CO, to Ca^{2+} ions, and to methylene blue (all of which block mitochondrial ATP synthesis) indicates that important differences exist between the 2 systems. The reasons for the differences and the routes of electron transport in thymus and other nuclei remain subjects for further investigation.

The question of the generality of nuclear oxidative phosphorylation is an important problem. Is aerobic ATP synthesis a property of all nuclear types, or is it restricted to thymus lymphocyte nuclei? One approach to this problem has involved the study of ATP synthesis in intact cells under conditions in which mitochondrial oxidative phosphorylation was selectively inhibited, e.g., by Janus Green B and by calcium ions. On the assumption that aerobic ATP synthesis under such conditions reflects the activity of the cell nucleus, it has been concluded that oxidative phosphorylation occurs in the nuclei of thymocytes, lymphocytes, bone marrow cells, ascites lymphoma, and cultured L and T cells, but no net ATP synthesis was observed in rhabdosarcoma, osteosarcoma, and kidney cells (48). It follows that differentiation of the nucleus in different cell types may lead to wide variations in nuclear capacity for aerobic phosphorylations.

It is an interesting fact that aerobic phosphorylation in the nucleus is very sensitive to X-irradiation. Creasey and Stocken have proposed that this is the reason for the differences in radiosensitivity observed between lymphoid tissues and liver (18b).

In summary, (1) many types of nuclei have a capacity for glycolysis. There are 2 substrate-linked oxidative phosphorylations in the glycolytic pathway which provide a source of nuclear ATP. The importance of this pathway is indicated by the great increase in nuclear glycolysis at times of increased nuclear activity, as in liver regeneration. (2) Isolated calf thymus nuclei have a complete citric acid cycle and can convert ^{14}C-glucose, pyruvate and lactate to $^{14}CO_2$. Several enzymes and substrates of the citric acid cycle have also been identified in liver, kidney, and brain cell nuclei. (3) Electrons produced in the catabolism of carbohydrates in isolated nuclei are passed on to NAD and NADP. The reduction of the pyridine nucleotide coenzymes can be observed directly by microfluorimetry of the nuclei in single cells immediately after the addition of glucose to the culture medium. The cell nucleus not only contains high concentrations of the pyridine nucleotide coenzymes, it also is the main site of NAD biosynthesis (see below). (4) In thymus nuclei, electrons are further transferred from NADH (or NADPH) to flavin coenzymes. This reaction appears to take place in other nuclear types as well, judging by their content of peptide-bound flavin. (5) Further steps in electron transport are not known for most types of cell nuclei, but thymus nuclei appear to contain a complete cytochrome sequence and are also capable of aerobic phosphorylation. Many types of nuclei contain cytochrome c and their outer membranes contain cytochrome b_5. The role of these cytochromes in nuclear electron transport is still unknown.

7. The Hexose Monophosphate Shunt

A direct oxidation of glucose phosphate, leading to the production of CO_2 and 5-carbon sugars is a key reaction sequence, because it offers a mechanism for the production of ribose 5-phosphate needed for RNA synthesis. Many nuclei have been found to contain enzymes of this pathway, including glucose-6-phosphate dehydrogenase, 6-phosphogluconate dehydrogenase, phosphoribose isomerase, and phosphoribose mutase. Isolated thymus nuclei have the ability to incorporate free purines into RNA and it can be presumed that they can synthesize ribose 5-phosphate and its pyrophosphoryl derivative (PRPP) and then form the required nucleotides. In this connection, it should be noted that liver nuclei isolated in nonaqueous media have many of the enzymes required for the conversion of orotic acid to the pyrimidine nucleotides, which also indicates the availability of ribosephosphate and its derivatives.

The shunt pathway for the direct oxidation of glucose can be followed by tracer techniques in isolated thymus nuclei. Such nuclei convert 4 times more glucose-1-^{14}C than glucose-6-^{14}C to $^{14}CO_2$, a finding which is usually interpreted to mean that the glucose shunt is operating very efficiently in comparison to glycolysis and the citric acid cycle.

VI. BIOSYNTHETIC REACTIONS IN THE NUCLEUS

A. NUCLEOTIDE COENZYME SYNTHESIS IN THE NUCLEUS

Nuclei isolated in nonaqueous media often have a higher concentration of NAD (on a dry weight basis) than does the cytoplasm. They also contain NADP, the required coenzyme for isocitric dehydrogenase and for enzymes in the glucose shunt. The pyridine nucleotide coenzymes are known to participate in electron transport in the nucleus, and their function is, of course, essential to dehydrogenations and electron transport elsewhere in the cell. It is a matter of some interest, therefore, that an enzyme responsible for the synthesis of NAD should be localized mainly, and perhaps exclusively, in the cell nucleus.

The mechanism of synthesis of nicotinamide-adenine dinucleotide proceeds by a condensation reaction involving ATP. The reaction can occur directly by the condensation of nicotinamide mononucleotide with ATP to yield NAD and inorganic pyrophosphate. (Alternatively, nicotinic acid mononucleotide could react with ATP to form desamido-NAD, followed by amidation to the corresponding coenzyme, NAD.) The condensation of ATP and the pyridine mononucleotide was first described by Kornberg, and the enzyme responsible was subsequently shown by Hogeboom and Schneider to be localized in liver nuclei isolated in sucrose solutions. Its presence in the nucleus has been verified by the use of the nonaqueous isolation technique which showed more than 90% of the total condensing enzyme activity of the cell to be localized in the nucleus.

It has also been reported that mammalian erythrocytes and reticulocytes (which lack nuclei) also lack the NAD-synthesizing enzyme, while nucleated chicken erythrocytes contain the enzyme which can be recovered in the nuclear fraction. Within the nucleus, much of the activity may be associated with the nucleolus, since high activities have been reported for nucleoli isolated from rat liver and from starfish oocytes.

The importance of the enzyme to the overall metabolism of the cell is indicated by the observation that the NAD content of amoeba falls markedly following removal of the nucleus by microdissection. Obviously, if the levels of NAD in the cytoplasm and mitochondria can be regulated by an enzyme which is present in the nucleus, the system confers upon the nucleus great potentialities for the control of glycolysis and oxidative reactions in the rest of the cell.

B. Kinase Activities in the Nucleus

The synthesis of polynucleotides requires the presence of nucleoside triphosphates as the substrates for the appropriate polymerizing enzymes. The RNA and DNA polymerase reactions occurring in nuclei will be discussed later, but it should be pointed out here that mechanisms do exist in the nucleus for the production of their substrates.

Nuclear ATP synthesis, as mediated by glycolysis or oxidative phosphorylations, has already been discussed. Nuclei also contain high concentrations of adenylate kinase, an enzyme which mediates the reversible conversion of 2 molecules of ADP to 1 molecule each of ATP and AMP. Adenylate kinase has been shown to occur in nuclei isolated from liver, from a mammary carcinoma, and from calf thymus. Its presence in the thymus nuclei was verified by isolations in nonaqueous media, thus ruling out adsorption artifacts.

Guanosine triphosphate (GTP) is known to occur in acid extracts of isolated nuclei. Its production in the thymus nucleus is indicated in a number of ways. First, it is produced in the course of the citric acid cycle (when α-ketoglutarate passes through succinyl-CoA to succinate, hydrolysis of succinyl-CoA is linked to the phosphorylation of GDP to GTP). Second, isolated thymus nuclei will incorporate isotopically labeled guanine and guanosine into RNA, thus indicating their potential for the required phosphoryl group transfers leading to GTP.

A similar approach has been applied in considering kinase activities for cytidine and uridine. Both of these isotopically labeled pyrimidine nucleosides are effectively utilized for RNA synthesis by isolated thymus nuclei, and it can be assumed that they were converted to CTP and UTP, respectively. This conclusion is supported by the results of analysis of liver nuclei isolated in nonaqueous media, in which it was shown that the nuclei surpassed the cytoplasmic fraction in their capacity to catalyze the phosphorylation of UMP to UDP and UTP when ATP was added to the incubation medium (79).

Thus, there is good evidence that nuclei possess the kinase activities required for the production of the 4 major ribonucleoside triphosphates required for RNA synthesis. These kinase activities will be heavily dependent upon the ATP concentration of the nucleus, and agents which interfere with nuclear ATP synthesis will also impair the production of the other required RNA precursors.

C. Protein Synthesis in the Nucleus

The great majority of cell types synthesize most of their proteins in the cytoplasm; but the machinery which makes protein synthesis possible is made or assembled in the nucleus. The nucleus is the prime source of the ribosomes, the amino acid tRNA's, the messenger RNA, and other RNA's (e.g., 5 S RNA) essential to ribosomal structure and function.

There is good evidence that within the nuclei of many cell types, protein synthesis proceeds at rates fully comparable to those measured in the cytoplasm. Moreover, the mechanism of amino acid incorporation into nuclear proteins can be studied in suspensions of isolated nuclei and in nuclear subfractions. The methods, and some results of this type of study, will be described below.

It should be stressed at the outset that a nuclear capacity for protein synthesis is an especially important consideration because of the probability that many of the proteins synthesized there may be involved in the regulation of gene activity and may serve as repressors or initiators of RNA synthesis. It should also be pointed out that some of the proteins known to be synthesized in the nucleus are utilized there; the lysine-rich histones, for example, are important elements of the structure of the chromatin; some enzymes, such as lactic dehydrogenase, have been shown to be synthesized in the nucleus and can be detected in association with nuclear ribosomes. LDH is an essential component of the nuclear glycolytic pathway. Other proteins are known to leave the nucleus and enter the cytoplasm, and some pass from one nucleus to another in a multinucleated cell. Proteins can pass as well in the opposite direction, from the cytoplasm to the nucleus, and this type of exchange is regulated precisely at different times in the life of the cell. In what follows, both processes must be considered—but particular emphasis will be placed on the synthetic aspects of nuclear protein metabolism and on the evidence for the nuclear localization of the enzymes, nucleic acids, and nucleoprotein complexes necessary for amino acid incorporation into polypeptide chains.

1. Evidence for Protein Synthesis in the Nucleus: Cell Fractionation Studies

Early experiments using ^{15}N-glycine as a precursor of proteins in the liver established that nuclear proteins were labeled *in vivo* (*12*). This conclusion was supported by kinetic studies of the incorporation of ^{15}N-labeled amino acids in different tissues of the rat, in bird erythrocytes, and *Arbacia* sperm (*21*). The cells were fractionated and their nuclei isolated in citric acid. Measurements of the extent of ^{15}N-glycine uptake made it clear that some proteins of the nucleus were labeled rapidly, but that the histones of nondividing cells were not being synthesized at comparable rates. In general, the synthesis of protein in the nucleus reflected the anabolic activity of the cell of origin; it was high in active tissues, such as liver and pancreas and vanishingly low in inert cells such as sperm and nucleated erythrocytes. The level of isotope uptake into nuclear proteins was approximately the same as for the mixed cytoplasmic proteins in liver and kidney. The results were confirmed and extended in studies of the incorporation of ^{14}C-formate, ^{35}S-methionine, and ^{15}N-glycine in the liver; it was again found that the isotope incorporations into nuclear proteins and into the cytoplasmic proteins were of the same order of magnitude.

Tracer studies of nuclear protein synthesis and metabolism have been carried out in 2 ways. The first, mentioned above, concentrates on the rate of uptake of isotopic amino acids into newly synthesized proteins. This depends, of course, on amino acid pool sizes, activating enzyme concentrations, tRNA concentrations, and other variables. A second approach, which is relatively independent of such considerations, allows the isotopic amino acid to enter the proteins and then measures the retention of previously incorporated amino acids as a function of time. Presumably, in the steady state, the loss of isotopic protein will be compensated by its replacement with unlabeled protein molecules. The rate of loss of the isotope is an indication of protein turnover and replacement, and the kinetics of such turnover can be particularly revealing in studies of nuclear protein metabolism.

It was shown, for example, that the uptake and retention of amino acids by proteins in the nucleus reflect the physiological state of the cell. In pancreatic cells, and in liver, both rates can be altered by feeding and fasting (*4*). This responsiveness of nuclear protein synthesis to changes in the environment is of particular interest because it indicates directly a pattern of continuing interaction between the cytoplasm and the chromosomes. It also suggests the existence of feedback mechanisms which regulate the involvement of the genetic material in the momentary activities of the cell.

The experiments outlined above were carried out by labeling cells *in vivo* and then fractionating the tissue homogenates to obtain nuclei and cytoplasmic components. It might be objected that some cytoplasmic proteins moved into the nucleus during the isolation procedure. (Actually, citric acid isolations tend to extract much of the protein from the nucleus, and it is more likely that uptakes reported for nuclear proteins isolated by such methods are much too low.) Such objections do not apply to studies of nuclear protein synthesis in which cell fractionations are carried out in nonaqueous media. Artifacts due to solubilization and migration of proteins are thus avoided. Kuehl (*50, 50a*) has shown that liver nuclear proteins are labeled 2 minutes after injection of isotopic amino acids. It was subsequently found that nuclear protein synthesis *in vivo* was much less susceptible to inhibition by Puromycin than was protein synthesis in the cytoplasm. As a result of this selective inhibition, the nuclear protein was identifiable as the most radioactive protein in the cell, a result which strongly indicates the presence of an independent protein-synthesizing system within the nucleus (*50a*). This conclusion is supported by the observation that isolated liver nuclei may preserve the ability to incorporate amino acids into their proteins, even after perinuclear ribosomes have been removed by detergent treatments (*90b*). However, more direct approaches to the study of nuclear protein synthesis exist. The most direct method is to demonstrate and analyze amino acid incorporation by isolated cell nuclei, a procedure which has proven very informative since its introduction in 1954 (*11*). This will be considered in some detail below, following a brief consideration of another relatively direct technique—autoradiography of cells and tissue sections.

2. Autoradiographic Evidence for Nuclear Protein Synthesis

Revealing insights into protein synthesis in the cell nucleus and cytoplasm are obtained when one compares grain densities over different areas of the cell following the incorporation of amino acids labeled with ^{14}C, ^{3}H, and ^{35}S. In this way one can arrive at important conclusions about intracellular localization of newly synthesized proteins, relative rates of synthesis, and the migration of proteins between the nucleus and the cytoplasm.

Early studies by Ficq and Errera in Brachet's laboratory indicated a rapid incorporation of ^{14}C-phenylalanine into the nuclear proteins of liver, pancreas, muscle, heart, kidney, lung, and intestinal epithelium [see (15)]. In oocytes and in developing embryos, very high nuclear uptakes of labeled amino acids were observed—a result in keeping with the expectation that dividing cells must synthesize more nuclear proteins, including ribosomal proteins, and lysine-rich histones.

Protein synthesis, or the appearance of newly synthesized protein in the nucleus, has been studied in many cell types by autoradiographic techniques. Amino acid uptake has sometimes been detected in subnuclear structures, such as the nucleolus, or along the chromosomes. Although extensive labeling of nuclear proteins is commonly observed, it has been objected that such studies leave unanswered the question of the site of protein synthesis, because a migration of labeled protein between nucleus and cytoplasm can often be demonstrated. For example, one of the most striking instances of migratory behavior of some nuclear proteins is seen in *Amoeba proteus*. This organism contains a class of proteins which has been shown to exist in much higher concentrations in the nucleus than in the cytoplasm, but these proteins shuttle back and forth between the nucleus and the cytoplasm, while other nuclear proteins remain relatively fixed (33). The peculiar affinity of such proteins for the nucleus, and their ability to leave one nucleus, pass through the cytoplasm, and enter another nucleus was dramatically shown in nuclear transplantation experiments. A nucleus was taken from a cell grown in the presence of ^{14}C-labeled amino acids and placed in an unlabeled host cell which retained its own nucleus. A large fraction of the radioactive protein left the transplanted nucleus and was found preferentially concentrated in the nucleus of the host cell (33). Indeed, the concentration of such cytonucleoproteins in the nucleus is approximately 75–80 times as great as their cytoplasmic concentration. Apart from their intrinsic interest with respect to nucleocytoplasmic interactions, these results clearly illustrate one of the pitfalls in autoradiographic studies of nuclear protein synthesis inside the cell—the extensive protein migration between nuclear and cytoplasmic compartments.

Fortunately, careful kinetic studies of the rates of labeling of nuclear and cytoplasmic proteins can often rule out the possibility that all radioactive proteins found in the nucleus were synthesized in the cytoplasm. Some of the most precise and informative studies are those of Schultze and co-workers (84) who followed the incorporation of 14 different amino acids (labeled with ^{3}H, ^{35}S, and ^{14}C) in 26 different tissues of the mouse, rabbit, guinea pig, and pigeon. A quantitative evaluation of nuclear protein synthesis in these diverse cell types was based on grain counts over equal areas of nucleus and cytoplasm at different times after administration of the radioactive amino acid. Grain densities over liver nuclei, for example, were found to be the same as over the cytoplasm. Moreover, the ratio of nuclear grains to cytoplasmic grains became constant almost immediately after administration of the amino acid. In adrenal cortex, for example, the ratio was constant between 5 and 315 minutes. This indicates that, after an initial brief period of equilibration of the free amino acid pools in nucleus and cytoplasm, the increase in protein radioactivity with time is the same in the nucleus and the cytoplasm, strongly suggesting that the rates of protein synthesis in the 2 compartments are equal. Of course, it might still be objected that these data could also be explained on the basis that all the newly synthesized protein was made in the cytoplasm, and that it trickled into the nucleus at a rate dependent upon cytoplasmic synthesis. This viewpoint is very unlikely because the ratio of nuclear to cytoplasmic grain densities increases greatly at times of increased nuclear activity, as in the course of liver regeneration. The migration theory was made still less likely by 2 other findings. First, the ratio of total grains over the nucleus to total grains over the cytoplasm was surprisingly constant in a wide variety of different cell types. This was true for muscle, where the nucleus is only 1–4% of the cell volume, for liver, where the average nuclear volume is about 10% of the cell, and for plasmocytes, in which the nucleus comprises about 30% of the cell volume. This result makes it highly improbable that the grain counts measured over nuclei are the final result of migration of labeled protein from the cytoplasm to the nucleus, because the same ratio for cells which differ so much in their relative volumes of nucleus and cytoplasm would be quite incomprehensible.

These studies also showed that the amount of radioactive protein appearing in the nucleus was proportional to nuclear volume. In all the different cell types examined there was an approximately equal rate of protein synthesis per unit nuclear volume, and one could make the generalization that large nuclei are most active in amino acid incorporation. Of course, this generalization about uniform nuclear activity has its limitations because the synthetic activity of the nucleus is known to vary with changes in the physiological state of the cell; e.g., in fasting or feeding, in response to hormones or mitogenic agents. Autoradiographic studies of ^{3}H-arginine uptake into *Allium* root meristematic cells show that the nucleolar proteins are very rapidly labeled. After 5 minutes' exposure to the isotopic amino acid many grains are located over the nucleolus, a few over the rest of the nucleus, and a few over the surrounding cytoplasm (*17a*).

A final and convincing autoradiographic demonstration of intranuclear protein synthesis is provided by studies of amino acid uptake in very large cells, such as amphibian oocytes. O. L. Miller has exposed *Xenopus* oocytes to tritium-labeled amino acids for short periods. The presence of radioactive protein in the interior of the nucleus was evident within seconds. Considering the dimensions of this nucleus (400–700 μ) it is unlikely that these results are to be explained in terms of a migration of proteins from the cytoplasm.

Thus, the autoradiographic evidence for intranuclear protein synthesis in intact cells is impressive, and it is considerably strengthened by tracer studies of protein synthesis in isolated nuclei where high-resolution autoradiography proves that amino acid uptake is an intranuclear phenomenon (see below). The early labeling of nuclear proteins and the rapid entry of amino acids into an intranuclear pool has also been verified in cell fractionation experiments, in which liver nuclei were isolated in nonaqueous media shortly after administration of radioactive amino acids (see below).

The resolution attainable by autoradiographic techniques has been considerably improved in recent years by the use of the electron microscope for the study of grain distributions. Some of the applications of this technique in the study of protein and RNA synthesis in the nucleus will be discussed below, but even ordinary light microscopy has permitted the study of protein synthesis in subnuclear organelles; for example, a heavy labeling of nucleolar proteins has been observed in embryonic tissues. and the accumulation of newly synthesized proteins in active regions of the large chromosomes of Dipteran salivary glands can also be readily detected by autoradiographic techniques. Of particular interest in this regard is the recent demonstration by Lezzi (51a) that chromosomes isolated from *Chironomus* salivary glands incorporate radioactive amino acids *in vitro*. Isotope uptake is most pronounced at the chromosomal regions ("puffs") at which protein synthesis can be detected *in vivo*.

3. *Amino Acid Incorporation in Isolated Cell Nuclei*

The most direct approach to nuclear protein synthesis is to study the phenomenon in isolated nuclei. The first experiments were carried out on calf thymus nuclei in 1954 (11). The selection of the thymus as a source of nuclei for experiments on protein synthesis was based on earlier observations that thymus nuclei isolated in isotonic sucrose managed to retain most of the enzymes and cofactors found in nuclei isolated in nonaqueous media; this made it likely that they would also retain the enzyme systems required for amino acid uptake into nuclear proteins.

Suspensions of nuclei of a high degree of purity (about 90–95% pure) can be obtained rapidly by differential centrifugation of homogenates of calf thymus tissue. The suspending medium is 0.25 M sucrose containing 3 mM Ca^{2+} (or Mg^{2+}), and the sedimented nuclei have been found to retain most of their nucleic acids, soluble enzymes, and low molecular weight compounds such as mononucleotides and free amino acids. The isolated lymphocyte nuclei also retain a surprising number of biosynthetic capabilities, including ATP synthesis, purine and pyrimidine nucleoside incorporation into RNA, and isotopic amino acid utilization for protein synthesis.

In studies of protein synthesis, the activity of the isolated nucleus can be followed by the uptake of tritiated or ^{14}C-labeled amino acids added to the nuclear suspension. After incubation at 37°C for various times, the total nuclear proteins (or particular protein fractions) are prepared for measurement of the incorporated radioactivity. The characteristic kinetics of amino acid uptake are shown in Fig. 8. After a brief lag phase, the significance of which will be discussed later, there is a steady and appreciable incorporation of amino acids into peptide bonds. No uptake is observed at 0°C or when ATP synthesis in the nucleus is impaired (see below).

It is important to establish that the uptake observed is actually an intranuclear event and is not due to the presence of whole cells or to cytoplasmic fragments or ribosomes adhering to the outer nuclear membrane. Many different types of evidence have been obtained which attest to the intranuclear localization of the newly synthesized protein, but the most convincing and graphic proof is provided by high-resolution autoradiography under the electron microscope. Isolated calf thymus nuclei were incubated in the presence of ^{3}H-leucine, purified again by centrifugation through a dense Ficoll solution, and then sectioned with the

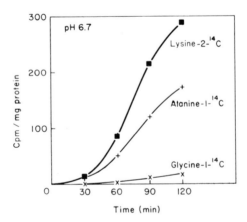

FIG. 8. Time courses of incorporation of different ^{14}C-labeled amino acids into the proteins of isolated calf thymus nuclei. Note the characteristic lag phase in amino acid incorporation at the outset of the incubation (*11*).

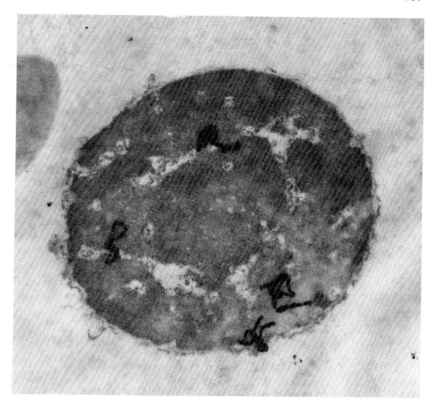

Fig. 9. Electron microscope autoradiograph showing grain distribution over an isolated calf thymus nucleus following incubation *in vitro* in the presence of ³H-leucine. Note that the grains are localized over the interior of the nuclear section and are not in a ring around the periphery (which would be the expected distribution if protein synthesis were limited to ribosomes on the outer nuclear membrane) (*5*) (30,000 ×).

ultramicrotome. The resulting nuclear sections (about 600 Å thick) were coated with a film of liquid photographic emulsion, stored in the dark, and then developed. Examination of the sections in the electron microscope revealed many grains over the nuclei. A typical nuclear cross section showing the localization of the newly synthesized protein is shown in Fig. 9. It is important to stress that the great majority of the grains in such sections occur within the nucleus and not in a ring on the nuclear periphery, a result which rules out the interpretation that the amino acid incorporation observed is due to ribosomes or cytoplasmic tabs on the outer nuclear membrane (*5*).

Chemical evidence that the presence of ³H-leucine in the nucleus is a true indication of protein synthesis was obtained at the same time, the most convincing evidence being the release by tryptic digestion of at least 8 different peptides containing radioactive leucine.

What fraction of the total protein made in intact lymphocytes is synthesized in the nucleus? How efficiently does an isolated nucleus function compared to a nucleus in an intact cell? Information bearing on these questions was obtained by counting grains over isolated nuclei and over different regions of intact cells incubated under the same conditions. In the whole thymocyte, an average of 10.9 grains per cell was observed, and 7.1 grains appeared over the nucleus. Thus, it appears that most of the protein synthetic activity in these small lymphocytes is localized in their nuclei. The free nuclei incubated under the same conditions contained an average of 3.7 grains per nucleus, considerably less than the 7.1 grain average over nuclei in intact cells, but still an impressive degree of retention of function in an isolated subcellular organelle.

Similar evidence for intranuclear protein synthesis in thymus nuclei was obtained with ³H-glycine as a protein precursor. In one of the earliest autoradiographic studies of protein synthesis by isolated nuclei, the presence of incorporated ¹⁴C-phenylalanine in the nucleus was detected by light microscope autoradiography (see Brachet, 15).

Autoradiography has also been employed to demonstrate protein synthesis in nuclei isolated from tissues other than the thymus. Vorbrodt established that nuclei isolated from a Novikoff hepatoma could incorporate ¹⁴C-phenylalanine almost as well as thymus nuclei, but he could not demonstrate amino acid uptake in liver nuclei isolated in isotonic sucrose (95). [Of course, it is known that many enzymes, nucleotides, and free amino acids are lost when liver nuclei are isolated in this way, and it should be pointed out that when liver nuclei are isolated in more complex media they are capable of leucine incorporation into protein (42, 90b).]

Amino acid uptake by suspensions of isolated nuclei has now been reported for a variety of nuclear types from both normal and cancer cells, and from plant and animal tissues. Isolated nucleoli from rat liver (55) and a Novikoff hepatoma also seem to be capable of amino acid incorporation into protein, as are nucleoli from pea seedlings (13), from starfish oocytes (93), and from HeLa cells (99a). In some cases, the uptake of labeled amino acids can only be demonstrated when the isolated nuclei or subnuclear fractions are supplemented with an energy-generating system (ATP), activating enzymes, and transfer RNA's (see below). Isolated nucleoli of HeLa cells are capable of amino acid incorporation (99a). An analysis of the nucleolar proteins by polyacrylamide gel electrophoresis showed that only two of the many protein bands were radioactive. The minor band was basic and may represent a histone, while the major radioactive protein has not yet been characterized. The selectivity of the labeling indicates that contamination by cytoplasmic proteins or bacterial contamination is unlikely, since many types of protein would have been labeled by cytoplasmic or bacterial incorporating systems.

4. Characteristics of Nuclear Amino Acid Incorporation

The uptake of amino acids by suspensions of isolated thymus nuclei is ATP-dependent, as already mentioned, and it can be inhibited by agents which inhibit nuclear ATP synthesis by blocking glycolysis or electron transport. A further test of the ATP dependence of nuclear protein synthesis became possible when it was shown that the nucleotide pools of isolated thymus nuclei could be removed by selective extraction with certain buffers at pH 5.1. Acetate buffer removed nuclear ATP; cacodylate or citrate buffers at the same pH did not (70). Nuclei after extraction with these buffers were incubated under neutral conditions with ^{14}C-alanine. Those exposed to the acetate buffer had lost ATP and the capacity for subsequent amino acid incorporation; nuclei exposed to citrate or cacodylate buffers at the same pH value retained their ATP and were active in protein synthesis. The ATP requirement is, of course, an indication of the presence and activity of amino acid activating enzymes.

One of the strongest arguments for intranuclear protein synthesis is the fact that amino acid uptake in isolated nuclei from calf thymus is not inhibited even in the presence of high concentrations of ribonuclease, unlike cytoplasmic incorporating systems which are inactivated even by trace amounts of the enzyme. It was subsequently shown that ribosomes within the nucleus are not degraded by the added nuclease although ribosomes outside the nucleus are broken down. A similar protective effect has also been observed in liver nuclei (60).

A peculiar aspect of amino acid uptake in isolated nuclei is its susceptibility to deoxyribonuclease (11), a finding which at first raised a question about the utilization of DNA itself as a messenger for nuclear protein synthesis. However, this possibility was found unlikely when it was shown that any large, negatively charged molecule (e.g., polyethylene sulfonate or polyacrylic acid) could substitute for DNA and restore function to DNase-treated nuclei. Without further details, these effects have now been explained in terms of histone release when the DNA is degraded, and histones have been shown to inhibit a wide spectrum of nuclear activities, including ATP, protein, and RNA synthesis (9). The addition of polyanions, such as polyethylene sulfonate, restores function because the histones are effectively removed by complex formation. [There are, in addition, some interesting effects of DNA itself on protein synthesis. DNA–ribosome complexes form readily (68), and denatured DNA will accelerate amino acid uptake in isolated nuclear ribosomes, as has also been shown for DNA and bacterial ribosomes. These aspects of nuclear activity will be considered in more detail later.]

Early experiments made it clear that amino acid incorporation by isolated thymus nuclei was a sodium ion dependent process, unlike amino acid uptake by cytoplasmic systems which is potassium ion dependent. The sodium ion requirement for the incorporation of different amino acids has been studied in some detail (6). It differs for different amino acids and different monovalent cations. It has been found that lithium ions are partly effective as a sodium substitute, but neither potassium, cesium, nor rubidium ions have any stimulatory effect. The specificity of the sodium ion requirement, apart from its value as additional evidence for the intranuclear, as opposed to cytoplasmic localization of the protein synthesizing system, opened the way to the study of transport reactions which control amino acid pool sizes in the nucleus. This control over the entry of amino acids into cells and nuclei may properly be regarded as the first step in the mechanism of protein synthesis.

5. Steps in Nuclear Protein Synthesis

Amino Acid Transport across Membranes. The incorporation of radioactive amino acids into the proteins of isolated nuclei proceeds rapidly in a sodium-containing medium and slowly in the presence of equivalent amounts of potassium. The magnitude of the effect is demonstrated by the comparative data summarized in Fig. 10, which show the effects of increasing concentrations of different ions on the uptake of ^{14}C-alanine. Sodium ions are clearly superior to equivalent concentrations of K^+, Cs^+, Li^+, and Rb^+. A number of quaternary ammonium ions were also tested because of their capacity to serve as effective sodium substitutes in maintaining membrane potentials in isolated nerve. But neither hydrazinium, guanidinium, or tetraethyl ammonium ions could promote alanine incorporation into the proteins of sodium-depleted thymus nuclei.

The reason for the sodium dependence of the incorporation process was sought, but could not be found in studies of the individual reactions of the protein synthetic pathway. There was no specific Na^+ requirement for the nuclear amino acid activating enzymes, nor for amino acid transfer to nuclear tRNA's, nor for amino acid uptake into isolated nuclear ribosomes. This suggested the possibility that sodium ions might influence the

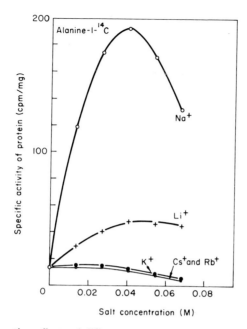

FIG. 10. Comparative effects of different monovalent cations on the incorporation of 1-¹⁴C-alanine into the proteins of isolated calf thymus nuclei. The specific activity of the proteins after a 60-minute incubation is plotted against the salt concentration of the medium (*6*).

rate at which amino acids enter the free amino acid pool within the nucleus. This approach has not only provided an explanation of the sodium ion effect, but it has revealed the presence of specific pumps or permease reactions which regulate transport across the nuclear membrane, and it has also supplied a basis for the understanding of the high intranuclear sodium ion concentration observed in so many different cell types.

The transport of amino acids across the nuclear membrane can be followed by measuring the radioactivity of the nuclear pool at different times after adding labeled amino acids to a nuclear suspension. By comparing the effects of different monovalent cations on this process, one can demonstrate the specific stimulation produced by sodium ions (see Fig. 10). The transport of amino acids into nuclei shows many properties of an enzymic reaction. First, it is highly specific. Only the L-isomers of the naturally occurring amino acids are involved; D-isomers are not acted upon. A further indication of specificity is provided by the results of competition experiments. The transport of ¹⁴C-valine into the nuclear amino acid pool is progressively diminished by increasing amounts of its structural analog, L-isoleucine, while D-isoleucine has no effect (*6*).

Similar competitive effects are observed between ^{14}C-alanine and un-labeled glycine, while the addition of dissimilar amino acids such as iso-leucine, lysine, or glutamic acid has little or no influence on ^{14}C-alanine transport into the nucleus. The results indicate that specific receptor sites are concerned with alanine or valine transport. These sites can readily discriminate between the L- and D-isomers of the appropriate amino acids, but close structural analogs, such as L-isoleucine or glycine, can competitively occupy the sites involved in valine or alanine trans-port, respectively. Such steric specificity is characteristic of enzymic processes, and it may be presumed that the transport of specific amino acids is mediated by specific proteins of the permease type. Specific sodium ion effects on amino acid transport have also been observed in isolated rat spleen nuclei (92f). The degree of sodium dependency varies for different amino acids. The incorporation of threonine, proline and alanine was found to be highly Na$^+$-dependent, but little effect was ob-served on the uptake of glycine, leucine, methionine, or phenylalanine (92f).

Other evidence for the enzymic nature of the process has been pre-sented (6). Amino acid transport has a well-defined pH optimum at neutrality; it has a temperature optimum in the physiological range (37°C) and the temperature coefficient ($Q_{10^\circ} = 2$) indicates that amino acid entry into the nucleus probably involves a chemical reaction and cannot be explained in terms of simple diffusion, since the temperature coefficient of the latter process would be small by comparison.

6. Evidence for Intranuclear Localization of the Amino Acid Pool

The presence of free amino acids in nuclei in vivo has been established by analysis of nuclei isolated in nonaqueous media. Such nuclei have also been used to establish the fact that amino acid entry into nuclear pools is a very rapid reaction in vivo. Within 30 seconds, the specific activity of the nuclear pool in the liver of rats injected with ^{14}C-leucine is equal to that of the cytoplasmic pool (50).

In the study of transport reactions in nuclei in vitro, it is important to minimize or eliminate contamination by intact cells, because whole cells have been shown to concentrate free amino acids to very high levels, often exceeding the concentration in the surrounding medium by several thousand-fold. In the experiments dealing with amino acid entry into isolated thymus nuclei and its sodium dependence, the results were checked in several ways. Nuclei were incubated in the presence of radio-active amino acids in an Na$^+$- or K$^+$-containing medium, then lyophilized and separated from the few cells present by density-gradient centrifuga-tion in organic solvents. Alternatively, the nuclei were purified by den-sity barrier centrifugation in dense sucrose or Ficoll solutions. Both types of purification showed that nuclei in a Na^+-containing medium had concentrated isotopic amino acids far more effectively than did nuclei in a K^+ medium.

However, the most convincing evidence for nuclear localization of amino acid transport reactions is provided by high resolution autoradiography. Nuclei were incubated briefly with ^3H-glycine in the presence of Na^+ or K^+. The distribution of radioactivity was then studied by counting the grains in autoradiographs viewed under the electron microscope. In the Na^+-containing medium, 296 out of 758 nuclei were labeled (39.1%) while in the K^+ medium only 38 out of 409 nuclei were labeled (9.3%). The results ($Na^+/K^+ = 4.2$) clearly indicated the sodium ion dependence of the process and also established the intranuclear localization of the isotope.

7. Amino Acid Accumulation as a Prelude to Protein Synthesis

The ability of the nucleus to accumulate and retain amino acids made it possible to test whether sodium ions must always be present in the medium in order for nuclear protein synthesis to occur. In these experiments, nuclei were placed in a sodium-containing medium and the transport of ^{14}C-alanine was allowed to proceed in the cold (thus preventing amino acid incorporation into protein, negligible below 30°C). The nuclei were then washed to remove the amino acid and the sodium from the suspending medium. They were then resuspended in either Na^+ or K^+ media and incubated at 37°C. A comparison of the time courses of ^{14}C incorporation into the nuclear proteins showed no major differences, establishing that once an amino acid had entered the nuclear pool its uptake into protein can take place in the absence of added sodium ions.

This experiment clarifies the role of sodium ions in nuclear metabolism. Protein synthesis in the nucleus is sodium dependent largely because Na^+ ions are needed to get amino acids to the site of synthesis. Once transport has occurred, the continued presence of sodium ions in the medium is not necessary for incorporation. This result is in full accord with other observations that amino acid activation and transfer to tRNA are not sodium-dependent processes.

8. Sodium Effects on the Transport of Nucleic Acid Precursors

Some degree of sodium ion dependence has also been established for the entry of bases and nucleosides into the nuclear pool (6). This can be followed by the same techniques employed in the study of amino acid transport, i.e., measuring the specific activities of the pools after exposure of isolated nuclei to isotopic RNA and DNA precursors, or by allowing incorporation of the precursor into the product nucleic acid. Not all precursors are affected to the same extent by altering the Na^+ and K^+ concentrations of the suspending medium. Thymidine-2-^{14}C transport was markedly Na^+ dependent, while orotic acid-6-^{14}C uptake was not (6). Adenosine, adenine, and uridine transport into the nucleus show a slight sodium ion dependence.

9. Sodium Ion Concentration in the Nucleus

The experiments on isolated nuclei have shown that protein and nucleic acid synthesis are sodium ion dependent processes, and that this dependence reflects the operation of specific transport systems, presumably localized on the nuclear membrane. These mechanisms are apparently enzymic, and they require sodium ions in order to transfer or admit amino acids and nucleosides to the intranuclear pool.

The conclusion that sodium plays such an important role in the regulation of nuclear metabolism is supported by studies of the intracellular distribution of sodium ions. It is usually claimed that the internal cellular fluid is potassium rich, unlike the plasma, which is rich in sodium. This is true for the cytoplasm, but the cell nucleus is a sodium-rich system. High nuclear sodium ion concentrations have been established in a variety of different cell types. In frog oocytes, for example, the accumulation of sodium ions by the nucleus has been shown by direct analysis of nuclei taken from frozen cells by microdissection (69). High concentrations of nuclear sodium have also been detected by autoradiography of cells exposed for brief periods to media containing ^{24}NaCl and ^{22}NaCl (69). Nuclei isolated in nonaqueous media from avian erythrocytes, calf thymus, and rat liver also have high sodium contents, and the entry of sodium ions into liver nuclei in vivo has been revealed as a very rapid process (89).

Thus, in vitro and in vivo, there is good evidence for a special role of sodium ions in nuclear metabolism, and this is reflected in high intranuclear sodium concentrations. For protein synthesis, the sodium plays some still unknown role in the transport of amino acids.

10. Amino Acid Activation in the Nucleus

In the nucleus, as in cytoplasmic and bacterial systems, the first step in the sequence of reactions leading to the formation of peptide bonds is an activation of amino acids requiring the participation of ATP. The reaction is catalyzed by enzymes similar to those originally described by M. Hoagland in 1955 in liver cytoplasm (36a). This step is described by the equation:

$$\text{Amino acid} + \text{ATP} + \text{activating enzyme} \rightleftharpoons \text{enzyme} - \text{AMP} \sim \text{amino acid} + \text{PP}$$

A convenient method of assay for this activity is based on its reversibility; in the presence of amino acids, ^{32}P-pyrophosphate is incorporated into ATP. By this technique it was shown that extracts of isolated thymus nuclei would form ^{32}P-ATP, but only if amino acids are present in the incubation medium (39). Alternatively, the activation of the carboxy groups of amino acids can be detected by their capacity to react with hydroxylamine (NH$_2$OH) to form the corresponding hydroxamates. When isotopically labeled amino acids are added to nuclear extracts, followed by hydroxylamine, radioactive amino acyl hydroxamates are formed which can be separated from the free amino acids by electrophoresis on paper (60).

It was shown by Hopkins (39) that the nuclear ATP–pyrophosphate exchange reaction is promoted by mixtures of amino acids and by individual amino acids tested separately. Neutral extracts of thymus nuclei contain activating enzymes for at least 15 L-amino acids: alanine, aspartic acid, cysteine, glutamic acid, histidine, isoleucine, leucine, lysine, methionine, proline, serine, threonine, tryptophan, tyrosine, and valine. A low order of activation was found for arginine, glycine, and phenylalanine. Only the natural L-isomers were activated; the corresponding D-amino acids were not able to promote ATP–pyrophosphate exchange. Most of the activating enzyme activity could be precipitated by lowering the pH of the nuclear extract to 5.2, as was first observed for the corresponding enzymes of liver cytoplasm, the well-known "pH 5 fraction."

Many tests have been carried out to make certain that the presence of activating enzymes in isolated nuclei is not an artifact caused by the adsorption of enzymes from the cytoplasm. The most reliable evidence is that provided by nonaqueous nuclear isolations, because such methods preclude any exchange of water-soluble enzymes between nucleus and cytoplasm. Two types of nonaqueous nuclei were selected for analysis because their purity had been established by chemical, enzymic, and immunological tests for the absence of cytoplasmic contamination; these were the nuclei prepared from lyophilized calf thymus and chicken kidney tissues. It was found that the nonaqueous thymus nuclei contained slightly more activating activity (in enzyme units per milligram dry weight) than the equivalent weight of whole tissue. In kidney nuclei the enzyme concentration was about half that observed in whole tissue. In both cases, the high purity of the preparations and their mode of isolation made it certain that the amino acid activating enzymes were of nuclear origin and were not due to cytoplasmic contamination.

Activating enzymes have also been detected in other nuclear types isolated in dense sucrose solutions. The activation of tyrosine has been reported in calf liver nuclei, and an alanine activating enzyme has been partially purified from pig liver nuclei. In recent work on nuclear protein synthesis in liver (60) it was found that the nuclear pH 5 enzyme fraction could activate phenylalanine just as well as the corresponding cytoplasmic fraction could; both enzymes initiated the formation of polyphenylalanine when polyuridylic acid was used as a template in a nuclear or cytoplasmic ribosome system (see below). The activation of valine by liver nuclear enzymes was also shown by the formation of the corresponding hydroxamate.

Considering the variety of nuclear types successfully tested for activating enzyme activity, the occurrence of such enzymes in the cell nucleus promises to be a widespread phenomenon. Ultracentrifugation experiments carried out on nuclear extracts have shown that these enzymes tend to remain associated with the nuclear ribosome fraction, and, as a result, they can be sedimented under conditions which bring down the ribonucleoprotein particles of the nucleus.

11. Amino Acid Transfer to Nuclear Ribonucleic Acids

Earlier studies of cytoplasmic and bacterial systems have shown that the sequel to amino acid activation is a transfer reaction in which the amino acid is coupled to a specific low molecular weight RNA (tRNA) according to the equation:

$$\text{Enzyme–amino acyl} \sim \text{AMP} + \text{tRNA} \rightleftharpoons \text{amino acyl–tRNA} + \text{enzyme} + \text{AMP}$$

The same type of reaction occurs in isolated cell nuclei.

A transfer of amino acids to RNA's in the nucleus can be shown directly by isolating the complexes. Nuclei are incubated with radioactive amino acids in the presence of chloramphenicol or puromycin. (These antibiotics block amino acid incorporation into nuclear proteins, but do not interfere with the formation of the amino acyl–tRNA complex.) The tRNA's can then be isolated by the phenol procedure without risk of contamination by radioactive proteins. Experiments using ^{14}C-leucine and isolated thymus nuclei indicated the rapid formation of a leucyl–RNA intermediate which could be isolated and further characterized. Neither puromycin nor chloramphenicol prevented the formation of this amino acyl–tRNA complex.

The fact that puromycin is an effective inhibitor of nuclear protein synthesis is itself of some interest, because the effectiveness of this antibiotic as an inhibitor of protein synthesis can be interpreted as due to a close structural resemblance between the antibiotic and a natural amino acyl–tRNA complex (98). It follows that puromycin inhibition can be taken as strong supporting evidence for the view that virtually all amino acid incorporation into nuclear proteins proceeds through an amino acyl–tRNA pathway.

The tRNA's of the nucleus, as in the cytoplasm, are low molecular weight RNA's with sedimentation coefficients of about 4 S. However, much of the tRNA in thymus nuclear extracts occurs in close association with ribonucleoprotein particles and can be sedimented rather easily. The leucyl–RNA complex of the nucleus has been studied in more detail than the other tRNA's. It has many properties which show its homology or identity with the amino acyl–tRNA complexes of the cytoplasm. It is nondialyzable and stable to dilute acids. On the other hand, a brief exposure to dilute alkali (e.g., 10^{-3} N NaOH) removes all the radioactivity from ^{14}C-leucyl–RNA, and subsequent chromatography on paper separates the released ^{14}C-leucine as the free amino acid.

The treatment of amino acyl–tRNA with ribonuclease is known to release the amino acid still bound to the terminal nucleoside of the RNA chain (99). A short ribonuclease digestion of the nuclear leucyl–RNA also releases the radioactivity in an acid-soluble form; and the resulting aminoacyl nucleoside can be separated by ionophoresis or chromatographic techniques. In both cases, treatment of the isolated leucyl nucleoside with dilute alkali releases all of the radioactivity as free leucine-^{14}C and leaves the nucleoside. These components are readily separable by chromatography or electrophoresis. The nucleoside has been identified as adenosine (2), and it follows that the receptor group in nuclear tRNA is a terminal adenylic acid, as is also the case for the tRNA's of cytoplasmic and bacterial systems.

To date, extensive purifications and nucleotide sequence studies have not been carried out on tRNA's derived from the cell nucleus, nor is it known whether any differences exist between tRNA's in the nucleus and the cytoplasm. Since the tRNA's of the cell are made in the nucleus, there is no reason to presume that differences do exist. However, the possibility remains that nuclear and cytoplasmic compartments may differ with respect to the concentration of tRNA's for particular amino acids (just as they differ in relative concentrations of glycolytic enzymes and their substrates). A multiplicity of tRNA's for particular amino acids is known to occur in other systems, and accounts, in part, for the apparent degeneracy of the code. It will be well worth investigating to see whether there is a preferential localization of different tRNA's specific for the same amino acid in different subcellular organelles.

Transfer RNA's are commonly characterized by their high content of methylated bases. Nuclear tRNA's are also methylated. The transfer of radioactive methyl groups from methionine or S-adenosyl methionine can be followed both by autoradiographic techniques *in vivo* and during *in vitro* incubations of isolated nuclei. It is not known whether nuclear and cytoplasmic tRNA's differ in their degree of methylation, but the methylating enzymes appear to be nuclear in origin.

12. *The Role of Ribonucleoprotein Particles in Nuclear Protein Synthesis*

So far, the pathway of amino acid incorporation into the proteins of the nucleus has been shown to resemble that in the cytoplasm, in the sense that both involve activating enzymes and specific tRNA's. In the cytoplasm, the next steps require the participation of ribosomes, GTP, associated enzymes or binding proteins, and messenger-RNA's which direct the sequence of amino acids in the growing polypeptide chain. There is good evidence that the nuclear mechanism for protein synthesis has much in common with that of the cytoplasm.

The most direct support for the involvement of ribosomes in nuclear protein synthesis comes from the study of 2 types of nuclei—thymus and liver. Early attempts to fractionate thymus nuclear proteins following the incorporation of ^{14}C-alanine *in vitro* showed that neutral extracts of the nucleus contained a ribonucleoprotein fraction of high specific activity (*11*). With successive salt extractions and differential centrifugation, many ribonucleoprotein fractions could be separated (*28*), a result in keeping with morphological evidence for the complexity of RNP particles in other nuclear types and along Dipteran chromosomes, and also with the expectation that nuclei should contain polysomes of varying size classes. Some of the RNP fractions from thymus nuclei contained particles with properties of composition, morphology, and sedimentation similar to those of cytoplasmic ribosomes.

Tracer studies in several laboratories have established that proteins associated with such particles are rapidly labeled after exposure of nuclei to isotopic amino acids and it has been concluded that they are the major sites of intranuclear protein synthesis (*2, 96, 99a*). A critical point to establish is that the uptake observed in these experiments took place on ribosomal particles inside the nucleus and not on ribosomes on the outer nuclear membrane. Autoradiographic evidence for intranuclear protein synthesis has already been described, but the biochemical evidence offers additional support for the conclusion that amino acid uptake by RNP particles is an intranuclear event. The intranuclear localization of the active ribosomes (i.e., RNP particles containing newly synthesized protein) seems certain because amino acid uptake is not halted when nuclei are exposed to ribonuclease; and it has been shown that RNP particles within isolated thymus nuclei and liver nuclei are not degraded by RNase under these conditions (*28, 60*). Moreover, the labeling of proteins associated with thymus nuclear ribosomes is a sodium-dependent process—a fact which indicates a dependence on amino acid transport across the nuclear membrane as discussed. Finally, amino acid uptake into the proteins associated with nuclear ribosomes is blocked if the nuclei are pretreated with deoxyribonuclease—yet this enzyme does not affect protein synthesis in cytoplasmic ribosomal fractions. The effect of DNase on nuclei is largely due to the inhibitory effects of the histones released when DNA is degraded. More than 92% of the histones still remain in nuclei after exposure to DNase under these conditions, which makes it unlikely that they act on extranuclear synthesizing systems. Moreover, it has been shown that the direct addition of histones to RNP particles isolated from the nucleus does impair their capacity for peptide bond synthesis *in vitro* (*9*).

Several laboratories have described the purification and properties of nuclear ribosomes. Thymus nuclear ribosomes were separated and characterized, both chemically and physically, and photographed under the electron microscope (*75*). It was found that the isolated particles had sedimentation coefficients of about 78 S. The structure and sedimentation properties of the purified nuclear ribosomes depend on the Mg^{2+} ion concentration of the suspending medium: At $5 \times 10^{-4} M$, the particles occur mainly as a 78 S component (monosomes). Removal of the Mg^{2+} (by dialysis against solutions of the chelating agent EDTA) resulted in extensive disruption of the 78 S component and the appearance of smaller particles (50 and 33 S) corresponding to the large and small ribosomal subunits, respectively. The 78 S particles were viewed under the electron microscope—they were roughly spherical and appeared to be about 200 Å in diameter in shadow-cast preparations.

Nuclear ribosomes have also been identified in extracts of rat liver nuclei. It is important to stress that these particles were prepared from nuclei after removal of the outer nuclear membrane (*82*), or after the ribosomes on the membrane were selectively destroyed by exposure to chelating agents and ribonuclease (*60*). Both monosomes and disomes have been reported in liver nuclear extracts, and the isolation of nuclear polysomes has also been described. Recently Zimmerman and co-workers described the isolation of ribosomal particles (i.e., particles with the buoyant density of ribosomes) from HeLa cell nuclei after the nuclear membrane had been removed by detergent treatment. The proteins synthesized *in vitro* by isolated HeLa nuclei and nucleoli were associated with these particles (*99a*). Protein synthesis in isolated HeLa nuclei was also observed by Gallwitz and Mueller (*31a*). It follows that earlier reports of the absence of ribosomes from HeLa nuclei (e.g., *71*) need not

be regarded as conclusive, although these reports are essentially correct
in concluding that by far the greater part of the ribosomes of the HeLa
cell occur in the cytoplasm (*96a, 71*).

13. *The Association of Nascent Protein with Nuclear Ribosomes*

What is the evidence that the labeling of nuclear ribosomes represents
the synthesis of nuclear proteins other than those which are part of the
structure of the ribosome itself? This is an important distinction because the nucleus, as will be shown later, is the site of ribosome synthesis
and assembly. One answer to this question is provided by the results
of "cold chase" experiments in which isolated thymus nuclei were incubated with a radioactive amino acid for very brief periods, washed, and
resuspended in a medium containing a great excess of the same amino
acid in the unlabeled (^{12}C) form. The distribution of radioactivity within
the nucleus then undergoes some very revealing changes.

The ribosomal fraction initially contains most of the newly synthesized
radioactive protein in short-term labeling experiments (1 to 3 minutes).
However, further incubation in a nonradioactive medium results in a
shift of radioactivity from the ribosomal fraction to the soluble phase
of the nucleus. The result is fully in accord with the expectation that
some of the protein associated with the ribosomes at early times is
nascent protein and is the precursor of other components of the nucleus.

A more detailed analysis of this precursor–product relationship became
possible when it was shown that treatment of the ribosomal fraction with
detergents would release loosely associated protein molecules but not
nascent polypeptide chains (*2*). For example, thymus nuclear ribosomes
prepared in sucrose density gradients contain 63% protein and 37%
RNA. Much of this protein is released from the ribosomes by treatment
with 0.5% deoxycholate (DOC), leaving the ribosomes with a correspondingly higher RNA content. The effects of the detergent fractionation
can be visualized if one pictures the ribosomes as containing a DOC-
resistant "core" and a removable protein "coat." [The terms core and
coat should not be taken in the strict sense of spatial localization, since
this has not been established for nuclear ribosomes, but recent electron
microscopic evidence for the existence of a channel in the large subunit
of cytoplasmic ribosomes (*81*) does suggest the importance of spatial
position in protein biosynthesis.]

Cold chase experiments have revealed some interesting shifts of
radioactivity from the core to the coat proteins and out into the soluble
phase of the nucleus. A typical experiment is summarized in Fig. 11,
which plots the specific activity of each protein fraction following a 1-
minute pulse of ^{14}C-leucine and a subsequent cold chase in a nonradio-

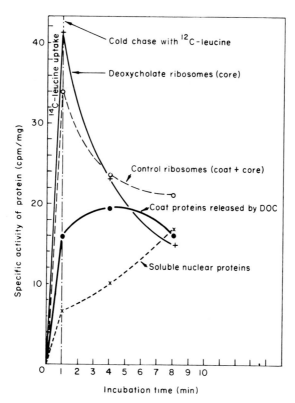

FIG. 11. Changes in the specific activity of nuclear ribosomal protein fractions following a brief incorporation of 1-¹⁴C-leucine and a cold chase. Isolated thymus nuclei were incubated with radioactive leucine for 1 minute, washed to remove the radioactive amino acid, and then incubated in a medium containing a great excess of unlabeled leucine. The ribosomes were prepared and fractionated by deoxycholate treatment to yield the coat proteins and the DOC-resistant core fraction. The specific activities of the coat, core, and supernatant proteins are plotted as a function of time during the cold chase. Note that the core protein is most radioactive at early times, but loses its activity quickly while the activity of the coat proteins rises and then declines. The specific activity of the soluble proteins of the nucleus gradually comes to exceed that of the ribosomal fractions (2).

active medium. Nuclear samples were withdrawn and analyzed at the 1-minute time point and at short intervals thereafter. The ribosomes were separated on a sucrose density gradient and then treated with DOC to release the associated coat proteins. The specific activity of this fraction was compared with that of the tightly associated proteins of the core and with the proteins in a soluble phase of the nuclear extract. It is clear that, at the outset, the proteins of the ribosomal core are much more radioactive than any other protein fraction in the nucleus. As time proceeds, the specific activity of the core protein falls, while that of the coat proteins rises and then falls. The activity of the soluble protein fraction continues to increase as the ribosomal proteins decline in activity.

The experimental data fit a hypothetical sequence of events which seems very plausible. The core proteins include polypeptide chains in the process of formation which are so tightly bound to the ribosome–tRNA–messenger RNA complex that they are not extractable by detergent treatment. As the synthesis of polypeptide chains proceeds to completion, the finished protein is released from the core and makes its way into the coat fraction, still weakly associated with the ribosome, but now removable with the aid of detergents. Within a few minutes, the protein leaves the vicinity of the ribosome and makes its way into the soluble phase of the nucleus (2). This sequence of events can be followed with comparative ease in a nucleus such as the thymus nucleus, in which the synthesis and replacement of ribosomal structural proteins is a relatively slow event; if replacement were rapid, much of the newly synthesized protein might remain attached to newly formed ribosomal subunits. The tight binding observed between nuclear ribosomes and nascent protein has a counterpart in cytoplasmic systems, in which it has been shown that nascent hemoglobin in reticulocyte ribosomes is resistant to protease digestion (57).

A study of the proteins associated with thymus nuclear ribosomes has revealed the presence of lactic dehydrogenase (96), an enzyme which has also been shown to be synthesized within the liver nucleus (50). Ribosomal particles prepared from HeLa cell nuclei after the incorporation of ^{14}C-labeled amino acid in vitro have been shown to contain two types of radioactive protein, one of which has the solubility properties of a histone (99a).

14. Amino Acid Incorporation by Ribosomal Fractions Isolated from Nuclei

Several studies have been carried out in which ribosomes prepared from calf thymus or rat liver nuclei were used to mediate peptide bond formation in vitro. The results have further clarified the mechanism of amino acid incorporation into nuclear proteins and have again emphasized the essential similarity of nuclear and cytoplasmic protein synthesis. For example, it could be shown that isolated nuclear ribosomes, like cytoplasmic ribosomes, require GTP for amino acid incorporation [see Allfrey (2)]. Both systems are magnesium ion dependent, and both can be inhibited by the addition of puromycin or ribonuclease.

Nuclear and cytoplasmic ribosomes from the liver have been compared according to their capacity to utilize cytoplasmic and nuclear pH 5 enzyme fractions. Ribosomes from both parts of the cell accepted amino acids equally well after activation by either pH 5 fraction (60).

Nuclear ribosomes could also be shown to accept synthetic polynucleotides as messenger RNA's. Polyuridylic acid promoted the synthesis of polyphenylalanine in vitro, while polycytidylic acid and polyadenylic acid directed the synthesis of polyproline and polylysine, respectively (60). Thymus nuclear ribosomes respond to the addition of adenylic acid–uridylic acid-rich RNA's by increased amino acid uptake (9). Since this class of RNA is isolated from nuclei (see below), their role as natural messengers is suggested.

One of the surprising results obtained in the isolated nuclear ribosome system is the stimulation of amino acid incorporation induced by the addition of DNA [see Allfrey (2)], an effect that is not obtained by adding other polyanions. It has been shown by Naora (68) that single-stranded DNA is far more effective than double-stranded DNA, a result which makes it likely that nuclear ribosomes (like bacterial ribosomes, e.g., ref. 38), can utilize single-stranded DNA directly as a messenger. Whether this ever happens in the nucleus is not known; nor is it known whether DNA-associated factors similar to those which are required for the translation of natural messenger RNA's in E. coli (80) occur in mammalian systems. There is good evidence, however, that protein synthesis in the nucleus depends upon a continuing synthesis of messenger-type RNA's.

15. The Role of "Messenger" RNA's in Nuclear Protein Synthesis

Studies of RNA synthesis in isolated thymus nuclei (85) and in many other types of nuclei in vivo and in vitro (see Georgiev, 32) have established the fact that RNA's resembling DNA in overall base composition exist in the nucleus and are synthesized there at very rapid rates. The present state of knowledge regarding the properties and function of such DNA-like RNA's leaves much to be desired (see below), but there is little reason to doubt that some of them serve as messengers for protein synthesis.

Are such "messengers" utilized within the nucleus? One approach to this fundamental question has involved the study of the linkage between protein and RNA synthesis in isolated cell nuclei. It was observed in 1955 that amino acid uptake in isolated nuclei has a characteristic lag phase which precedes the period of most active protein synthesis (see Fig. 8). This delay in amino acid incorporation can be abolished by a prior incubation of the nuclei under conditions which permit RNA synthesis, but not if RNA synthesis is blocked (11). If RNA synthesis is inhibited during the lag phase, the rate of protein synthesis falls progressively as the incubation continues. However, if one waits 30 minutes before adding inhibitors of RNA synthesis [both Actinomycin D and DRB (5,6-dichloro-β-D-ribofuranosyl benzimidazole) have been used successfully], the rate of protein synthesis remains high and shows no signs of inhibition for another 30–60 minutes although RNA synthesis stops immediately. Similarly, the addition of DRB at 60 minutes has no effect on amino acid uptake for the next 30 minutes. (See Fig. 12.) Eventually, once RNA synthesis is stopped, protein synthesis also declines, but not precipitously. It should be pointed out that RNA synthesis in the isolated thymus nucleus is not dominated by ribosomal RNA synthesis, since much of the newly synthesized RNA's appear to be AU-rich in base composition (85) unlike ribosomal RNA's which have high guanylic acid + cytidylic acid contents. These results fit the hypothesis that the uptake of amino acids by nuclear ribosomes *in situ* requires a prior synthesis of messenger RNA's (9, 11). Once such RNA's have been formed, protein synthesis can continue for upward of 30 minutes. Since

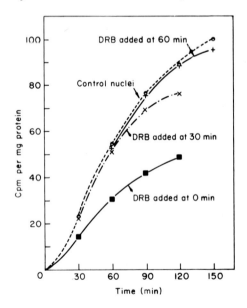

FIG. 12. The effect of an inhibitor of RNA synthesis (DRB) on the incorporation of 1-^{14}C-alanine into the proteins of isolated calf thymus nuclei. The time course of incorporation of the amino acid in the presence of DRB is compared with that observed in control nuclei (upper curve). Note that DRB (which stops RNA synthesis quickly) also inhibits protein synthesis when it is added before the lag phase in uptake ($t = 0$). However, addition of DRB at later times ($t = 30$ or 60 minutes) is much less effective in blocking amino acid uptake (11).

amino acid uptake does begin to decline 30–60 minutes after RNA synthesis is blocked, it seems probable that messenger RNA's of limited lifetimes direct much of the protein synthesis on intranuclear sites. In this connection it should be noted that a large part of the RNA made in the nucleus stays in the nucleus and is degraded there without ever going out to the cytoplasm. (Evidence for this will be summarized below.) Presumably, some of this RNA turnover within the nucleus reflects the continuous synthesis and replacement of messengers for intranuclear protein synthesis.

D. Ribonucleic Acid Synthesis in the Nucleus

In the past few years a number of excellent reviews have appeared on the subject of ribonucleic acid synthesis in the nuclei of unicellular and higher organisms (21b, 32, 43, 73, 77, 90, 94). These will be cited frequently in the following discussion as an alternative to a more detailed bibliography of this popular field.

In what follows, emphasis will be placed on the evidence that (1) clearly establishes the nucleus as the major site of RNA synthesis in the cell, (2) implicates both the DNA of the chromatin and DNA associated with the nucleolus in the synthesis of ribonucleic acids, (3) places the enzymic mechanism of nucleotide polymerization (RNA polymerase activity) within the cell nucleus, (4) proves the nucleolus to be the site of synthesis and assembly of ribosomes, (5) demonstrates that only a small fraction of the DNA in the differentiated cells of higher organisms is functional as a template for RNA synthesis, while most of it is inert, and (6) indicates that the function of DNA in RNA synthesis is subject to hormonal and other types of control.

The latter topic—the control of genetic activity—has many facets, some of which will not be considered in this review; but attention will be paid to the relationship between the fine structure of the chromatin and its function, and to the problem of the role of DNA-associated proteins (such as the histones and nuclear phosphoproteins) in the organization and activity of the genetic material.

1. A Brief Classification of RNA Types

Most of the ribonucleic acid of the cell is ribosomal RNA. In parenchymal cells of the liver, for example, estimates of ribosomal RNA content range from 60 to 90%. Ribosomal RNA's like the ribosomal subunits from which they are derived, can be separated according to their sizes and densities by centrifugation techniques.

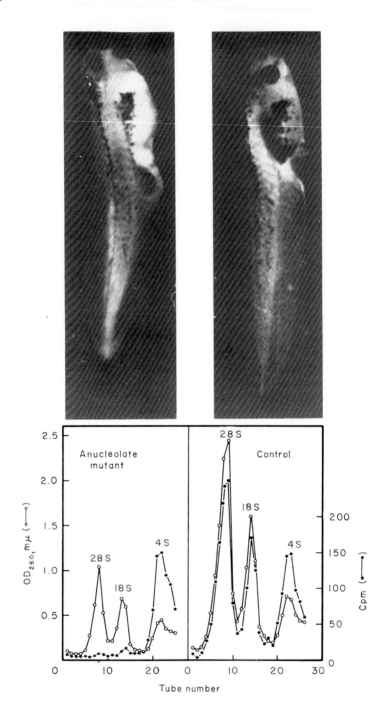

Intact liver ribosomes have sedimentation coefficients of 78 S, corresponding to a particle weight of about 5 million. In the absence of Mg^{2+} ions, ribosomes dissociate into equal numbers of large subunits (47 S; MW 3.3×10^6) and small subunits (32 S; MW 1.6×10^6). About 60% of the mass of the mammalian cell ribosome is protein; the rest is RNA which can be isolated by displacing and denaturing the protein with phenol. The RNA remains in the aqueous phase and can be collected by precipitation with ethanol.

The analysis of ribosomal RNA's by centrifugation in sucrose density gradients reveals 2 well-defined peaks (see Fig. 13). In HeLa cells, the larger component has a sedimentation coefficient of 28 S, corresponding to a molecular weight of 1.9×10^6, while the smaller component appears as an 18 S peak with a molecular weight of 0.71×10^6 (60a). These RNA's arise from the different ribosomal subunits, and they differ in overall base composition. [In liver ribosomes, the 18 S component is 20.7% A, 21.4% U, 32.2% G, and 25.7% C, while the 28 S component is 17.2% A, 19.2% U, 34.9% G, and 28.7% C (32).] A relatively high content of guanylic and cytidylic acids (G + C/A + U = 1.6) is typical of ribosomal RNA's. The 2 ribosomal RNA peaks also differ with respect to the sequence of nucleotides in the polynucleotide chain, and it has been shown by hybridization experiments that 18 and 28 S RNA's are complementary to different regions of the DNA in the cells of origin. The 28 S RNA appears to be complex, and to consist of a large RNA molecule joined by hydrogen-bonding to a small 7 S component of molecular weight about 40,000 [see (60a, 21b)].

Ribosomes also contain a relatively low molecular weight RNA with a sedimentation coefficient of about 5 S. This is not an amino acid tRNA, since it also differs from the latter in its low proportion of unusual or methylated bases. The 5 S RNA is found in the large ribosomal subunit.

Isotopic studies of RNA synthesis also reveal the presence of huge molecules with sedimentation coefficients of 45 and 32–36 S. These are ribosomal RNA precursors and their formation will be discussed in more detail below.

The amino acid tRNA's comprise about 10–15% of the total cellular

FIG. 13. Absence of ribosomal RNA synthesis in the anucleolate mutant of *Xenopus laevis*. The sucrose density-gradient analyses of the total RNA's isolated from ^{14}C-labeled embryos containing 0-nucleoli and 2-nucleoli are shown below the photographs of the corresponding mutant and wild-type embryos. Note that although the mutant does contain some 18 S and 28 S RNA (of maternal origin) this RNA is not radioactive, while the wild-type RNA's show radioactivity beneath both ribosomal RNA peaks: ○—○, optical density at 260 mμ; ●—●, radioactivity (from Brown and Gurdon, *15a*).

RNA and are a more heterogeneous population. Their sedimentation coefficients (4 S) indicate their relatively small size, with molecular weights of 25,000–30,000 and 75–100 nucleotides in length. Like ribosomal RNA's they are synthesized in the cell nucleus. Isolated cell nuclei contain tRNA's in readily detectable amounts. The multiplicity of tRNA's reflects the requirement for at least 1 tRNA for each of the naturally occurring amino acids. (See Chapter III by von Ehrenstein.)

In addition to the above major classes of RNA, there are many other RNA populations in the nucleus. Since these are usually characterized by their relatively high contents of adenylic and uridylic acids they are referred to as "AU-rich" RNA's or "DNA-like" RNA's. (Their resemblance to DNA is based on the equivalence of uracil in RNA and thymine in DNA for complementary base pairing to adenine in the formation of a Watson-Crick double helix.) In part, the heterogeneity of this type of RNA reflects the multiplicity of the structural messengers required for the synthesis of the many proteins coded for by the functional DNA in a particular cell type. In addition, however, there appear to be other "DNA-like" RNA's whose function in protein synthesis has not been established. Some of these have enormous molecular weights and sediment as 60–80 S peaks.

At the other extreme are very low molecular weight RNA's of unusual base composition which have been obtained from the nuclei of plant cells (14) and from chick embryos (40a). Their function is unknown, though it has been suggested that they play some role in genetic regulation (11a, 14, 40a).

A further complexity is the presence in higher cells of small amounts of homopolymers, such as polyadenylic acid, and of enzymes, such as poly A synthetase and poly C synthetase, which catalyze their formation. Again, the biological function of these RNA-like molecules is unknown, though it has been proposed that poly A formation would offer a storage mechanism for ATP in which some of the potential chemical energy of ATP would be preserved (see Allfrey, 2). A low molecular weight uridylic acid-rich RNA has been prepared from nuclei of a Novikoff hepatoma (65b). It is one of many nuclear RNA's in the 4–7 S size range for which the function has not yet been established. One of the most unusual recent developments in the study of nuclear polynucleotides is the discovery of poly-ADP-ribose. This compound is formed from nicotinamide–adenine dinucleotide by the elimination of nicotinamide; the resulting ADP-ribose subunits are enzymically linked together by the formation of glycosidic linkages between the C1'-position of the free ribose (which had previously been joined to nicotinamide) and the 2'-carbon of the adenylic acid ribose in the next ADP-ribose subunit (29a, 77c). The enzyme which carries out this reaction occurs in the chromatin (92h). A large proportion of the product appears to be joined to histones (69a).

These, then, are the multiple classes of RNA molecules found in the cell. Most of them play some part in directing the synthesis of proteins. Since the genetic information specifying amino acid sequences in proteins is encoded in the DNA, it is not surprising that the synthesis of RNA's should be so largely relegated to the subcellular organelles containing DNA: the nucleus, the mitochondria, and the chloroplast.

2. Evidence for Nuclear Synthesis of Ribonucleic Acids

Since the early experiments of Marshak (see Brachet, 15) and of Bergstrand and co-workers (12), many investigators have confirmed the observation that, whatever the chosen precursor, incorporation always proceeds much faster into nuclear RNA than into RNA of the cytoplasm. So uniformly was this observed that it was assumed from the outset that the cell nucleus is the primary site of RNA synthesis in the cell. Kinetic studies were carried out in an attempt to demonstrate a simple relationship between the nuclear RNA as precursor and cytoplasmic RNA as product. The approach had very limited success; the kinetic experiments did not yield simple solutions but indicated only that some of the RNA in the cytoplasm could have been derived from RNA's previously synthesized in the nucleus. In retrospect, the limited success of the kinetic approach is hardly surprising, because simple kinetic equations would not be expected to apply to complex mixtures of different RNA's which not only differ in base composition and nucleotide sequence, but also differ in their relative proportions in nucleus and cytoplasm, in their rates of synthesis and decay, and in the time required for subsequent modifications (such as methylation) and for movement to the cytoplasm. Earlier workers could not have surmised that much of the nuclear RNA would be degraded *in situ* without ever entering the cytoplasm, nor that some cytoplasmic RNA's would be made independently in DNA-containing subcellular organelles such as mitochondria and chloroplasts. Still, nuclear RNA was much more radioactive than cytoplasmic RNA, particularly at early times after injection of the isotopic precursors.

Compelling and graphic evidence for the central role of the nucleus in cellular RNA synthesis came from autoradiographic studies of the type shown in Fig. 14(A). Short-term exposure of cells to isotopically labeled RNA precursors always leads to a clear-cut localization of radioactive RNA in the nucleus. When the labeled cells are subsequently transferred to a nonradioactive medium, the radioactive RNA is lost from the nucleus and accumulates in the cytoplasm. This transition is particularly evident in Prescott's autoradiograph of *Tetrahymena* cells which were labeled

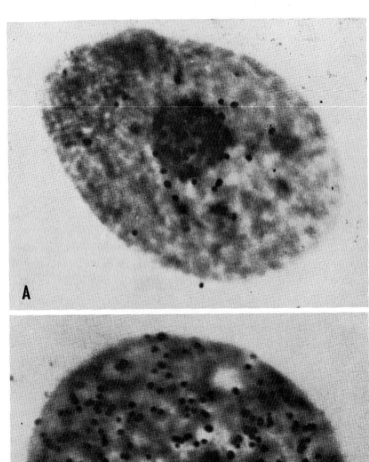

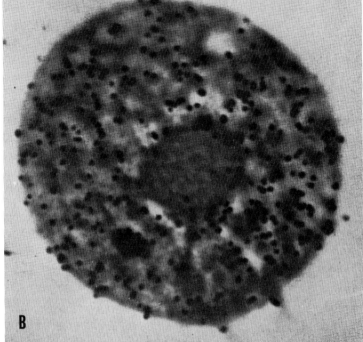

with ^{3}H-cytidine for 15 minutes and then given a cold chase for 88 minutes [see Fig. 14(B)].

The time lapse between labeling of the nucleus and appearance of radioactive RNA in the cytoplasm varies from one cell type to another. In *Neurospora*, the delay is 1–4 minutes, while in HeLa cells, the lag is of the order of 10–30 minutes. In very large cells, such as those of *Triturus*, the lag can be much longer.

The transit of RNA from nucleus to cytoplasm has been observed by autoradiographic techniques. Goldstein and Plaut transplanted the nucleus of an ameba that had been heavily labeled with ^{32}P-orthophosphate into an unlabeled host cell. Radioactivity gradually moved from the radioactive nucleus into the cytoplasm of the host, but no radioactive RNA was observed to move into the unlabeled nucleus of the recipient cell. It was concluded that RNA is transported from the nucleus to the cytoplasm but not in the reverse direction (see Prescott, 77). However, recent tracer studies of RNA migrations following nuclear transplantations in ameba suggest that radioactive RNA's can enter a nonradioactive nucleus from the surrounding cytoplasm. The nature and extent of this phenomenon and its relevance to the massive flow of RNA's in the reverse direction remain to be determined.

Compelling arguments for nuclear RNA synthesis come from studies of enucleated cells. In enucleated *Tetrahymena*, labeling of RNA is completely absent (77). Similarly, HeLa cells in culture, *Acanthamoeba*, and amphibian cells fail to incorporate nucleosides into RNA after the removal of their nuclei. In the differentiation of red blood cells in mammals, the nucleus is normally ejected during the maturation of the normoblast to form a reticulocyte, and when this occurs, the cell loses its capacity for RNA synthesis. In avian reticulocytes and erythrocytes, the nucleus is not ejected, and such nuclei can be induced to resume RNA synthesis when they are placed in the proper environment (34).

An ingenious experiment by Zalokar offers still more evidence for the central role of the nucleus in RNA synthesis. He centrifuged the tubular hyphae of *Neurospora* under conditions in which the various subcellular organelles were stratified into distinct and recognizable layers. The organelles were viable after this treatment and could incorporate both RNA and protein precursors. In short-term labeling experiments with ^{3}H-uridine, the isotopic RNA appeared only in the nuclear zone of the centri-

---

Fig. 14. Changes in the intracellular distribution of radioactive RNA in *Tetrahymena*, following: (A) a brief labeling with ^{3}H-cytidine, (B) followed by a cold chase in nonradioactive medium. Note that the nucleus has lost all its label while radioactive RNA has accumulated in the surrounding cytoplasm (from Prescott, 77).

fuged hyphae, while at later times it made its appearance in other segments of the hyphae. Grain counts indicated that over 99% of the RNA synthesized was made in the nuclear zone, and it was concluded that virtually all the cytoplasmic RNA must have had its origin there.

Impressive as the autoradiographic evidence is, objections have been raised about the complexities of unequal pool sizes in nucleus and cytoplasm, and the possibility that some nuclear RNA might be degraded to yield radioactive bases or nucleosides which could move to the cytoplasm and be used there for independent cytoplasmic RNA synthesis. Critical analysis of nuclear RNA turnover by Henry Harris (34, 34a) did, in fact, offer strong evidence that some nuclear RNA is degraded while still inside the nucleus. It has become increasingly clear that this is a major aspect of nuclear RNA metabolism. Studies of nuclear RNA "turnover" in a variety of cell types support the view that most of the newly synthesized RNA is subject to rapid degradation soon after synthesis, and that this RNA does not leave the nucleus. In HeLa cells, for example, at least 90% of the heterogeneous nuclear RNA's turn over within the cell nucleus with extraordinary rapidity. The average half-life of such RNA's is only about 3 minutes (90a).

A more direct approach became possible with the availability of isolated nuclei that are capable of ATP synthesis. In 1955 it was observed that isolated calf thymus nuclei could incorporate ^{32}P-orthophosphate and a variety of isotopic purine bases, purine and pyrimidine nucleosides, and orotic acid into RNA (see Allfrey and Mirsky, 9). (The subsequent conversion of these precursors into the required nucleoside triphosphates was made possible by the presence in thymus nuclei of both ATP and the required enzymes.)

An important point made in these early experiments was that RNA synthesis is inhibited if the nuclei are first treated with deoxyribonuclease and that complete removal of the DNA left the nucleus incapable of nucleoside, ^{32}P, or orotic acid uptake.

With the discovery and characterization of RNA polymerase in rat liver nuclei by S. B. Weiss in 1960 (32), not only the site, but the mechanism of RNA synthesis was elucidated. RNA is synthesized from the ribonucleoside triphosphates by an enzymatic reaction which also requires DNA as a template. By supplying the nucleoside triphosphates as precursors, the requirement for ATP synthesis by isolated nuclei is bypassed. Both the enzyme RNA polymerase and the DNA remain in most types of nuclei during their isolation in aqueous media. As a result it has been possible to test for RNA polymerase activity in a wide variety of nuclei from normal and tumor cells and from plant tissues. This has settled the question of the localization of the enzymic machinery for RNA synthesis.

3. DNA Function as a Template for RNA Synthesis

Experiments on isolated nuclei indicated that RNA synthesis was sensitive to deoxyribonuclease, and that nuclei completely lacking in DNA could not incorporate RNA precursors. It had been shown, moreover, that when nuclei have lost nearly all their DNA, the amount of DNA which remains determines their capacity for RNA synthesis (8). Though not all the DNA in the nuclei of higher organisms is used as a template for RNA synthesis, that part which is functional can combine (by hybridization) with the many classes of RNA found in the cell.

Reich and co-workers established that the antibiotic Actinomycin D would combine specifically with double-stranded DNA and prevent further RNA synthesis on the DNA template (78). Similar results were obtained with the bacterial RNA polymerases and Actinomycin D-treated DNA (see Hurwitz and August, 43). These results not only establish the DNA dependence of RNA synthesis, but they also indicate that the natural template is double-stranded DNA (with which Actinomycin D specifically combines).

This does not mean that both strands of the DNA double helix are copied during transcription by RNA polymerase. On the contrary, it is more likely that only 1 DNA strand is transcribed. Apart from the elegant experiments that have been carried out in viral systems, showing that bacteriophage SP 8 messenger RNA's will hybridize with only one of the viral DNA strands (e.g., 58), the microanalysis of the base compositions of RNA and DNA in specific segments of insect chromosomes did not show the equivalence expected if all the DNA were copied. It is probable that asymmetric transcription is the rule in the chromosomes of higher organisms, as it is in simpler forms. However, it is likely that crossing-over, inversions, and other modifications of chromosome structure in higher organisms may have resulted in more complex patterns of transcription involving both DNA strands of the double-helix, using one or the other at different genetic loci, and perhaps synthesizing RNA's in different directions, depending upon the 5'–3' polarity of the DNA template.

The use of Actinomycin D is the chemical equivalent of enucleation, and it has permitted a direct answer to the question of whether all species of RNA are synthesized on DNA templates. Tamaoki and Mueller reported that low levels of this antibiotic block the incorporation of RNA precursors into all RNA classes in HeLa cells (except 4 S RNA, in which some slight terminal labeling, end group turnover, persisted). Inhibitions of 99% are easily achieved, and the synthesis of ribosomal RNA's (both 18 S and 28 S) seems especially sensitive to low concentrations of the antibiotic (see Perry, 73).

A similar form of chemical enucleation is provided by another anti-biotic, mitomycin C, which depolymerizes DNA in some way and also inhibits all RNA synthesis.

Different RNA's are made in different parts of the nucleus. Ribosomal RNA, for example, is synthesized on DNA in that part of the chromatin associated with the nucleolus (see Vincent and Miller, 94). Since most of the RNA of the cell is ribosomal RNA, synthetic activity in the nucleolus is intense, especially in rapidly growing and dividing cell populations. In accord with expectations, it has been found that the RNA polymerase activity of isolated liver nucleoli is over 20 times more concentrated than that of the nucleus as a whole (94). In amphibian oocytes, hundreds of independent nucleoli, each with its own DNA complement specific for ribosomal RNA, carry out a mass synthesis of ribosomes needed for further development of the egg and the embryo. Hybridization studies by Davidson and co-workers have established that over 90% of the RNA synthesized at the "lampbrush" stage is ribosomal RNA (21c). This synthesis is also actinomycin D sensitive, indicating its DNA dependence. In amphibian oocytes at the diplotene stage of meiosis, there is a great deal of RNA synthetic activity along the loops of the lampbrush chromosomes (31). These loops contain DNA which serves as the template for the synthesis of messenger-RNA's required for further development. It is an interesting fact that many of the messengers are not immediately utilized but are stored—assembled into nonfunctional polysomes that will not be activated until after fertilization (44). In any case, RNA synthesis along the loops is inhibited by Actinomycin D, which not only stops nucleoside incorporation into RNA, but, also causes the loops to contract (10).

Thus, there can be no question about the localization of the RNA polymerases and the required template DNA's in the nucleus; more specifically, it can be said that they occur along the chromosomes and in association with the nucleolus.

4. Ribosomal RNA Synthesis in the Nucleolus

There are few aspects of the chemistry of the cell nucleus that have achieved the certainty of this topic. Indeed, the evidence relating the nucleolus to the synthesis of ribosomes has an experimental basis with few peers in the whole of molecular biology (see Perry, 73; Vincent and Miller, 94; and Darnell, 21b).

A relationship between nucleolar function and the RNA content of the cell was suggested by early microanalyses. Of particular interest were the observations of Edstrom that the RNA content of supraoptic neurons was directly proportional to the nucleolar volume. The development of quantitative microelectrophoresis led to analyses of the RNA in the nucleolus and in other parts of the cell separated by microdissection techniques (*23*). Some of the results are summarized in Fig. 15, which shows the electrophoretic patterns of the RNA hydrolysates from nucleoli, nucleoplasm, and cytoplasm of spider oocytes. Nucleolar RNA, unlike RNA in

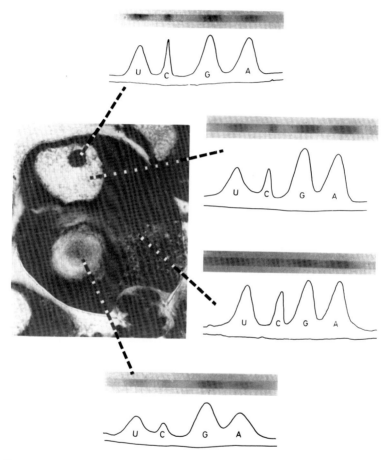

Fig. 15. Microelectrophoretic analyses of RNA's in different regions of the spider oocyte. The oocyte, stained with methylene blue, is shown on the left. Dark areas indicate high concentrations of RNA. The cell nucleus with its nucleolus is seen in the upper part of the cell; the yolk nucleus is seen below it. The microelectrophoretic patterns showing the positions of (A) adenine, (G) guanine, (C) cytidylic acid, and (U) uridylic acid are shown for RNA's prepared from parts of the cell separated by microdissection. The photographs were taken in ultraviolet light at 257 mμ and then scanned in a photometer. Densitometry indicates that nucleolar RNA and cytoplasmic RNA have very similar compositions, while RNA of the nucleoplasm has a higher A + U content (from Edstrom, *23*).

other regions of the nucleus, was found to have a base composition similar
to that of cytoplasmic RNA (which is largely ribosomal).

In an ingenious series of experiments, Perry and co-workers irradiated
the nucleolus of HeLa cells with a microbeam of ultraviolet light. The
result was startling—not only was the incorporation of RNA precursors
suppressed in the target nucleolus, but the subsequent accumulation of
radioactive RNA in the cytoplasm was reduced by 70%. RNA synthesis
in other parts of the nucleus was not markedly reduced by this treatment (94).

Strong confirmatory evidence for nucleolar synthesis of ribosomal
RNA's was provided by the demonstration that low doses of Actinomycin
D, which preferentially inhibit RNA synthesis in the nucleolus, also
prevent labeling of the 45 S precursor of the ribosomal RNA's.

The most remarkable results, however, are those obtained in a study
of a mutation in the African clawed toad *Xenopus laevis*. Mutants occur
in which a deletion has occurred in the DNA of the nucleolar-organizing
region. The result (in homozygous animals) is an embryo without nucleoli
which lives only until the tail bud stage (see Fig. 13). It was shown by
Brown and Gurdon that such anucleolate mutants were incapable of synthesizing ribosomal RNA's (Fig. 13). Direct evidence that DNA in the
nucleolar-organizing region codes for ribosomal RNA's was obtained by
the technique of DNA–RNA hybridization. In an impressively controlled
set of hybridization experiments, Birnstiel and co-workers established
that 28 S and 18 S RNA's combine with 0.04–0.07 and 0.025–0.04%, respectively, of denatured (i.e., single-stranded) wild-type *Xenopus* DNA.
(The RNA–DNA hybrids were characterized by their high $G + C$ content, ribonuclease resistance, thermostability, and buoyant densities.)
The saturation levels for hybridization indicate that there are 500–800
ribosomal DNA complements for each of the 28 S and 18 S RNA's in the
haploid genome of the wild-type cell. In short, there is a high degree of
redundancy in the genetic information for ribosomal RNA. The anucleolate mutant of *Xenopus*, in the homozygous condition, has few, if any,
ribosomal cistrons, while the heterozygote has only half the number of
ribosomal cistrons found in the wild-type DNA (73).

Similar results have been obtained in *Drosophila melanogaster* by
Ritossa and Spiegelman. They studied RNA–DNA hybridization in
stocks of *Drosophila* containing increasing numbers of nucleolar organizers. It was shown that the DNA from these stocks would hybridize with
ribosomal RNA's to an extent that was directly proportional to the number of organizers present (from one to four). Moreover, a mutation in a
heterochromatic region of the X-chromosome in *Drosophila* (bobbed)
leads to a decrease in the number of ribosomal RNA cistrons. The muta-

tion is characterized by short bristles, and it is likely that a deficiency in ribosomes has limited the synthesis of the bristle proteins.

An actual isolation of the DNA cistrons coding for ribosomal RNA has been achieved in *Xenopus*. When wild-type *Xenopus* DNA is banded by ultracentrifugation in CsCl density gradients, a heavy satellite band separates. This comprises about 0.2% of the total DNA of the wild-type, but it is not found in the anucleolate mutant. It is far more active in the formation of hybrids with ribosomal RNA's than the major DNA band is. The size of the satellite DNA and its binding capacity for ribosomal RNA's also indicate that ribosomal cistrons occur in repeating units or clusters on the DNA (*73, 94*).

This isolation of a fragment of the genome with a specific function is impressive, but the *Xenopus* system has also provided an answer to the question of whether the nucleolus is a site of synthesis of other types of RNA. Work in other animals had suggested that amino acid tRNA's are synthesized and methylated in the nucleolus. However, analysis of the RNA's in the anucleolate *Xenopus* mutant does not show a deficiency of tRNA's, and it has been concluded that the synthesis of tRNA's is extra-nucleolar in this organism. In *Drosophila*, too, there seems to be a separation of the cistrons for ribosomal and tRNA's, since tRNA cistrons lie outside of the region of the chromosome map which contains the nucleolar organizer. In other insect species, such as *Smittia*, and in starfish oocytes, there is evidence for the synthesis of tRNA's in the nucleolus; and this suggests that organisms differ with respect to the proximity of the genes for ribosomal RNA's and tRNA's. This view is supported by the observation that in some microorganisms (e.g., *B. subtilis*) these cistrons map very close to each other on the bacterial chromosome.

5. *High Molecular Weight Precursors of Ribosomal RNA's*

Tracer studies of RNA synthesis in a variety of cell types reveal that, at an early time, isotopic precursors are preferentially incorporated into components of very high molecular weights. These can be separated from ribosomal RNA's by zonal centrifugation on sucrose gradients, and the radioactivity of the various RNA's can be compared by sampling different regions of the centrifuge tube.

Perry observed that low concentrations of Actinomycin D, which selectively block the synthesis of ribosomal RNA's also prevent the labeling of a 45 S component in the gradient (see Perry, *73*). This led to the proposal that the 45 S RNA is a precursor of the 28 S and 18 S ribosomal RNA's. This interpretation of events is supported by "Actinomycin D chase" experiments, one of the best known of which is shown in Fig. 16. Scherrer and co-workers labeled HeLa cells with ^{14}C-uridine and found

V. G. ALLFREY

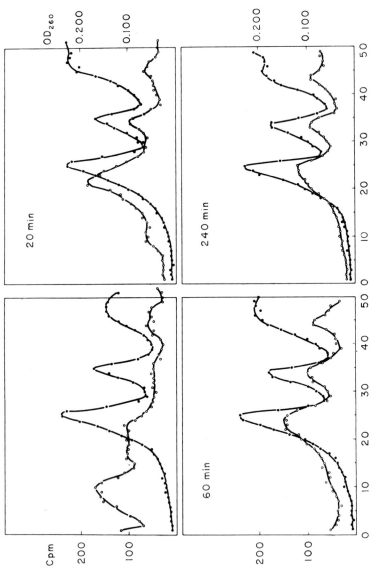

FIG. 16. Evidence for the conversion of 45 S RNA to 28 and 18 S ribosomal RNA's. HeLa cells were labeled with 2-¹⁴C-uridine for 30 minutes. RNA was prepared from a sample of the culture and subjected to sucrose density-gradient analysis (upper left). The remainder of the culture was treated with Actinomycin D to prevent further RNA synthesis. Samples were withdrawn at 20, 60, and 240 minutes for RNA gradient analysis. Note that the radioactivity seen in heavier regions of the gradient (45 S) at early times gradually shifts over and appears under the 28 and 18 S ribosomal peaks: ● — ●, optical density at 260 mμ; ○ — ○, radioactivity in cpm (from Scherrer *et al.*, 83).

that, after 30 minutes, the radioactivity was predominantly in the heavy RNA peaks. Actinomycin D was then added to prevent the further synthesis of RNA, and samples were withdrawn from the culture 20, 60, and 240 minutes later. As shown in Fig. 16, gradient analyses indicated that the label progressively disappears from the heavy peaks and accumulates in those regions of the gradient occupied by 28 and 18 S ribosomal RNA's (*83*). Apart from the kinetic evidence for the conversion of 45 S RNA into ribosomal RNA's, there is strong evidence pointing to the chemical relatedness of these RNA molecules. Base composition studies indicate that the 45 S component has the composition expected for a 1:1 mixture of 28 S and 18 S RNA's. The nucleotide sequences in the precursor RNA and in the ribosomal peaks are presumably identical, because both compete in the formation of hybrids with DNA. As expected, there is a diminished synthesis of the 45 S component in the anucleolate mutant of *Xenopus* which, as has been shown above, lacks the cistrons for the synthesis of ribosomal RNA. (Some radioactivity is still seen in the 45 S region of the gradient, but this is due to the synthesis of other heavy RNA's in the anucleolate mutant, some of which overlap the 45 S zone.)

An estimation of the size of the 45 S RNA places its molecular weight at 4.5×10^6. A single covalently-linked polynucleotide chain of this size could contain more than one molecule each of 28 S RNA (MW 1.9×10^6) and 18 S RNA (MW 0.7×10^6). However, tracer studies of the maturation of the 45 S ribosomal RNA precursor indicate that 1/3–1/2 of the RNA is lost during subsequent processing and is not incorporated into 18 S or 28 S ribosomal RNA's. Obviously this macromolecule must be cleaved in a very specific way to produce the ribosomal products. A brief outline of the further steps in ribosomal RNA synthesis will be presented below, but it should be noted here that the structure of the 45 S precursor is modified very soon after its synthesis. Methylating enzymes transfer methyl groups from *S*-adenosyl methionine to purine and pyrimidine bases and to the 2'-position of the ribose in the polynucleotide chains. This alteration of RNA structure seems to occur prior to cleavage of the macromolecule, and it may possibly play a role in determining the precise sites of cleavage of the 45 S component.

6. *Cleavage of 45 S RNA and the Assembly of Ribosomes*

Experiments on a variety of cell types have shown that 45 S RNA is cleaved to produce 18 S ribosomal RNA and a 32–36 S component (see Perry, *73*). The latter is then further modified to produce 28 S ribosomal RNA. These maturation steps are believed to happen in the nucleolus because sequential labeling of the 45 and 32–36 S fractions could be observed in isolated nucleolar fractions prepared from cells after short-term labeling and "Actinomycin D chase" experiments (see Perry, *73*).

There is good evidence that further processing of the ribosomal RNA's takes place along different pathways. The 18 S component is linked to proteins and is transferred to the cytoplasm as the small (40 S) ribosomal subunit. Since many kinetic studies indicate that 40 S particles bearing radioactive 18 S RNA are the first ribosomal subunits to enter the cytoplasm, the assembly and export of the small subunit from the nucleus must be a rapid process.

Back in the nucleolus, the 36 S RNA is further modified, presumably by cleavage reactions which produce 28 S RNA and small RNA fragments of unknown function. The 28 S RNA is then joined to proteins to produce the large (60 S) ribosomal subunit. Many of these particles remain in the nucleolus, or near the rim of the nucleolus, for some time, but eventually, by some unknown mechanism, they are transported into the cytoplasm as 60 S ribosomal subunits. Thus, the overall sequence of reactions in the maturation of ribosomal RNA precursors is a complex phenomenon involving a number of intermediate stages [see (21b)]. The process is further complicated by the fact that these operations also involve RNA-associated proteins. The production of the 60 S ribosomal subunit in L cells, for example, proceeds through ribonucleoprotein particles of progressively smaller size. A 110 S particle contains the 45 S ribosomal RNA precursor, while a 78 S particle contains the 32–36 S RNA intermediate, and a 62 S particle contains 32 S RNA and some 28 S RNA (51b).

It should be stressed that free 18 S and 28 S RNA's are never observed in the cytoplasm and that they occur only as components of the ribosomal subunits. Their incorporation into ribonucleoprotein complexes takes place in the nucleolus and is known to involve a very complex sequence of attachments to protein.

Some inkling of the complexity of ribosome assembly is provided by recent studies of ribosomal proteins. A detailed discussion of the ribosomal proteins is beyond the scope of this review, but it should be noted briefly that the large subunit contains about 40 different protein components and the small subunit contains about 21, most of which differ from those present in the large subunit. The function of the ribosome in protein synthesis depends on these structural proteins [as has been well documented by studies of the loss and restoration of function accompanying the dissociation and reassociation of ribosomal subunits in bacteria (see Part B, the chapter on "The Structure and Function of Ribosomes")].

The attachment of ribosomal RNA's to ribosomal proteins is believed to take place in the nucleolus soon after the RNA is synthesized. Indeed, recent experiments by Warner indicate that a substantial pool of preformed ribosomal proteins exists in the nucleolus for this purpose and that different proteins are added to ribosomal RNA's in a programmed sequence. The evidence for a preformed ribosomal protein pool is quite convincing because newly synthesized RNA's make their appearance as ribosomal subunits in the cytoplasm even under conditions in which protein synthesis is 99% inhibited (by cycloheximide). It follows that ribosome assembly must have involved the combination of ribosomal RNA's with proteins which were synthesized before the addition of the antibiotic and which were stored in, or available to, the nucleolus.

The site of synthesis of ribosomal structural proteins is not known, though it is clear that, wherever they are synthesized, they must eventually accumulate in the nucleolus. In this connection, it should be noted that there is a great deal of autoradiographic evidence for localization of radioactive protein in the nucleoli of growing cells, and that nucleolar fractions isolated from rat liver, plant cells, and starfish oocytes have a capacity for amino acid incorporation into protein (see Maggio, *55*; and Vincent *et al.*, *93*).

A recent aspect of research on the assembly of the ribosome centers on observations that newly synthesized or newly emergent ribosomal subunits have more associated proteins than are found in older particles present in the cytoplasm. The function of these extra associated proteins and the means of their removal is not yet known (see Perry, *73*).

7. *Subunit Utilization for Polysome Formation and Ribosome Turnover*

After their assembly, the ribosomal subunits appear in the cytoplasm, and, as pointed out earlier, they are released from the nucleus at different rates, the 40 S subunit appearing first, and the larger 60 S particle, (which presumably takes longer to complete) enters the cytoplasm later. The 40 S and 60 S subunits are soon used in the formation of polysomes, and it appears that their combination to form the complete 80 S ribosome (monosome) takes place during the creation of the polysome. A dissociation of polysomes after short-term labeling experiments with RNA precursors shows that the 18 S RNA in the small subunit is far more radioactive than the 28 S RNA in the large subunit, as might be predicted from the different kinetics of release of the subunits into the cytoplasm. The radioactivity subsequently appears in the monosome peak which is presumably derived from the dissociation of polysomes. Most of the details of polysome formation, attachment to membranes, function, and dissolution remain to be worked out. What is certain is that the subunits are eventually degraded. Wilson and Hoagland have reported that the average half-life for a ribosome in the liver cell is only 5 days. The cell must make up for this loss by a continuing synthesis of new ribosomes. Moreover, the ribosomal RNA content in fasted animals may drop to half the normal value, and then be restored 10–12 hours after refeeding. It follows that the synthesis and turnover of ribosomal RNA's and associated structural proteins are subject to physiological control mechanisms—a conclusion which is in accord with a number of other findings, particularly the fact that heterozygous stocks of *Xenopus,* which have only 1 nucleolar organizer, nevertheless synthesize an amount of ribosomal RNA equivalent to that found in the wild-type animal, which has 2 nucleolar organizers.

For further information on the synthesis of ribosomal RNA's and on the function of the nucleolus, the reader is referred to recent excellent reviews of this subject (*21b, 73, 77*).

8. *Some Aspects of Nuclear RNA Polymerase Activity*

The polymerization of the ribonucleoside triphosphates, as pointed out before, is an enzymically catalyzed reaction which is DNA-dependent (see Georgiev, *32*; and Hurwitz and August, *43*). In higher forms this activity is largely localized in the cell nucleus, though some polymerase activity is also found in the DNA-containing organelles of the cytoplasm, such as the mitochondria and chloroplasts.

RNA polymerase (RNA nucleoside triphosphate nucleotidyl transferase) was first discovered in rat liver nuclei by S. Weiss, who demonstrated the requirement for all 4 ribonucleoside triphosphates and a DNA template. The occurrence of similar enzymes in bacteria, yeasts, and plant cells was established by other workers at about the same time. It is now known that the enzyme is essentially a polymerizing enzyme and that the sequence of nucleotides in the RNA product is rigorously determined by the complementary sequence in the DNA primer (*43*).

In most cells, the enzyme is very closely associated with DNA. In higher organisms, the enzyme–DNA complex remains stable during the isolation of cell nuclei in sucrose solutions (or other aqueous media of low ionic strength). This has made it possible to test for changes in nuclear RNA polymerase activity induced by hormones (*91*), ions (*76*), and other physiological control mechanisms (*32, 76, 91*).

Results of such studies have established that the RNA synthetic capacity of the nucleus depends upon at least 3 major variables: (1) the concentration of the necessary substrates, (2) the amount of enzyme available, and (3) the number of DNA sites available to the enzyme for transcription. The latter is an extremely important variable, and, as will be shown below, not all of the DNA of higher organisms is functional in any given cell type; i.e., different cells utilize different parts of the totality of genetic information they carry in their chromosomes. One of the great problems in cell biology is to understand how this pattern of variable gene activity is directed and controlled during development, later life, and senescence.

Before proceeding to this aspect of nuclear function, mention should be made of current views of RNA polymerase structure and function in bacterial cells (from which a soluble enzyme can be prepared), for it is likely that the mammalian enzyme will have much in common with that of microorganisms. The RNA polymerase of *E. coli* is a complex composed of three principal types of polypeptide chain: α, β, and β', which have molecular weights of about 40,000, 155,000, and 165,000, respectively. They occur in the minimal enzyme in the proportion $2\alpha:1\beta:1\beta'$. The enzyme is responsible for the stepwise elongation of the polynucleotide chain, following the sequence directed by complementary bases in the DNA template, but the initiation of transcription requires the interaction of the enzyme with another factor [called the sigma factor (*92e*)] which catalyzes the initiation of new RNA chains. The sigma factor has a molecular weight of about 95,000. It appears to interact reversibly with the enzyme, being released from the polymerase–factor complex and from the DNA template soon after initiation; it may then be reused by another molecule of minimal enzyme to initiate a new RNA chain. It is likely that similar factors regulate RNA synthesis in higher organisms, as well. Growing RNA chains remain attached to the polymerase–DNA complex. The accumulation of the product in the vicinity of the template can sometimes be visualized directly, as in the "puffs" in Dipteran salivary gland chromosomes and the "loops" of the lampbrush chromosomes in amphibian oocytes. The attachment of newly synthesized RNA's to the DNA–enzyme complex makes it difficult to extract the most radioactive RNA's from mammalian cell nuclei after short-term labeling experiments. Recourse has been made to heat, detergents, proteases, phenol, and other agents to facilitate the isolation of DNA-like RNA's (see Hurwitz and August, *43*; Georgiev, *32*). Complete extractions are rarely achieved.

Since all the RNA products obtained so far are single-stranded polynucleotides which do not hybridize with themselves, it follows that cellular RNA's do not contain chains of complementary nucleotide sequence. This is a strong argument for transcription of only 1 DNA strand, because transcription of both strands would be expected to yield complementary RNA molecules.

The activity of RNA polymerases in different regions of the cell nucleus can be altered by changing the ionic strength of the medium and the concentration and nature of the divalent cations present. This was strikingly demonstrated in studies of isolated rat liver nuclei. Such nuclei, as isolated in sucrose containing Mg^{2+} ions, synthesized largely ribosomal RNA's (U/G ratio = 0.53), and high resolution autoradiography showed the grains to be localized over the nucleolus. An apparent activation of genetic loci other than those associated with ribosomal RNA synthesis was achieved by adding Mn^{2+} ions and increasing the salt concentration slightly. The U/G ratio then shifted upward to 0.97, indicating that more and different DNA loci were activated for transcription (76). This conclusion was supported by autoradiographic evidence which showed that grain densities over the chromatin (i.e., over extranucleolar regions of the nucleus) were correspondingly increased. The observation is of more than technical interest because it is now known that the ionic environment within the nucleus is subject to physiological controls; nuclear membrane permeability in insect cells, for example, is altered by the administration of the steroid hormone ecdysone (54). This type of ionic control may well be one of the important variables underlying hormonal effects on gene activity.

9. Evidence for Selective Gene Activity in Nuclei of Differentiated Cells

In considering the regulation of nuclear activity, particularly with respect to RNA synthesis, much can be learned from the study of the differentiated cells of higher organisms. We begin with a realization that the various somatic cells of the body, despite their differences, carry equivalent genetic endowments. Their nuclei may differ in size and shape, but their chromosome numbers are, by and large, equivalent. This equivalence is evident not only in number, but in the fact that different cells in different regions of the fruit fly *Drosophila* display traits (such as eye color, bristle number, and wing form) that can be correlated with map positions along the chromosomes of the salivary gland.

Direct chemical evidence for the view that nuclei in different cell types carry equivalent genetic endowments came from the work of Vendrely and Vendrely (92i) and Mirsky and Ris (65), who showed that nuclei in different somatic cells of an organism carry equal amounts of the genetic material DNA while sperm cells contain half that amount.

It follows that, for all its specialization, each cell carries the totality of genetic information characteristic of the organism; but this potential is only partially expressed. Red blood cells make hemoglobin, but not digestive enzymes; while the acinar cells of the pancreas make trypsin, ribonuclease and amylase, but not hemoglobin. What types of control mechanisms are employed to impose this high degree of specialization upon cells in different somatic tissues? The subsequent discussion will concentrate on the evidence that this control involves selection processes which determine which portions of the genome are called into activity at different times during the life of the cell. In this view, the great differences in chemistry, morphology, and behavior of various cell types can be traced ultimately to the selection mechanisms that decide which genetic loci will be active and which will not.

The molecular basis of this selection is the control over transcription of different DNA molecules. As pointed out earlier, the synthesis of RNA's in the nucleus is DNA dependent, but it has become increasingly clear that not all the DNA in a differentiated cell is actually functioning as a template for RNA synthesis. This fact has been demonstrated directly by experiments on isolated nuclei (8) in which the DNA was progressively removed by exposing the nuclei to increasing concentrations of deoxyribonuclease. The treated nuclei were then tested for the effects of DNA removal on nuclear function. It was found that 70–80% (and possibly more) of the DNA could be removed from isolated lymphocyte nuclei with no apparent effect on the amount of RNA synthesized. It was concluded that the DNA which had been removed must have been nonfunctional as far as RNA synthesis is concerned, i.e., it was not available as a template for the RNA polymerase reaction.

However, when all the remaining DNA was removed, RNA synthesis came to a halt, indicating the complete DNA dependency of the nuclear RNA polymerase reaction. If only a portion of the remaining DNA was removed, the rate of RNA synthesis decreased proportionately, and what was observed was a linear correspondence between the amount of DNA remaining in the nucleus and its capacity for RNA synthesis (8). It follows that the DNA of the interphase nucleus must be sequestered in ways that permit a small portion of the total genome to be functional as a template for RNA synthesis, while most of the genetic information is maintained in an inactive condition.

This point of view is supported by hybridization experiments that test the capacity of RNA's from different tissues to combine with complementary nucleotide sequences in DNA. Although the techniques of RNA–DNA hybridization are not yet under full control, they do offer strong support for selective transciption mechanisms. For example, it has been shown by McCarthy and Hoyer (59) that ^{32}P-labeled RNA's from mouse L cells in culture will hybridize effectively with DNA from any murine tissue. In this binding to the complementary regions of the DNA, nonradioactive RNA's from L cells will compete with their radioactive forms, as would be expected. However, RNA's from other cell types, such as liver and spleen, are much less effective in this type of competition, and it can be concluded that they represent a population of RNA molecules of different nucleotide sequence, i.e., they are complementary to regions of the genome other than those transcribed in the L cell. Of course, L cells, liver, and spleen may be expected to have common characteristics, and common sequences in DNA are probably used for transcription in all cell types (such as the cistron concerned with the synthesis of RNA polymerase itself).

Another approach to selective transcription has been employed recently by Paul and Gilmour, who used chromatin from different tissues as a template for the RNA polymerase reaction. Subsequent tests of the product RNA's by DNA hybridization showed differences between the competitive properties of bone marrow and thymus RNA's (70b).

The hybridization technique has great potential promise, but it requires careful controls. When properly applied, it can be used to follow stages in differentiation. Patterns of gene activation and repression during development of amphibians have been extensively studied by this technique (see Gurdon, 33a).

While it is obvious that different tissues of the body differ in their enzyme and protein composition and must, therefore, differ in their messenger RNA's which specify amino acid sequence in the protein, it should also be pointed out that particular cell types can also vary their patterns of RNA synthesis in response to environmental influences. The change in template activity of the DNA can be followed by both RNA polymerase assays and by hybridization experiments, as has been done for liver regeneration. For example, Pogo and co-workers (74) were able to show that nuclei isolated from regenerating liver had much more DNA template activity for RNA synthesis than did the corresponding nuclei from normal or sham-operated animals. Hybridization studies of the liver RNA's produced at different times after partial hepatectomy have also shown the presence of new molecules of RNA which are not present in normal or sham-operated animals. Moreover, it was shown by Church and McCarthy that some of these molecules appear 1 hour after partial hepatectomy and that their synthesis is turned off in a few hours (17b).

What are the mechanisms by which cells choose which genetic loci will be functional and which loci will remain inactive? This problem involves structural considerations combined with specific control mechanisms involving nuclear proteins.

10. Active and Inactive Chromatin

The lymphocyte nucleus offers some interesting structural corollaries to the biochemical evidence for active and inactive DNA. Electron microscopy of such nuclei in tissue sections, or following their isolation, reveals that the chromatin is not homogeneously spread throughout the nucleus, but is heterogeneously distributed into dense clumps of compacted fibrils and diffuse regions of extended filaments about 100–150 Å in average diameter. Finer filaments (about 50 Å) are also evident (*52*).

The RNA synthetic activity in different regions of the nucleus can be visualized directly by the techniques of high resolution autoradiography under the electron microscope. Figure 17 shows the distribution of radioactive RNA in thymus nuclei following a 30-minute incorporation of uridine-³H. Note that the vast majority of the grains occur over or near the diffuse chromatin regions, while few grains are seen over the large clumps of condensed chromatin. Since most of the DNA in the thymus lymphocyte nucleus is present in the dense masses of compact chromatin, it follows that most of the DNA is nonfunctional in promoting RNA synthesis (*52*). Such observations are not limited to thymus nuclei. Hsu found that chromocenters in mouse cell nuclei (strain H4c) are relatively inert in uridine uptake, while other workers have reported that the diffuse chromatin is the preferred site of RNA synthesis in kidney cells and in plant tissues (*Trillium* microspores).

The relationship between chromatin structure in the nucleus and RNA synthesis can be visualized in other systems as well. The lampbrush chromosomes of amphibian oocytes are characterized by long loops of a DNA-containing fibril on which RNA synthesis takes place (*31*). The puff in insect salivary gland chromosomes is also a site of intense RNA synthesis. On the other hand, extreme condensation of the chromatin can be correlated with inactivity. Metaphase chromosomes, for example, have been shown by autoradiographic studies to be relatively inert, and Johnson and Holland have observed that HeLa cells isolated at metaphase synthesize very little RNA. They further established that the chromatin of the metaphase cells was a very poor primer for the RNA polymerase reaction, while the DNA alone was fully active. It may be concluded that DNA function as a template for transcription is seriously repressed during cell division at the time the chromosomes assume their compact configuration. [A somewhat related phenomenon is seen in the unicellular organism *Euplotes* in which DNA synthesis occurs in a wavelike fashion, traversing the long macronucleus. The synthesis of DNA is restricted to a narrow replication band which is readily recognized. RNA synthesis is turned off in the region of this band, indicating the mutual exclusiveness of transcription and DNA replication (*77*).]

The lack of genetic activity in compact chromatin applies to DNA synthesis as well as RNA synthesis. High resolution autoradiographs of proliferating cells of the salamander after ³H-thymidine incorporation showed that isotope uptake was largely limited to regions of diffuse chromatin (*35*). This observation raises some interesting questions, be-

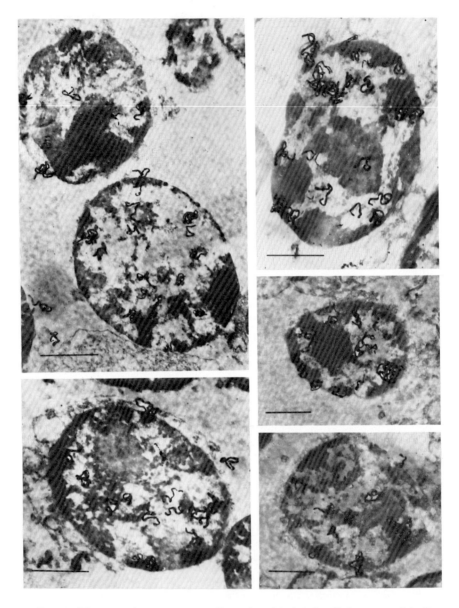

Fig. 17. Electron microscope autoradiographs of isolated calf thymus nuclei after short-term incubations in ³H-uridine. Note that the radioactive RNA is localized almost exclusively in the diffuse regions of the chromatin and that few grains are seen in the condensed chromatin areas. The line on the lower left is 1 μ (52).

cause all the DNA in the nucleus must be replicated, including the DNA in the clumped chromatin. It follows that all the chromatin in a dividing cell must spend some time in the diffuse state, which, in turn, suggests that the physical state of the chromatin is subject to transformation, presumably by active mechanisms that spool and unravel the deoxyribonucleoprotein fibrils as replication proceeds.

How is the physical state of DNA in chromatin determined? This is the question which will now be considered.

11. *Histones in Chromatin: Their Influence on Structure and Genetic Activity*

Much of the mass of the nucleus is made up of basic proteins called histones. In the thymus nucleus, the histones comprise nearly a third of the nuclear dry weight. There are many types of histone associated with DNA in the nucleus, and a full discussion of their fractionation and properties is beyond the scope of this review. [The reader is referred to several recent reviews of the chemistry and function of histones and other chromosomal proteins (*15b, 35a, 67, 90c*).]

What will be discussed here is the role of histone in the structural organization of chromatin, the effects of histones on transcription, and some structural modifications of these remarkable proteins which may be a clue to their role in nuclear metabolism.

Histones are basic proteins, i.e., they have a preponderance of the basic amino acids lysine, arginine, and histidine. As a result, their isoelectric points lie in the alkaline range, they combine readily with negative molecules such as DNA, and they can be extracted from nuclei in dilute acid solutions.

Histones exist in the nucleus and in the chromosomes complexed with DNA, as was clearly indicated in the early work of Mirsky and Ris (*64*) who also showed that the complex could be dissociated by raising the salt concentration to $1 M$. Histones can be detected and measured quantitatively when insect and animal chromosomes *in situ* are stained with Fast Green and subjected to microspectrophotometry. They remain attached to DNA during the isolation of metaphase chromosomes from HeLa cells and other cell types (*41, 56*).

Histones are usually extracted from nuclei in acid, and chromatography of the extracts on ion-exchange columns usually yields major fractions which differ in amino acid composition, particularly with respect to their lysine-to-arginine ratios. Although histones differ in other respects as well, the Lys/Arg ratio has become a convenient standard upon which to base their classification. In the very lysine-rich histones, this ratio is about 14/1, while in the arginine-rich histones it is only 0.7/

1. This nomenclature has proven useful even though many of the histone fractions isolated to date are themselves mixtures, as judged by electrophoretic analysis on starch gels or polyacrylamide gels (*15b, 35a, 45a, 90c*). Depending upon the conditions of analysis, it would appear that about 16–30 histone bands are separable, but precise estimates of the total number are made difficult by their tendency to aggregate and by structural modifications which alter their shapes and charges. There are about 5 major types of histones.

In the interest of simplifying the further discussion of histone chemistry, they are conveniently grouped into 3 classes: lysine-rich, arginine-rich, and intermediate. Many of the histones are further modified by substitution reactions–acetylation, methylation, and phosphorylation. (These will be described in some detail below.)

The lysine-rich histones differ in several respects from other histone fractions. For example, they are soluble in 5% perchloric acid (which precipitates most other proteins), and this has become a commonly used technique for their selective extraction from nuclei. The lysine-rich histones are displaced relatively easily from their sites of attachment to DNA in the nucleus. This can be accomplished by lowering the pH (especially in the presence of chelating agents) or by raising the salt concentration. Daly and Mirsky (*21a*) employed both methods: They found that 0.1 M citric acid would extract lysine-rich but not arginine-rich histones from calf thymus nuclei. Alternatively, much of the lysine-rich histone could be displaced at salt concentrations which left the arginine-rich fraction intact. Similar differences in extractability have been observed by many workers in both plant and animal nuclei.

Such differences in ease of extraction of different histone fractions suggest differences in the nature and extent of their binding to DNA in the chromatin, and there is now good evidence that the lysine-rich histones play a special role in the organization of the condensed state of the chromatin. For example, it was shown by Littau and co-workers (*53*) that removing the lysine-rich histones (which comprise 20% of the total histone), and only those histones, loosened the structure of the dense chromatin in thymus nuclei. Selective removal of the arginine-rich histones did not have this effect. The special role of lysine-rich histones in binding chromatin threads together was shown by restoring histones to histone-depleted nuclei, in which the chromatin masses had broken down into a loose fibrous network. Only the lysine-rich histones caused chromatin masses to reappear, although both arginine-rich and lysine-rich histones could combine with the nuclear DNA. The results suggest that the lysine-rich histone fraction may combine with the phosphoric acid residues of adjacent DNA molecules, thus cross-linking many DNA double helices to form the condensed chromatin masses. The possibility that histones may link adjacent DNA helices has also been suggested by X-ray diffraction studies (see Allfrey and Mirsky, *10*).

In considering the role of the histones in nuclear function, it was suggested years ago by Edgar Stedman that histones may play a role in genetic regulation. Experimental support for this view has been obtained only relatively recently. One approach to the problem has been to consider the effects of histones on biosynthetic reactions in isolated thymus nuclei and, particularly, their effects on RNA synthesis.

In 1961 it was observed that the addition of histones to suspensions of isolated calf thymus nuclei caused an immediate inhibition of RNA synthesis (*3, 9, 10*). The effect is illustrated by the experiments summarized in Fig. 18(A), which shows that adding increasing amounts of total thymus histone or histone subfractions leads to a progressive depression of ^{14}C-guanosine or ^{14}C-adenosine uptakes into RNA.

However, it was known at the time that this was not necessarily a specific target effect, because histones had previously been shown to inhibit other synthetic reactions in nuclei, in mitochondria, and in isolated enzyme systems. For example, it was shown that histones inhibit nuclear ATP synthesis and protein synthesis, as well as cytochrome c oxidase activity and ATP synthesis in mitochondria. Histones were also known to inhibit amino acid transport into nuclei and to block amino acid incorporation by isolated nuclear ribosomes.

For these reasons, the experimental approach was changed, and emphasis was placed on the effects of removing histones from nuclei. This can be done selectively by tryptic digestion, because histones are more susceptible to attack by this enzyme than are the generality of nuclear proteins. Controlled tryptic digestion of nuclear suspensions leads to a removal of 70% of the total histone—without loss of RNA polymerase activity and without hydrolyzing more than 5% of the nonhistone proteins (*10*). The effects of this treatment are summarized in Fig. 18(B), which shows that RNA synthesis by the treated nuclei may be 300–400% higher than that observed in untrypsinized controls. The lower curves in Fig. 18(B) show the time course of RNA labeling in control nuclei; the upper curves show the corresponding time courses in nuclei from which 70% of the histones had been removed. In some experiments, the tryptic digestion was moderated by the simultaneous addition of the soybean trypsin inhibitor. The increase in RNA synthesis is less marked, but still evident (curve 2a). Moreover, the addition of more histone to such nuclei again leads to a suppression of RNA synthesis (curve 2b).

The newly synthesized RNA's produced in histone-depleted nuclei presumably represent a population of RNA's which would not otherwise

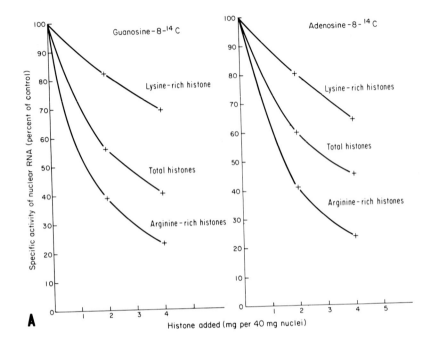

A

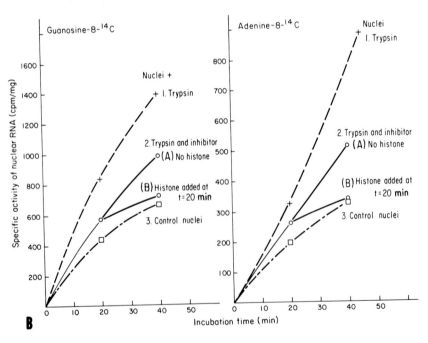

B

have been synthesized in the thymus nucleus, because tryptic digestion leads to a loosening or extension of the previously repressed chromatin in the clumps. The synthesis of new RNA's was suggested by base composition studies of the RNA formed before and after tryptic digestion. Differences in overall nucleotide composition were observed, but it remains to be demonstrated that removal of the histones did, in fact, turn on gene loci that were previously repressed.

In this connection, it should be noted that autoradiographic studies of polytene chromosomes in insect cells also show that tryptic digestion leads to an increase in the rate of chromosomal RNA synthesis. Moreover, Robert and Kroeger observed that uptake is increased at the puffs which were already engaged in RNA synthesis. This is an especially interesting result because it implies that active loci of the chromosomes still contain histone proteins, a conclusion which would argue against the simple concept that gene activation demands the complete removal of histones from the vicinity of the DNA template. In addition, the conclusion that histones at the puffs are more susceptible to proteolytic attack than are histones at inactive regions of the chromosome, suggests that they are not complexed with DNA in the same way. This raises the possibility that the DNA–histone linkages are not permanently fixed, but can be modified in the transition from inactive to active chromatin. Some direct chemical evidence supporting these views will be presented briefly.

A major breakthrough in the study of histone effects on RNA synthesis took place in 1962, when Huang and Bonner (40) examined the RNA polymerase activity of chromatin fractions prepared from pea-seedling nuclei. They observed that the plant chromatin (containing DNA plus associated histones) was relatively inert as a primer or template for RNA synthesis, as compared to the template activity of an equivalent amount of free DNA. Similar observations on the inhibition of RNA polymerases from bacterial, animal, and plant sources have since been

---

FIG. 18. Evidence for histone suppression of nuclear RNA synthesis. (A) Effects of adding histones on 8-^{14}C-adenosine and 8-^{14}C-guanosine uptake by isolated calf thymus nuclei. The amount of histone added is plotted against the relative specific activity of the total nuclear RNA after a 30-minute incubation. (B) Effects of histone removal by selective tryptic digestion on the uptake of 8-^{14}C-guanosine and 8-^{14}C-adenine into the RNA's of isolated calf thymus nuclei. The lower curve (3) shows the time course of RNA synthesis in control nuclei (no histone added or removed). The upper curve (1) shows the uptake in nuclei which had lost about 70% of their histone content. Curve 2A is for nuclei treated with a mixture of trypsin + soybean inhibitor (to moderate enzyme action and preserve nuclear structure). Curve 2B shows that addition of more histone to histone-depleted nuclei again represses RNA synthesis (3).

made in many laboratories. The methods have sometimes been criticized for the fact that the nucleohistone complexes are insoluble and tend to precipitate from the assay solution, but this is not the case for all the experiments which have been carried out. On considering all the experiments in which the removal of the histones from chromatin fractions, using acid, or salts, or tryptic digestion, did lead to an increased capacity for RNA synthesis, it is hard to avoid the conclusion that combination with histones does effectively limit DNA template activity in the RNA polymerase reaction. (Combination with histones also appears to limit DNA template activity in the DNA polymerase reaction, but this aspect of histone activity will not be considered here.)

12. *Enzymic Modifications of Histone Structure*

Although the primary structure of histones, like other proteins, involves amino acids joined in peptide linkage, it also includes a number of substituent groups attached to different residues in the polypeptide chain.

At least 3 structural modifications are known: (1) the presence of N-terminal acetyl groups was first detected by Phillips (see Murray, *67*) and ϵ-N-acetyl lysine has recently been detected in certain arginine-rich histone fractions (*32a*). (2) The occurrence of ϵ-N-methyl lysine in histones was first described by Murray (*67*). Dimethyl lysine is also present: Paik and Kim (*70a*) recently showed that its concentration in thymus histones exceeds that of the monomethyl derivative by a factor of two. (3) Histones are also phosphorylated, and several laboratories have described the isolation of ^{32}P-labeled serine phosphate from histone digests after incubating nuclei with ^{32}P-phosphate.

All the above reactions have in common the facts that they are enzymically catalyzed and that they take place after completion of the polypeptide chain. Such group addition reactions can be readily distinguished from histone synthesis by the use of puromycin, since this antibiotic blocks amino acid incorporation into the proteins of the nucleus, but it has no effect on either acetylation or methylation of the lysine residues in histones (see Allfrey, *3*). Studies of histone phosphorylation in isolated nuclei also show that puromycin fails to block the histone phosphorylation reaction. In this connection, it should be noted that Stevely and Stocken have reported that increasing levels of phosphorylation of the lysine-rich histones correlate with a decreased capacity to inhibit the RNA polymerase reaction.

Something is known of the mechanisms underlying these very interesting histone modifications. In acetylation, the group donor is acetyl-coenzyme A, which transfers its acetyl residues to amino groups on the histones. There is also evidence for the enzymic acetylation in *in vitro* systems of serine hydroxyl groups, particularly in the *f3* fraction of the arginine-rich histones.

In methylation, the group donor is *S*-adenosyl methionine; its methyl groups are transferred to the epsilon amino position of lysine residues in the polypeptide chain. It is of interest that the lysine-rich histones are not methylated at all, while arginine-rich histone fractions contain both mono- and dimethyl lysines, with the latter in excess. This is another fundamental distinction between these 2 histone types.

Like the above reactions, the phosphorylation of hydroxyl groups in histones is enzymically catalyzed. It requires the participation of ATP as the phosphoryl group donor and, in thymus nuclei, phosphorylation occurs preferentially on the lysine-rich histones. A very suggestive observation relevant to the role of histone phosphorylation is the recent finding that some histones (including an arginine-rich histone) are extensively phosphorylated during the maturation of fish spermatocytes. This occurs when the histone proteins are removed from the DNA and replaced by protamines.

An additional structural complication is provided by the fact that some arginine-rich histones contain 1 cysteine residue per molecule, i.e., they possess thiol groups which can be oxidized. Ord and Stocken have recently reported that the state of oxidation varies in different states of the chromatin, with more of the thiol form in diffuse chromatin, and more of the disulfide form in the condensed chromatin (*69b,c*).

What is the functional meaning of all these structural modifications of the basic proteins attached to DNA? Why are some histones phosphorylated but not methylated? Why do acetyl groups preferentially attach to the epsilon-amino groups of lysine residues in some histones but not in others? What is the significance of one or more methyl groups on lysine residues in arginine-rich histones?

In the hope that these remarkable structural modifications of histones may offer significant clues to their role in the regulation of chromatin function, studies have been made comparing the histones in different fractions of chromatin, in resting nuclei, and in nuclei triggered into new biosynthetic activities. To date, the most promising correlations have been obtained in studies of histone acetylation.

13. *Histone Acetylation and Chromosomal Activity*

When isolated nuclei are incubated in the presence of radioactive acetate or ^{14}C-acetyl-coenzyme A, and the histones are subsequently fractionated by column chromatography or electrophoretic techniques, it is observed that the highest specific activity is found in the arginine-rich histone fraction (*3*). Similar results are obtained *in vivo*.

Since arginine-rich histones were known to inhibit the RNA polymerase reaction, tests were carried out to see whether further acetylation would influence histone suppression of RNA synthesis. Three different RNA polymerase systems were tested (from calf thymus nuclei, *E. coli*, and *Azotobacter vinelandii*). All systems were inhibited by the arginine-rich histones of calf thymus, but, in all cases, the acetylated histones were much less effective than the parent proteins in blocking UTP or ATP utilization for RNA synthesis. Since the acetylated histones can combine with DNA, it was tentatively concluded that comparatively subtle changes in histone structure could influence DNA–histone interactions in ways that modify the template activity of the complex. Despite the apparent success of this direct approach, the results must be regarded cautiously because many factors can influence the *in vitro* RNA polymerase assay system. However, these results did prompt further studies of acetylation *in vivo*, in an attempt to answer the question: Does the enzymic acetylation of histones—a process which proceeds independently of histone synthesis—affect the structure and function of chromatin?

The question is an important one because the number of different histone fractions in different somatic tissues, or in the same tissue at different stages in development, is not likely to be able to account for the great complexity and selectivity of gene activation. At present there is no convincing evidence that puffs in insect chromosomes lose histones as they become active in RNA synthesis. [However, it has been reported that the heterochromatic (dense and synthetically inert) regions of mealy bug chromatin have an increased concentration of lysine-rich histones.] Similarly, there are no major differences in the histone complements of active and inactive thymus chromatin fractions after their isolation. The problem arises as to how histone–DNA interactions can be influenced without necessarily removing or destroying the basic proteins. One possibility is by altering the structure of the histone. Acetylation of lysine residues, for example, would be expected to decrease histone charge and affinity for DNA.

To answer the question of the relationship between histone acetylation and chromatin structure, recourse was made to procedures for chromatin fractionation. Nuclei were incubated in the presence of 2-^{14}C-uridine or 6-^{14}C-orotic acid as RNA precursors and acetate-2-^{14}C as a precursor of the histone acetyl groups. Following incubation, the nuclei were washed, broken by sonication, and fractionated (as described earlier in this review) to yield the active and inactive chromatin subfractions. The specific activity of the histones and of the RNA in each fraction was then determined. A very strong proportionality was observed: Histone acetylation, like RNA synthesis, was much more active in the diffuse chromatin fraction (3).

There is good reason to believe that acetylation takes place in the chromatin and not elsewhere in the nucleus. High resolution autoradiographs of thymus nuclei labeled with ³H-acetate show labeling over the chromatin, and light microscope autoradiographs of the salivary gland chromosomes of *Chironomus thummi* also show histone labeling along the chromosome (3).

Another type of correlation between histone acetylation and chromatin function is indicated by tracer studies of cells, such as avian erythrocytes, in which the genetic apparatus exists in a highly repressed state. The mature nucleated red blood cells of the duck, for example, incorporate only negligible amounts of RNA precursors. They also fail to acetylate histones at rates comparable to those observed in cells which can synthesize RNA, such as lymphocytes. The results support the view that cells which are not actively engaged in RNA synthesis are not actively acetylating their histones either.

More direct evidence relating histone acetylation to gene activation has been obtained in the study of cells in culture, using procedures which activate or depress gene function. Perhaps the most striking results are those of Pogo and co-workers on changing patterns of acetylation during the response of human lymphocytes to phytohemagglutinin stimulation (see Allfrey, 3). It is now well known that lymphocytes, which do not ordinarily divide in culture, can be stimulated to enlarge and divide by the addition of mitogenic agents, of which phytohemagglutinin (a protein derived from the common bean *Phaseolus vulgaris*) is the best known example. Under the proper conditions, nearly all the cells are transformed—they soon begin to synthesize new proteins and to increase their rates of RNA synthesis—they enlarge, and, at a relatively late stage in their response to phytohemagglutinin (PHA), they synthesize DNA and histones. After 48–72 hours, 70–90% of the cells go on to divide.

Tracer studies of this lymphocyte transformation process show that new RNA's and new proteins are made soon after PHA is added to the cell suspension. What happens to the histones when lymphocyte genetic activity is triggered in this way? The most striking effect is a great increase in the rate of acetylation of the arginine-rich histone fractions, detectable within minutes after PHA is added to the cell suspension. A comparison of the kinetics of RNA synthesis and histone acetylation in PHA-stimulated cells shows very clearly that the increase in acetylation rate precedes the increase in the rate of RNA synthesis.

It was concluded that a change in the structure of the chromatin—

brought about by, or coincident with histone acetylation—might be a necessary prerequisite for the synthesis of new RNA's at previously repressed gene loci (3). This conclusion is supported by recent evidence that the DNA of the "transformed" lymphocyte has become more accessible to staining by acridine orange dye, and to binding by Actinomycin D. The kinetics of these changes in DNA reactivity are very similar to those observed for histone acetylation.

An interesting extension of these experiments became possible when it was found that RNA synthesis in equine granulocytes is depressed when phytohemagglutinin is added to the cell culture. Histone acetylation is suppressed at the same time. Moreover, if acetate uptake is allowed to proceed in normal cells for 15 minutes and PHA is then added, previously incorporated acetyl groups are discarded, while no loss of radioactive amino acids from histones can be detected in this time interval.

Further evidence for the modification of histone structure during periods of gene activation comes from a comparison of histone acetylation in normal and regenerating liver. In normal rat liver, radioactive acetate is rapidly incorporated into the arginine-rich histones. The pattern of acetylation of histones in regenerating liver is complex, because in dividing cells, new histone is synthesized. The synthesis of certain histones (the f2a fraction) involves an acetylation of the NH_2-terminal serine residues. At the same time there is an independent, puromycin-resistant acetylation of the lysine residues in the protein. Most of the acetate incorporated into the arginine-rich histone fraction occurs there as ϵ-N-acetyl lysine (32a). The acetylation of normal liver histones appears to be largely reversible, because 70% of the counts present at 15 minutes are lost in the next 45-minute interval.

After partial hepatectomy, the remaining parenchymal cells grow and divide, and, as mentioned earlier, this response involves an increase in DNA template activity and the synthesis of new types of RNA. A change in histone acetylation is also observed, and this is one of the earliest events seen in regenerating liver. One of the most striking differences between regenerating and normal liver is in the turnover of acetyl groups on arginine-rich histone fractions. Throughout the period of maximum gene activation for RNA synthesis, this turnover is drastically curtailed. Later, after DNA template activity begins to level off, acetyl group turnover on the histones resumes and reverts to the normal pattern. This occurs at a precise time interval after partial hepatectomy (between 4 and 5 hours) and before DNA template activity has reached its peak.

The results certainly suggest an involvement of histone acetylation in changing patterns of gene function, and this conclusion finds further support in studies of hormone effects on nuclear activity. It is now well established, as a result of studies in many laboratories, that RNA synthesis in the mammalian liver can be affected by steroid hormones, hydrocortisone in particular (see Tata, *91*). Particularly incisive were the studies of Kenney and co-workers (*96b*) who showed that hydrocortisone increases the rate of labeling of many types of RNA without changing the rates of side reactions, such as terminal nucleotide turnover in amino acid tRNA's. It is also known, from the work of Kidson and Kirby, that cortisol administration induces the synthesis of new types of RNA which can be recognized by their patterns of separation in countercurrent distribution systems. Hydrocortisone also stimulates histone acetylation in the livers of normal or adrenalectomized animals. Not only is there an increase in the extent of labeling of the arginine-rich fraction, but acetyl group turnover remains suppressed during the period of gene activation. This stimulation of RNA synthesis and histone acetylation is only one aspect of hormonal effects on chromatin function. It is well known that different tissues respond differently to the same hormone. With cortisol, liver function is stimulated while lymphoid tissues are adversely affected. Lymphocyte cultures, for example, show a rapid decline in their capacity for RNA synthesis within 5 minutes after cortisol is added. A decreased acetylation of the histones is also evident when isolated lymphocyte nuclei are treated with soluble cortisol derivatives.

The above examples all point to a correlation between gene activation for RNA synthesis and the degree of acetylation of the arginine-rich histones. It should be observed that this may be coincidental, and that present evidence does not constitute a proof that histone acetylation is causally related to the activation of particular genetic loci, but the correlation with RNA synthesis is strong and it would appear that the phenomenon warrants further investigation.

In this connection, it should be noted that no such correlations have yet been seen for histone methylation. On the contrary, in regenerating liver, the methylation of the arginine-rich histones is a relatively late event. It occurs long after the early activation of RNA synthesis and reaches its maximal rate at 30–36 hours after partial hepatectomy—at a time when histone, DNA, and RNA synthesis have already begun to decline. This is a clear distinction between histone acetylation and methylation in the same tissue under the same physiological conditions.

There is some evidence that phosphorylation of the histones is also a signal of increased genetic activity. In regenerating liver, the phosphorylation of the lysine-rich fraction is increased, but the kinetics of the increase remain to be investigated.

As mentioned earlier, the phosphorylation of histones involves the transfer of phosphoryl groups from ATP to hydroxyl groups on serine residues in the protein. This reaction has an important counterpart in another recently discovered class of nuclear components—the nuclear phosphoproteins.

14. *Intranuclear Location and Metabolism of Phosphoproteins*

It has been known for some time that cells exposed to ^{32}P-orthophosphate incorporate ^{32}P into phosphoprotein linkages. What is new is that this is largely an intranuclear phenomenon. Recent experiments by Langan and Lipmann (*50b*) have provided convincing evidence for the nuclear localization of a highly phosphorylated protein fraction which appears to contain tracts of adjacent serine phosphate residues and phosphothreonine as well. Such proteins have been found in the nuclei of calf thymus lymphocytes and in rat liver nuclei, where they occur in amounts that suggest a major role in nuclear metabolism—up to 45% of the DNA mass. A striking illustration of their nuclear localization is the fact that they are readily detected in the nucleated red blood cells of birds but not in the nonnucleated erythrocytes of mammals.

It is a matter of some interest that condensed and diffuse chromatin fractions, which differ in their rates of RNA synthesis, also differ in their content of phosphoproteins. The active chromatin has approximately 3.5 times more alkali-labile phosphate (per unit of DNA) than does the condensed chromatin. In this connection, it should be recalled that the 2 chromatin fractions, as isolated by the method of Frenster and co-workers, do not differ appreciably in their histone-to-DNA ratios (*29*). The presence of large amounts of phosphoprotein in the active chromatin raises the possibility that these acidic proteins, by complexing with histones, may change the state of the chromatin by altering the affinity between DNA and histones, and thus modify the template activity of the chromatin in RNA synthesis. This possibility has been tested by Langan and Smith who showed that phosphoproteins do combine with histones *in vitro* and that this diminishes the inhibitory effects of the histones on the RNA polymerase reaction (*50b*). Kleinsmith has recently studied the effects of adding phosphoproteins to isolated nuclei and observed that the state of the condensed chromatin is altered to form a more diffuse network, a result which lends support to an earlier observation by Frenster that nuclear phosphoproteins added to condensed chromatin fractions increase their RNA polymerase activity.

The phosphorylation of the nuclear proteins is an energy-dependent reaction which has been studied in some detail by Kleinsmith and co-workers (*47a,b*). Not only ATP, but other nucleoside triphosphates, will donate their terminal phosphoryl groups to serine and threonine residues in the nuclear protein. The reaction is catalyzed by a phosphoprotein kinase which also occurs in the nucleus. In the nucleus, phosphorylation is a remarkably active exchange reaction; 60–70% of previously incorporated ^{32}P is lost from thymus nuclear phosphoproteins in a 2 hour incubation under conditions in which there is no loss of previously incorporated ^{14}C-serine.

What is the significance of this rapid turnover of phosphate groups linked to proteins of the chromatin? The possibility that regions of high negative charge density on the phosphoproteins might influence DNA–histone interactions and modify the structure of the chromatin has already been mentioned, together with the evidence in its support. Since the phosphoprotein concentrations of diffuse chromatin greatly exceed those in condensed chromatin, and since the phosphate groups of these proteins turn over rapidly, a mechanism is suggested for influencing DNA–histone interactions in a dynamic way, and for changing the condensed inactive state of the chromatin to a more extended functional state. Dephosphorylation of the diffuse chromatin, on the other hand, could again lead to a tighter coiling of the DNA–histone–protein complex. Such an active spooling mechanism for altering the state of the chromatin could begin to account for the autoradiographic evidence (35) that, in DNA synthesis, all the chromatin must spend part of its time in the diffuse state.

As mentioned earlier, RNA synthesis in many nuclear types appears to require that the template DNA exist in the diffuse state, and one would predict an increase in nuclear phosphorylation during periods of gene activation. This has been found to be the case in human lymphocytes stimulated by phytohemagglutinin (47c), in which ^{32}P-phosphate uptake into nuclear phosphoproteins, like histone acetylation, is increased soon after addition of the mitogenic agent. It is of interest that, at this time, when the phosphorylation of the nuclear proteins is augmented, acridine orange binding to DNA increases in the treated cells, again suggesting a change in the state of the chromatin.

15. Some Other Aspects of the Control of RNA Synthesis

In all the foregoing discussion of DNA–histone–phosphoprotein interactions, emphasis was placed on the importance of the physical state of the chromatin. On the premise that DNA must be placed in a diffuse chromatin region to be functional, such considerations are clearly relevant to the control of RNA synthesis.

It is equally obvious that the control of genetic activity in the cells of higher organisms must involve more than histone interactions with DNA, even if one accepts the view that histones do influence DNA function.

The problem of specifically inducing or repressing RNA synthesis at hundreds or thousands of different loci in the chromosomes of different cell types is not readily answerable in terms of small numbers of histones. Although it has been pointed out that lysine-rich and arginine-rich histones influence the structure of chromatin in different ways, and that histones are chemically modified (e.g., by acetylation) at times of gene activation, the question of specificity remains to be considered. How are particular cistrons turned on and off in different cell types, or in the same cell type, at different stages in its development? Whatever theory is advanced, it must account for the fact that nuclear activity is subject to feedback control from the cytoplasm and the environment.

In this connection, reference should be made to work on the hormonal induction of RNA synthesis and to recent isolations of gene repressors in bacterial systems. Both offer significant insights into control mechanisms.

The control exerted by hormones over nuclear activity has now been documented for mammalian, plant, and insect cells (see Tata, 91). Steroid hormones in particular, have been shown to influence the rate of RNA synthesis in target tissues. When cells increase their rates of RNA synthesis in response to hormonal stimulation, the hormone itself is often localized in the chromatin. Radioactive aldosterone, for example, was detected in kidney nuclei, and testosterone was shown by Wilson to be preferentially bound to the active chromatin fraction in the preen gland of the duck. Estrogens bind to a nuclear fraction of the uterus and also appear along the chromosomes of amphibian oocytes. In some of these cases, at least, the binding of steroid hormones to nuclei involves the mediation of specific proteins. It is presumed that such proteins are gene repressors or inducers and that the effect of the hormone is to induce an allosteric change in protein conformation, thus modifying its effect on the genome (see Tata, 91).

The existence of repressor proteins specific for a particular gene locus in the bacterial chromosome is now fairly well established. In *E. coli*, the repressor protein for the β-galactosidase cistron has been isolated by Gilbert. (32b). It combines both with the appropriate region of the DNA and with the inducer (isopropyl thiogalactoside). A similar repressor protein involved in the control of bacteriophage replication is also known. It not only recognizes a specific cistron in the phage genome, but it will combine only with DNA in the double-stranded form. These findings should offer significant clues in the search for repressor proteins in the chromatin of higher organisms.

At this time, mention should be made of another important aspect of nuclear RNA metabolism, namely, the fact that much of the RNA synthesized in the nucleus does not enter the cytoplasm. This intranuclear

RNA turnover was first observed in critical autoradiographic studies by Harris and associates (*34a*), and has since been verified by more direct biochemical methods in several laboratories. Particularly impressive are recent RNA–DNA hybridization studies by Shearer and McCarthy showing that all nucleotide sequences of RNA found in the cytoplasm of the L cell are also found in the nucleus, but that cytoplasmic RNA was unable to compete against a large fraction of the nuclear RNA molecules—indicating that the latter are not present in the cytoplasm. Estimates of the amount of nucleus-specific RNA indicated that approximately 80% of the nucleotide sequences being transcribed into RNA in L cells are in unstable molecules which are retained by the cell nucleus (*84a*).

Relevant to this problem are observations in Attardi's laboratory, and by Scherrer and co-workers, of very high molecular weight RNA molecules (about 20,000 nucleotides in length; molecular weights in the range of 2×10^6–10^7 daltons) which are found in the nuclei of HeLa cells and duck erythroblasts. These RNA's differ in base composition and size from the high molecular weight ribosomal RNA precursors. Though they are rapidly synthesized, they do not appear to be simple precursors of the messenger RNA's found on polysomes in the cytoplasm, although this possibility has not yet been excluded. The notion that they are polycistronic messenger RNA's, or that they are repeating units of the same message seems unlikely, especially in the case of the erythroblast, in which 90% of the newly synthesized protein is globin and in which genetic data rule out linkage or redundancy. Such RNA's have exceedingly rapid rates of synthesis and degradation (*90a*), and they are restricted to the cell nucleus (*84a*).

In considering the possible significance of such rapidly synthesized RNA's of unknown function, many of which are restricted to the cell nucleus, the problem of genetic regulation offers an obvious role. It can be assumed that in higher cells, which have many more nucleotide sequences in their DNA than would be required to code for known cytoplasmic proteins, most of the genome is concerned with regulation—telling other structural genes when to act and when to remain silent in the synthesis of ribosomal RNA's and messenger RNA's for the cytoplasm. This control, if mediated through proteins, must require an enormously complex set of structural genes for repressor proteins. It would follow that protein synthesis in the cell nucleus, and much of the RNA synthesis there, would be largely concerned with the mechanism of genetic regulation.

This aspect of the biochemistry of the nucleus is still in its early stages, and the constant influx of new and significant information is sure to alter our present concept of genetic transcription and its control. Yet the present state of our knowledge does permit an integrated view of chromatin activity which may help tie together many of the observations described earlier in this review: DNA is largely complexed with histones and the complex is effectively silenced as a template for RNA synthesis. At active loci, the state of the chromatin is transformed and histone–DNA interactions are modified, perhaps by structural changes in the histones themselves (e.g., acetylation) or by interactions with nuclear phosphoproteins (or other specific proteins). This is a necessary but not a sufficient condition for gene activation at a specific locus. This control requires, in addition, the binding or release of repressors or inducers which inhibit (or facilitate) transcription of DNA into RNA. The latter process is carried out in the chromatin by the tightly associated RNA polymerases. Repressors of transcription are probably proteins and their specificity is due to their capacity to recognize and bind to specific tracts of nucleotides in double-stranded DNA. The site of repressor protein synthesis is not known, but if it occurs in the nucleus, this would confer a function to RNA molecules of the type described above which are restricted to the cell nucleus. (An alternative proposal is that such RNA's are themselves regulators of the genetic transcription mechanism, a consideration which must be taken seriously because of the strength of the experimental evidence indicating that added RNA's and especially amino acid tRNA's, can inhibit the RNA polymerase reaction.) In either case, the presence of nuclear-specific RNA's could be explained in terms of their role in the control of DNA transcription.

Finally, it should be pointed out that the nucleus is subject to feedback control from the cytoplasm. This is dramatically illustrated by observations on avian erythrocyte nuclei, which have ceased RNA synthesis, but which will resume it when placed in the cytoplasm of the HeLa cell or macrophage (*34*). Such control is presumably mediated by proteins which move from cytoplasm to nucleus, and this may be the function of the cytonucleoproteins mentioned earlier in this review.

E. DNA AND CHROMOSOMAL REPLICATION

Until now, this review has concentrated largely on the role of the nucleus in the synthesis of ribonucleic acids and proteins, and on its energy metabolism, and has failed to consider the major attribute of the genetic material, namely, its capacity for self-replication. The dividing cell poses a whole new set of problems in molecular biology, some of which will be considered now.

How is DNA replicated? It is generally considered that the model which best explains the specificity of DNA replication is the Watson-Crick double helix, in which the bases in 1 strand pair specifically with complementary bases in the opposing strand—adenine with thymine and guanine with cytosine. In this way the order of bases in 1 strand fixes the base sequence in the complementary strand, and it is presumed that the double helix separates at the replication point to permit each strand to be copied. This produces 2 double helices, each of which contains 1 parental strand and 1 newly synthesized strand. This is the semiconservative model of DNA replication which has been proved by the experiments of Meselson and Stahl to be the pattern followed in bacterial systems. Autoradiographic studies of plant and animal chromosomes labeled with ^{3}H-thymidine are also best explained in terms of the semiconservative replication of the genetic material; i.e., the newly synthesized DNA strands and the unlabeled parental strands are distributed in the chromatids of successively dividing cells as would be predicted by the model (see Taylor, *92, 92a,b,c*). However, the autoradiographic evidence is sometimes complicated by other phenomena affecting chromosome structure in higher forms (such as crossing-over and translocation of chromosome segments) and the approach has been combined with biochemical techniques which also establish the semi conservative mode of replication. For example, when HeLa cells are grown in the presence of ^{14}C-bromodeoxyuridine in place of thymidine, the newly synthesized DNA strands contain bromouracil in place of thymine and are not only radioactive, but are also more dense than the parental DNA strands. As a result, the hybrid (T-DNA/BrU-DNA) can be separated by sucrose density-gradient centrifugation. The existence of the hybrid form, plus the fact that it can be denatured to produce 1 strand of labeled BrU-DNA and 1 strand of unlabeled T-DNA establishes that replication in higher cells does proceed by a strand separation mechanism and that each parental strand is paired with its complementary daughter strand.

This mechanism accounts for the specificity and virtually error-free precision in the copying of millions of DNA nucleotide sequences during cell division, and it would also appear to be a logical explanation for the mechanism of repair of single-strand deletions or breaks in damaged DNA molecules.

The enzymic mechanism of deoxynucleotide polymerization to form DNA has not been fully explored in higher organisms. There seems little doubt that the 5′-deoxyribonucleoside triphosphates are the required precursors, and that their polymerization involves the formation of 3′–5′ phosphodiester linkages between adjacent nucleotides in the growing chain. (The reaction proceeds with the elimination of inorganic pyrophosphate which is destroyed by a pyrophosphatase present in the nucleus.) DNA-synthesizing enzymes are known. Some are similar to the DNA polymerases originally described by Kornberg in bacterial extracts. However, the thymus DNA polymerase which has received most of the attention has not displayed the expected specificity for native double-stranded DNA but prefers instead heat-denatured (i.e., single-stranded) DNA as a primer. As a result, it has been suggested that this enzyme is really a repair enzyme, and that there ought to be a DNA replicase with specificity for native DNA. Studies of a DNA polymerase in the nuclei of HeLa cells by Mueller and co-workers indicate the presence of an enzyme, the activity of which is proportional to the degree of synchrony of the culture (i.e., to the number of cells engaged in DNA synthesis). This makes it likely that this polymerase is involved in DNA replication *in vivo*. Like other DNA-synthesizing enzymes, it requires all four deoxyribonucleoside triphosphates and a DNA template. Its activity also depends on factors derived from the cytoplasmic portion of the cell homogenate. This interesting observation is reminiscent of the findings (1) that avian erythrocyte nuclei (which are inactive in DNA synthesis) resume thymidine uptake when they are placed in HeLa cell cytoplasm (*34*); and (2) that transplantation of an ameba nucleus from a cell in the G2 phase to a cell in the S phase (period of DNA synthesis) may reinitiate synthesis of DNA in the transplanted nucleus (*77a,b*). Similar effects are seen in *in vitro* experiments (*92d*).

The duplication of DNA at the molecular level is only one aspect of the complex series of events occurring in chromosomal replication, but there is good evidence that chromosomal DNA is subdivided into units of DNA which replicate independently; these are referred to as replicons, by analogy with the single replicon of the bacterial chromosome. Autoradiographic studies of dividing cells have made it clear that different parts of the chromosome can replicate independently, and that some chromosomes do not synthesize DNA at all until most of the other chromosomes have already completed thymidine uptake. Usually the heterochromatic X chromosome is the last to be duplicated in somatic cells.

From autoradiographic studies, Taylor has estimated the time of replication of the long arm of the X chromosome in the Chinese hamster to be about 210 minutes. This arm of the X chromosome is 3 μ long and has been estimated to contain about a 41,000 μ length of DNA in a tightly coiled configuration. Estimates of the rate of DNA synthesis in cells of higher organisms vary from 1–60 μ of DNA/minute. If the lower estimates are correct, this means that about 200 replicons exist in the long arm (41,000 μ/1 μ per minute $\times$ 210 minutes = 195); higher rates of DNA synthesis would reduce this estimate proportionately. In any case, the long arm of the X is only 3 μ of a total chromosomal length of

124 μ, and it is obvious that the whole genome must contain several hundred independently replicating DNA subunits. The virtual autonomy of the replicons in different regions of the chromosome is strikingly demonstrated in recent experiments by Hsu, who found that translocated arms of X chromosomes replicate at a time when the remainder of the X chromosome is duplicated, and not when the DNA is synthesized in the somatic chromosome to which they are attached.

Two new aspects of DNA replication in the chromosomes of higher organisms should now be considered. The first concerns the mechanism and rate of growth of DNA chains. Pulse-labeling studies of replication in Chinese hamster cells show that the newly synthesized DNA first appears in short segments (less than 10 S, 30–40 μ in length). These are linked to longer chains (20–40 S) within 2 minutes, and within 2–3 hours the pulse-labeled segments have become connected to DNA chains about 200 μ long (70–75 S). The conclusion was drawn that DNA is replicated in short segments which are quickly joined to larger pieces and then later linked to form the long-strand structural units of the chromosome (92a,b,c). This conclusion is in accord with autoradiographic studies of chromosomal DNA synthesis. Autoradiography of single DNA fibers after pulse-labeling with ³H-thymidine shows that chromosomal DNA consists of long fibers which are subdivided into much shorter replication units (replicons). DNA replication begins in the interior of each replication unit and proceeds outward from two forklike growing points to the ends of the replication unit (41a).

For further considerations of DNA replication in higher organisms, the reader is referred to recent reviews (92b, 41a).

VII. CONCLUSION

There are many aspects of the biochemistry of the cell nucleus that have not been considered in this review, or which have been mentioned only fleetingly. Nothing has been said of phospholipid and lipid metabolism, lipoproteins, low molecular weight RNA's, nucleopeptides, and other substances known to exist in the nucleus or in the nuclear envelope.

The topics which have been considered do reflect a large part of current research on nuclear function—and many were selected to illustrate the important contributions of biology to research at the molecular level.

(Consider, for example, the value of the anucleolate mutant of *Xenopus* in the study of ribosomal RNA synthesis.) The complexity of the metazoan should be regarded not as an obstacle, but as an opportunity. Differentiated cells represent the ultimate in experimental design of genetic control mechanisms—and the implications of further research in this area will extend to the clinic as well as to the laboratory.

REFERENCES

1. V. G. Allfrey, The isolation of subcellular components. *In* "The Cell" (J. Brachet and A. E. Mirsky, eds.), Vol. 1, p. 193. Academic Press, New York, 1959.
2. V. G. Allfrey, Nuclear ribosomes, messenger-RNA and protein synthesis. *Exptl. Cell Res.* 9, 183–212 (1963).
3. V. G. Allfrey, Control mechanisms in RNA synthesis. *Cancer Res.* 26, 2026–2040 (1966).
4. V. G. Allfrey, M. M. Daly, and A. E. Mirsky, Some observations of protein metabolism in chromosomes of non-dividing cells. *J. Gen. Physiol.* 38, 415–424 (1955).
5. V. G. Allfrey, V. C. Littau, and A. E. Mirsky, Methods for the purification of thymus nuclei and their application to studies of nuclear protein synthesis. *J. Cell Biol.* 21, 213–231 (1964).
6. V. G. Allfrey, R. Meudt, J. W. Hopkins, and A. E. Mirsky, Na-dependent "transport" reactions in the cell nucleus and their role in protein and nucleic acid synthesis. *Proc. Natl. Acad. Sci. U. S.* 47, 907–932 (1961).
7. V. G. Allfrey and A. E. Mirsky, The role of DNA and other polynucleotides in ATP synthesis by isolated cell nuclei. *Proc. Natl. Acad. Sci. U. S.* 43, 589–598 (1957).
8. V. G. Allfrey and A. E. Mirsky. Evidence for the complete DNA-dependence of RNA synthesis in isolated thymus nuclei. *Proc. Natl. Acad. Sci. U. S.* 48, 1590–1596 (1962).
9. V. G. Allfrey and A. E. Mirsky, Mechanisms of synthesis and control of protein and RNA synthesis in the cell nucleus. *Cold Spring Harbor Symp. Quant. Biol.* 28, 247–262 (1963).
10. V. G. Allfrey and A. E. Mirsky, Role of histone in nuclear function. *In* "The Nucleohistones" (J. Bonner and P. O. P. Ts'o, eds.) pp. 267–288. Holden-Day San Francisco, California, 1964.
11. V. G. Allfrey, A. E. Mirsky, and S. Osawa, Protein synthesis in isolated cell nuclei. *J. Gen. Physiol.* 40, 451–490 (1957).
11a. I. Bekhor, G. M. Kung, and J. Bonner, Sequence-specific interaction of DNA and chromosomal protein. *J. Mol. Biol.* 39, 545–550 (1969).
12. A. Bergstrand, N. A. Eliasson, E. Hammersten, B. Norberg, P. Reichard, and H. von Ubisch, Experiments with N¹⁵ on purines from nuclei and cytoplasm of normal and regenerating liver. *Cold Spring Harbor Symp. Quant. Biol.* 13, 22–25 (1948).
13. M. L. Birnstiel and B. B. Hyde, Protein synthesis by isolated pea nucleoli. *J. Cell Biol.* 18, 41–50 (1963).
14. J. Bonner and R. C. Huang, Histones as specific repressors of chromosomal RNA synthesis. *Ciba Found. Study Group* 24, 18–33 (1966).

15. J. Brachet, "Biochemical Cytology." Academic Press, New York, 1957.
15a. D. D. Brown and J. B. Gurdon, Absence of ribosomal RNA synthesis in the anucleolate mutant of *Xenopus laevis*. *Proc. Natl. Acad. Sci. U. S.* **51**, 139–146 (1964).
15b. J. A. V. Butler, E. W. Johns, and D. M. P. Phillips. Recent investigations on histones and their functions. *Prog. Biophys. Mol. Biol.* **18**, 209–244 (1968).
16. H. Chantrenne, in "Modern Trends in Physiological Sciences," Vol. 14 Pergamon Press, Oxford, 1961.
17. J. Chauveau, Y. Moule, and C. Rouiller, Isolation of pure and unaltered liver nuclei—morphology and biochemical composition. *Exptl. Cell Res.* **11**, 317–321 (1956).
17a. L. A. Chinouard and C. P. Leblond, Sites of protein synthesis in nucleoli of root meristematic cells of *Allium cepa* as shown by autoradiography with ^{3}H-arginine. *J. Cell Sci.* **2**, 473–480 (1967).
17b. R. B. Church and B. J. McCarthy, Ribonucleic acid synthesis in regenerating and embryonic liver. I. The synthesis of new species of RNA during regeneration of mouse liver after partial hepatectomy. *J. Mol. Biol.* **23**, 459–475 (1967).
18. T. E. Conover and G. Siebert, On the occurrence of respiratory components in rat liver nuclei. *Biochim. Biophys. Acta* **99**, 1–12 (1956).
18a. T. E. Conover, Respiration and adenosine triphosphate synthesis in nuclei. *Current Topics in Bioenergetics* **2**, 235–267 (1967).
18b. W. A. Creasey and L. A. Stocken, Biochemical differentiation between radio-sensitive and non-sensitive tissues in the rat. *Biochem. J.* **69**, 17p (1958).
19. J. Cronshaw, L. Hoefert, and K. Esau, Ultrastructural features of Beta leaves infected with beet yellow virus. *J. Cell Biol.* **31**, 429–443 (1966).
20. W. D. Currie, N. M. Davidian, W. B. Elliott, N. F. Rodman, and R. Penniall, Respiratory activity of isolated mammalian nuclei. III. The reduced NAD oxidase of rat liver nuclei and nucleoli. *Arch. Biochem. Biophys.* **113**, 156–166 (1966).
21. M. M. Daly, V. G. Allfrey, and A. E. Mirsky, Uptake of N^{15}-glycine by components of cell nuclei. *J. Gen. Physiol.* **36**, 173–179 (1952).
21a. M. M. Daly and A. E. Mirsky, Histones with high lysine content. *J. Gen. Physiol.* **38**, 405–413 (1955).
21b. J. Darnell, Ribonucleic acids from animal cells. *Bact. Rev.* **32**, 262–290 (1968).
21c. E. H. Davidson, V. G. Allfrey, and A. E. Mirsky, On the RNA synthesized during the lampbrush phases of amphibian oogenesis. *Proc. Natl. Acad. Sci. U. S.* **52**, 501–507 (1964).
21d. H. Denis, Gene expression in amphibian development. *J. Mol. Biol.* **22**, 285–304 (1966).
22. C. W. Dingman and M. B. Sporn, Studies on chromatin-isolation and characterization of nuclear complexes of DNA, RNA and protein from embryonic and adult tissues of the chicken. *J. Biol. Chem.* **239**, 3483–3492 (1964).
23. J. E. Edstrom, Synthesis of RNA from various parts of spider oocytes. *J. Biophys. Biochem. Cytol.* **8**, 47–51 (1960).
24. D. M. Fambrough and J. Bonner, On the similarity of plant and animal histones. *Biochemistry* **5**, 2563–2569 (1966).
25. D. W. Fawcett, "The Cell: Its Organelles and Inclusions—an Atlas of Fine Structure." Saunders, Philadelphia, Pennsylvania, 1966.
26. F. Fischer, G. Siebert, and E. Adloff, Charakterisierung von zwei Adenosintriphosphatasen in Schweinenieren-Zellkernen. *Biochem. Z.* **332**, 131–150 (1959).

27. R. F. Fisher, D. J. Holbrook, and J. L. Irvin, Density-gradient isolation of rat liver nuclei with high DNA content. *J. Cell Biol.* **17**, 231–236 (1963).

27a. W. W. Franke, Isolated nuclear membranes. *J. Cell Biol.* **31**, 619–623 (1966).

28. J. H. Frenster, V. G. Allfrey, and A. E. Mirsky, Metabolism and morphology of ribonucleoprotein particles from the cell nucleus of lymphocytes. *Proc. Natl. Acad. Sci. U. S.* **46**, 432–444 (1960).

29. J. H. Frenster, V. G. Allfrey, and A. E. Mirsky, Repressed and active chromatin isolated from interphase lymphocytes. *Proc. Natl. Acad. Sci. U. S.* **50**, 1026–1032 (1963).

29a. T. Fujimura, S. Hasegawa, Y. Shimizu, and T. Sugimura, Polymerization of the adenosine-5'-diphosphate-ribose moiety of NAD by nuclear enzyme. *Biochim. Biophys. Acta* **145**, 247–252 (1967).

30. J. G. Gall, Octagonal nuclear pores. *J. Cell Biol.* **32**, 391–400 (1967).

31. J. G. Gall and H. G. Callan, Uridine-H^3 Incorporation into Lampbrush Chromosomes. *Proc. Natl. Acad. Sci. U. S.* **48**, 562–570 (1962).

31a. D. Gallwitz and G. C. Mueller, Protein synthesis in nuclei isolated from HeLa cells. *Eur. J. Biochem.* **9**, 431–438 (1969).

32. G. P. Georgiev, Nature and biosynthesis of nuclear RNAs. *Prog. Nucleic Acid Res. Mol. Biol.* **6**, 259–351 (1967).

32a. E. L. Gershey, G. Vidali, and V. G. Allfrey, The occurrence of epsilon-N-acetyl lysine in the *f2al* histone. *J. Biol. Chem.* **243**, 5018–5022 (1968).

32b. W. Gilbert and B. Müller-Hill, Isolation of the LAC repressor. *Proc. Natl. Acad. Sci. U. S.* **56**, 1891–1898 (1966).

33. L. Goldstein, Interchange of protein between nucleus and cytoplasm. *Symp. Intern. Soc. Cell Biol.* **4**, 79–94 (1965).

33a. J. B. Gurdon, Nucleic acid synthesis in embryos and'its bearing on cell differentiation. *Essays in Biochem.* **4**, 26–68 (1968).

34. H. Harris, The reactivation of the red cell nucleus. *J. Cell Sci.* **2**, 23–32 (1967).

34a. H. Harris and J. W. Watts, The relationship between nuclear and cytoplasmic ribonucleic acid. *Proc. Roy Soc. B.* **156**, 109–118 (1962).

35. E. D. Hay and J. P. Revel, The fine structure of the DNP component of the nucleus: An electron microscopic study utilizing autoradiography to localize DNA synthesis. *J. Cell Biol.* **16**, 29–51 (1963).

35a. L. S. Hnilica, Proteins of the cell nucleus. *Progr. Nucleic Acid Res. Mol. Biol.* **7**, 25–106 (1967).

36. M. B. Hoagland, M. L. Stephenson, J. F. Scott, L. I. Hecht, and P. C. Zamecnik, A soluble ribonucleic acid intermediate in protein synthesis. *J. Biol. Chem.* **231**, 241–257 (1958).

36a. M. B. Hoagland, An enzymic mechanism for amino acid activation in animal tissues. *Biochim. Biophys. Acta* **16**, 288–291 (1955).

37. G. H. Hogeboom, W. C. Schneider, and G. E. Palade, Cytochemical studies of mammalian tissues. I. Isolation of intact mitochondria from rat liver: Some biochemical properties of mitochondria and submicroscopic particulate material. *J. Biol. Chem.* **172**, 619–636 (1948).

38. J. J. Holland and B. J. McCarthy, Stimulation of protein synthesis *in vitro* by denatured DNA. *Proc. Natl. Acad. Sci. U. S.* **52**, 1554–1561 (1964).

39. J. W. Hopkins, Amino acid activation and transfer to ribonucleic acids in the cell nucleus. *Proc. Natl. Acad. Sci. U. S.* **45**, 1461–1470 (1959).

40. R. C. Huang and J. Bonner, Histone, a suppressor of chromosomal RNA synthesis. *Proc. Natl. Acad. Sci. U. S.* **48**, 1216–1222 (1962).

40a. R. C. C. Huang and P. C. Huang, Effect of protein-bound RNA associated with chick embryo chromatin on template specificity of the chromatin. *J. Mol. Biol.* **39,** 365–378 (1969).

41. J. Huberman and G. Attardi, Isolation of metaphase chromosomes from HeLa cells. *J. Cell Biol.* **31,** 95–105 (1966).

41a. J. A. Huberman, Mechanism of DNA replication in mammalian chromosomes. *In* "Exploitable Molecular Mechanisms and Neoplasia," pp. 337–356. Williams and Wilkins Co., Baltimore, Maryland, 1969.

42. N. T. Hubert, P. Favard, N. Carasso, R. Rozencwajg, and J. P. Zalta, Méthode d'isolement de noyaux cellulaires à partir de foie de rat. *J. Microscopie* **1,** 435–444 (1962).

43. J. Hurwitz and J. T. August, The role of DNA in RNA synthesis. *Progr. Nucleic Acid Res. Mol. Biol.* **1,** 59–92 (1963).

44. A. A. Infante and M. Nemer, Accumulation of newly-synthesized RNA templates in a unique class of polyribosomes during embryogenesis. *Proc. Natl. Acad. Sci. U. S.* **58,** 681–689 (1967).

45. V. M. Ingram, A specific chemical difference between the globins of normal human and sickle-cell anaemia haemoglobins. *Nature* **178,** 792–794 (1956).

45a. E. W. Johns, The electrophoresis of histones in polyacrylamide gel and their quantitative estimation. *Biochem. J.* **104,** 78–82 (1967).

46. H. M. Keir and J. N. Davidson, Purine and pyrimidine derivatives in the acid-soluble fraction of animal tissues and cell nuclei. *Arch. Biochem. Biophys.* **77,** 68–80 (1958).

47. H. M. Keir, R. M. S. Smellie, and G. Siebert, Intracellular localization of DNA-nucleotidyl transferase. *Nature* **196,** 752–754 (1962).

47a. L. J. Kleinsmith and V. G. Allfrey, Nuclear phosphoproteins. I. Isolation and characterization of a phosphoprotein fraction from calf thymus nuclei. *Biochim. Biophys. Acta* **175,** 123–135 (1969).

47b. L. J. Kleinsmith and V. G. Allfrey, Nuclear phosphoproteins. II. Metabolism of exogenous phosphoprotein by intact nuclei. *Biochim. Biophys. Acta* **175,** 136–141 (1969).

47c. L. J. Kleinsmith, V. G. Allfrey, and A. E. Mirsky, Phosphorylation of nuclear protein early in the course of gene activation in lymphocytes. *Science* **154,** 780–781 (1966).

48. H. M. Klouwen and A. W. M. Appelman, Synthesis of ATP in isolated nuclei and intact cells. *Biochem. J.* **102,** 878–884 (1967).

49. E. Kohen, G. Siebert, and C. Kohen, Metabolism of reduced pyridine nucleotides in ascites cell nuclei. *Histochemie* **3,** 477–483 (1964).

50. L. Kuehl, Evidence for nuclear synthesis of lactic dehydrogenase in rat liver. *J. Biol. Chem.* **242,** 2199–2206 (1967).

50a. L. Kuehl, Effects of various inhibitors on nuclear protein synthesis in rat liver. *J. Cell Biol.* **41,** 660–663 (1969).

50b. T. A. Langan, Phosphorylation of proteins in the cell nucleus. *In* "Regulatory Mechanisms for Protein Synthesis in Mammalian Cells" (A. San Pietro, M. Lamborg, and F. T. Kenney, eds.) pp. 101–118. Academic Press, New York, 1968.

51. G. M. Lehrer and R. F. Mathewson, Enzyme analysis of nuclei and cytoplasm of single supramedullary neurons in puffer fishes. *J. Cell Biol.* **23,** 53A (1964).

51a. M. Lezzi, RNS und Protein Synthese in Puffs Isolierte Speicheldrüsechromosomen von *Chironomus. Chromosoma* **21,** 72–88 (1967).

362   V. G. ALLFREY

51b. M. C. Liau and R. P. Perry, Ribosome precursor particles in nucleoli. *J. Cell Biol.* **42**, 272–283 (1969).

52. V. C. Littau, V. G. Allfrey, J. H. Frenster, and A. E. Mirsky, Active and inactive regions of nuclear chromatin as revealed by electron microscope autoradiography. *Proc. Natl. Acad. Sci. U. S.* **52**, 93–100 (1965).

53. V. C. Littau, C. J. Burdick, V. G. Allfrey, and A. E. Mirsky, The role of histones in the maintenance of chromatin structure. *Proc. Natl. Acad. Sci. U. S.* **54**, 1204–1212 (1965).

54. W. R. Lowenstein, Y. Kanno, and S. Ito, Permeability of nuclear membranes. *Ann. N. Y. Acad. Sci.* **137**, 708–716 (1966).

55. R. Maggio, Progress report on the characterization of nucleoli from guinea pig liver. *Natl. Cancer Inst. Monograph* **23**, 213–222 (1966).

56. J. Maio and C. Schildkraut, Isolated mammalian chromosomes. I. General characteristics of nucleic acids and proteins. *J. Mol. Biol.* **24**, 29–39 (1967).

57. L. Malkin and A. Rich, Partial resistance of nascent polypeptide chains to proteolytic digestion due to ribosomal shielding. *J. Mol. Biol.* **26**, 329–346 (1967).

58. J. Marmur and C. M. Greenspan, Transcription *in vivo* of DNA from bacteriophage SP8. *Science* **142**, 387–389 (1963).

59. B. J. McCarthy and B. H. Hoyer, Identity of DNA and diversity of messenger-RNA molecules in normal mouse tissues. *Proc. Natl. Acad. Sci. U. S.* **52**, 915–922 (1964).

60. K. S. McCarty, J. T. Parsons, W. A. Carter, and J. Laszlo, Protein synthetic capacities of liver nuclear subfractions. *J. Biol. Chem.* **241**, 5489–5499 (1966).

60a. E. H. McConkey and J. W. Hopkins, Molecular weights of some HeLa ribosomal RNA's. *J. Mol. Biol.* **39**, 545–550 (1969).

61. B. S. McEwen, V. G. Allfrey, and A. E. Mirsky, Studies of energy-yielding reactions in thymus nuclei. *J. Biol. Chem.* **238**, 758–766, 2571–2578, and 2579–2586 (1963).

61a. B. S. McEwen, V. G. Allfrey, and A. E. Mirsky, Dependence of RNA synthesis in isolated thymus nuclei on glycolysis, oxidative carbohydrate catabolism and a type of "oxidative phosphorylation." *Biochim. Biophys. Acta* **91**, 23–28 (1964).

62. O. L. Miller, Structure and composition of peripheral nucleoli of salamander oocytes. *Natl. Cancer Inst. Monograph* **23**, 53–66 (1966).

63. A. E. Mirsky and A. W. Pollister, The nucleoprotamine of trout sperm; chromosin, a DNP complex of the cell nucleus. *J. Gen. Physiol.* **30**, 101–148 (1946).

64. A. E. Mirsky and H. Ris, Isolated chromosomes—the chemical composition of isolated chromosomes. *J. Gen. Physiol.* **31**, 1–18 (1947).

65. A. E. Mirsky and H. Ris, Variable and constant components of chromosomes. *Nature* **163**, 666–667 (1949).

65a. A. Monneron and W. Bernhard, Fine structural organization of the interphase nucleus in some mammalian cells. *J. Ultrastruct. Res.* **27**, 266–288 (1969).

65b. Y. Moriyama, J. L. Hodnett, A. W. Prestayko, and H. Busch, Studies on the nuclear 4–7 S RNA of the Novikoff hepatoma. *J. Mol. Biol.* **39**, 335–349 (1969).

66. M. Muramatsu, J. L. Hodnett, W. J. Steele, and H. Busch, Synthesis of 28-S RNA in the nucleolus. *Biochim. Biophys. Acta* **123**, 116–125 (1966).

67. K. Murray, The basic proteins of the cell nucleus. *Ann. Rev. Biochem.* **34**, 209–246 (1965).

68. H. Naora, DNA-dependent protein synthesis in nuclear ribosomes *in vitro*. *Biochim. Biophys. Acta* **123**, 151–162 (1966).
69. H. Naora, H. Naora, M. Izawa, V. G. Allfrey, and A. E. Mirsky, Some observations on differences in composition between the nucleus and cytoplasm of the frog oocyte. *Proc. Natl. Acad. Sci. U. S.* **48**, 853–859 (1962).
69a. Y. Nishizuka, K. Ueda, T. Honjo, and O. Hayaishi, Enzymic adenosine-diphosphate ribosylation of histone and poly-adenosine-diphosphate-ribose synthesis in rat liver nuclei. *J. Biol. Chem.* **243**, 3765–3768 (1968).
69b. M. G. Ord and L. A. Stocken, Metabolic properties of histones from rat liver and thymus gland. *Biochem. J.* **98**, 888–897 (1966).
69c. M. G. Ord and L. A. Stocken, Variations in the phosphate content and thiol/disulfide ratio of histones during the cell cycle. *Biochem. J.* **107**, 403–410 (1968).
70. S. Osawa, V. G. Allfrey, and A. E. Mirsky, Mononucleotides of the cell nucleus. *J. Gen. Physiol.* **40**, 491–513 (1957).
70a. W. K. Paik and S. Kim, Epsilon-*N*-dimethyl lysine in histones. *Biochem. Biophys. Res. Commun.* **27**, 479–483 (1967).
70b. J. Paul and R. S. Gilmour, Organ-specific restriction of transcription in mammalian chromatin. *J. Mol. Biol.* **34**, 305–316 (1968).
71. S. Penman, RNA metabolism in the HeLa cell nucleus. *J. Mol. Biol.* **17**, 117–130 (1966).
72. S. Penman, I. Smith, and E. Holtzman, Ribosomal RNA synthesis and processing in a particulate site in the HeLa cell nucleus. *Science* **154**, 786–789 (1966).
73. R. P. Perry, The nucleolus and the synthesis of ribosomes. *Progr. Nucleic Acid Res. Mol. Biol.* **6**, 220–257 (1967).
74. A. O. Pogo, V. G. Allfrey, and A. E. Mirsky, Evidence for increased DNA-template activity in regenerating liver nuclei. *Proc. Natl. Acad. Sci. U. S.* **56**, 550–557 (1966).
75. A. O. Pogo, B. G. T. Pogo, V. C. Littau, V. G. Allfrey, A. E. Mirsky, and M. G. Hamilton, The purification and properties of ribosomes from the thymus nucleus. *Biochim. Biophys. Acta* **55**, 849–864 (1962).
76. A. O. Pogo, V. C. Littau, V. G. Allfrey, and A. E. Mirsky, Modification of RNA synthesis in nuclei isolated from normal and regenerating liver. The effects of salt and specific divalent cations. *Proc. Natl. Acad. Sci. U. S.* **57**, 743–750 (1967).
77. D. M. Prescott, Cellular sites of RNA synthesis. *Progr. Nucleic Acid Res. Mol. Biol.* **3**, 33–57 (1964).
77a. D. M. Prescott, Sequential events of the cell life cycle. *In* "Exploitable Molecular Mechanisms and Neoplasia," pp. 359–370. Williams and Wilkins Co., Baltimore, Maryland, 1969.
77b. D. M. Prescott and L. Goldstein, Nuclear–cytoplasmic interaction in DNA synthesis. *Science* **155**, 469–470 (1967).
77c. R. H. Reeder, K. Ueda, T. Honjo, Y. Nishizuka, and O. Hayaishi, Studies on the polymer of adenosine-diphosphate-ribose. II. Characterization of the polymer. *J. Biol. Chem.* **242**, 3172–3177 (1967).
78. E. Reich and I. H. Goldberg, Actinomycin and nucleic acid function. *Progr. Nucleic Acid Res. Mol. Biol.* **3**, 183–234 (1964).
79. E. Reid, A. B. A. El-Aaser, M. K. Turner, and G. Siebert, Enzymes of RNA and ribonucleotide metabolism in rat liver nuclei. *Z. Physiol. Chem.* **339**, 135–149 (1964).

80. M. Revel and F. Gros, A factor from E. coli required for the translation of natural messenger RNA. *Biochem. Biophys. Res. Commun.* **27**, 12–19 (1967).

81. D. Sabatini, Y. Tashiro, and G. E. Palade, On the attachment of ribosomes to microsomal membranes. *J. Mol. Biol.* **19**, 503–524 (1966).

82. P. D. Sadowski and J. Alcock-Howden, Two distinct classes of polysomes from a nuclear fraction of rat liver. *J. Cell Biol.* **37**, 163–181 (1967).

83. K. Scherrer, H. Latham, and J. E. Darnell, Demonstration of an unstable RNA and of a precursor to ribosomal RNA in HeLa cells. *Proc. Natl. Acad. Sci. U. S.* **49**, 240–248 (1963).

84. B. Schultze, P. Citoler, K. Hempel, K. Citoler, and W. Maurer, Cytoplasmic protein synthesis in cells of various types and its relation to nuclear protein synthesis. *Symp. Intern. Soc. Cell Biol.* **4**, 107–139 (1965).

84a. R. W. Shearer and B. J. McCarthy, Evidence for RNA molecules restricted to the cell nucleus. *Biochemistry* **6**, 283–289 (1967).

85. A. Sibatani, S. R. deKloet, V. G. Allfrey, and A. E. Mirsky, Isolation of a nuclear RNA fraction resembling DNA in its base composition. *Proc. Natl. Acad. Sci. U. S.* **48**, 471–477 (1962).

86. G. Siebert, Enzyme und Substrate der Glykolyse in Isolierten Zellkernen. *Biochem. Z.* **334**, 369–387 (1961).

87. G. Siebert, K. Dahm, and H. Breur, Activitäten von Steroidenzymen in Zellkernen der Rattenleber. *Naturwissenschaften* **53**, 615 (1966).

88. G. Siebert and G. B. Humphrey, Enzymology of the nucleus. *Advan. Enzymol.* **27**, 239–288 (1965).

89. G. Siebert, H. Langendorf, R. Hannover, D. Nitz-Litzow, B. C. Pressman, and C. Moore. Untersuchen zur Rolle des Natrium Stoffwechsels im Zellkern der Rattenleber. *Physiol. Chem.* **343**, 101–115 (1965).

90. R. M. S. Smellie, The biosynthesis of RNA in animal systems. *Progr. Nucleic Acid Res. Mol. Biol.* **1**, 27–58 (1962).

90a. R. Soiero, M. H. Vaughan, J. R. Warner, and J. E. Darnell, Jr., The turnover of nuclear DNA-like RNA in HeLa cells. *J. Cell Biol.* **39**, 112–118 (1968).

90b. C. E. Sripati, C. K. Pyne, and Y. Khouvine, Aspect morphologique et activité des noyaux de foie de rat isolés en utilisant des détergents nonioniques dans un milieu contenant du saccharose. *Compt. Rend. Acad. Sci. (Paris)* **268**, 2752–2755 (1969).

90c. R. H. Stellwagen and R. D. Cole, Chromosomal proteins. *Ann. Rev. Biochem.* **38**, 951–990 (1969).

91. J. R. Tata, Hormones and the synthesis and utilization of ribonucleic acids. *Progr. Nucleic Acid. Res. Mol. Biol.* **5**, 191–250 (1966).

92. J. H. Taylor, The replication of chromosomes. *In* "Molecular Genetics" Part 1, (J. H. Taylor, ed.), pp. 65–111. Academic Press, New York, 1963.

92a. J. H. Taylor, Rates of chain growth and units of replication in DNA of mammalian chromosomes. *J. Mol. Biol.* **31**, 579–594 (1968).

92b. J. H. Taylor and P. Miner, Units of DNA replication in mammalian chromosomes. *In* "Symposium on the Developmental Biology of Neoplasia" (H. Pitot, ed.), Williams and Wilkins Co., Baltimore, Maryland, 1968.

92c. J. H. Taylor, N. Straubing, and E. Schandl, Units and patterns of replication in mammalian chromosomes. *J. Cell Biol.* **39**, 134a (1968).

92d. L. R. Thompson and B. J. McCarthy, Stimulation of nuclear DNA and RNA synthesis by cytoplasmic extracts *in vitro*. *Biochem. Biophys. Res. Commun.* **30**, 166–170 (1968).

92e. A. A. Travers and R. R. Burgess, Cyclic reuse of the RNA polymerase Sigma factor. *Nature* **222**, 537–540 (1969).

92f. J. Tsuzuki, The effect of monovalent cations on the incorporation of various amino acids in protein synthesis by isolated rat spleen nuclei. *Biochim. Biophys. Acta* **182**, 580–582 (1969).

92g. K. Ueda, T. Matsuura, N. Date, and K. Kawai, The occurrence of cytochromes in the membranous structures of calf thymus nuclei. *Biochem. Biophys. Res. Commun.* **34**, 322–327 (1967).

92h. K. Ueda, R. H. Reeder, T. Honjo, Y. Nishizuka, and O. Hayaishi, Polyadeno-sine-diphosphate-ribose synthesis associated with chromatin. *Biochem. Biophys. Res. Commun.* **31**, 379–382 (1968).

92i. R. Vendrely and C. Vendrely, La teneur du noyau cellulaire en acide desoxy-ribonucléique à travers les organes, les individus, et les espèces animales. *Experientia* **4**, 434 (1948).

93. W. S. Vincent, E. Baltus, A. Lovlie, and R. E. Mundell, Proteins and nucleic acids of isolated starfish oocyte nucleoli and ribosomes. *Natl. Cancer Inst. Monograph* **23**, 235–253 (1966).

94. W. S. Vincent and O. L. Miller, The nucleolus: Its structure and function. *Natl. Cancer Inst. Monograph* **23** (1966).

95. A. Vorbrodt, The effect of ATP and RNA on the uptake of 2-^{14}C-phenyl-alanine and 6-^{14}C-orotic acid by nuclei isolated from rat thymus, liver and hepatoma. *Bull. Acad. Polon. Sci. Ser. Sci. Biol.* **8**, 489–492 (1960).

96. T. Y. Wang, Enzymes associated with nuclear ribosomes. *Biochim. Biophys. Acta* **55**, 392–395 (1962).

96a. R. A. Weinberg, U. Loening, M. Willems, and S. Penman, Acrylamide gel electrophoresis of HeLa cell nucleolar RNA. *Proc. Natl. Acad. Sci. U. S.* **58**, 1088–1095 (1967).

96b. W. D. Wicks, D. L. Greenman, and F. T. Kenney, Stimulation of RNA synthesis by steroid hormones. *J. Biol. Chem.* **240**, 4414–4426 (1965).

97. E. B. Wilson, "The Cell in Development and Heredity." Macmillan, New York, 1924.

98. M. Yarmolinsky and G. de la Haba, Inhibition by puromycin of amino acid incorporation into protein. *Proc. Natl. Acad. Sci. U. S.* **45**, 1721–1729 (1959).

99. H. G. Zachau, G. Acs, and F. Lipmann, Isolation of adenosine-amino acid esters from an RNase digest of soluble liver RNA. *Proc. Natl. Acad. Sci. U. S.* **44**, 885–889 (1958).

99a. E. F. Zimmerman, J. Hackney, P. Nelson, and I. M. Arias, Protein synthesis in isolated nuclei and nucleoli of HeLa cells. *Biochemistry* **8**, 2636–2644 (1969).

100. G. Zubay and P. Doty, The isolation and properties of deoxyribonucleo-protein particles containing single nucleic acid molecules. *J. Mol. Biol.* **1**, 1–20 (1959).

VI

The Three-Dimensional Structure and Evolution of Proteins

Charles J. Epstein

I. THE ACQUISITION OF THREE-DIMENSIONAL STRUCTURE *IN VITRO*

The functional characteristics of proteins are determined, directly or indirectly, by 2 fundamental properties: their amino acid sequences and their three-dimensional configurations. Other chapters in this book are concerned with the mechanisms by which proteins, as polypeptide chains with defined amino acid sequences (or primary structures), are synthesized. It will be the purpose of this chapter to discuss what may be considered the final stage of protein synthesis—the conversion of the polypeptide chain into a highly complex native protein. In addition, we shall be concerned with the influence of evolutionary forces on both the sequences and the conformations of proteins.

A. STRUCTURE OF NATIVE PROTEINS

Implicit in any discussion of protein conformation is the assumption that native proteins do have definite three-dimensional structures. While this assumption is generally regarded as valid, the evidence for it is largely circumstantial. The only method presently available for directly visualizing proteins in 3 dimensions is X-ray crystallography, and relatively few proteins have been satisfactorily analyzed by this technique. These include the 2 heme-containing proteins myoglobin and hemo-

globin, egg white lysozyme, and pancreatic ribonuclease (*6, 63, 89*) as
well as chymotrypsin, carboxypeptidase A, and papain. Several other
proteins (including carbonic anhydrase C, staphylococcal nuclease, and
cytochrome c) are in various stages of analysis, but the degree of resolution thus far obtained has not been sufficient for definitive structural
evaluation. For the purposes of the present part of the discussion, the
most important feature of the determination of structure by X-ray
crystallography is not the nature of the structure observed, but the fact
that a definite structure is found at all. For this to occur, it is necessary
that there be very little structural variation among the many molecules
of the protein in the crystal. A significant degree of conformational
variability would result in a blurring of the electron density maps and
would make the patterns uninterpretable. Therefore, with regard to the
proteins that have been studied, it may be stated that in the crystalline
form they have highly specific, perhaps even unique, conformations.
This concept of uniqueness must be somewhat loosely interpreted since,
even in the crystalline state, the various atoms and bonds will be in
constant states of vibration. Furthermore, the side chains of certain
amino acid residues may be so situated as to be capable of considerable
freedom of movement within the crystals.

Unfortunately, there are no methods presently available for structural
determinations on proteins in solution that in any way approach the
power of X-ray crystallography. However, information may be obtained
about the general aspects of three-dimensional structure by using various
physical–chemical techniques. These include titration, ultracentrifugation,
spectroscopy, and measurements of optical rotation, fluorescence, viscosity, light scattering, and diffusion. While these methods do not by themselves yield a detailed picture of conformation, they do make possible
the correlation of the properties of proteins in solution with the known
structures of crystalline proteins. Such comparisons have been carried out,
for example, with regard to the numbers of amino acid residues present
in an α-helical conformation as estimated from measurements of the
optical rotatory dispersions of solutions of myoglobin and lysozyme.
Within the technical and theoretical errors inherent in this method, the
results correspond reasonably well to the helical contents revealed by
crystallography (*107*). Another approach to the investigation of the
solution and crystal conformations of proteins has been the measurement
of the activity of enzymes in the crystalline state. Several crystalline
enzymes have been found to be highly active, and it has been inferred
that the active sites of the enzymes in this form are probably quite
similar to, if not identical with, the active sites of the enzymes in solution
(*22*). Despite the information derived by these means, it is still not

possible to answer definitively 2 fundamental questions; first: Does a protein in solution have exactly the same conformation as the protein in the crystal? and, second: Does a protein in solution have a single unique structure?

In any discussion of protein conformation, it is necessary to define the specific conditions which obtain in the system. Particular attention must be paid to the presence and state of small molecules which, because of their ability to produce profound effects on conformation, may interact with the protein. For example, the change in the state of oxidation of the iron in the heme group of cytochrome c, from $+2$ to $+3$, leads to a partial unfolding of the protein and makes it much more susceptible to denaturation by guanidine hydrochloride (107). Likewise, even in the crystalline state, the addition of oxygen to deoxyhemoglobin results in a significant decrease in the distance between the 2 β-chains of the molecule (82). Perhaps more relevant, if less dramatic, are the effects on the conformation of the enzyme phosphoglucomutase which result from its interaction with the substrate glucose 6-phosphate (116). Certain groups, such as the sulfhydryls, become more reactive to reagents which combine with them while others, such as lysine and methionine, become less reactive. In addition, there are alterations in the ultraviolet absorption and fluorescence spectra indicating changes in the environments of the tyrosine and tryptophan residues. And, finally, the presence of substrate makes the enzyme more susceptible to inactivation. All these effects can be explained by postulating substrate-induced (induced fit) conformational changes both adjacent to and distant from the substrate-binding site. The general term applied to these changes, as well as to the conformational alterations produced by the binding of small nonsubstrate molecules to proteins, is allosteric, and a general hypothesis to explain their origin and significance has recently been advanced (80). Quite often allosteric proteins are composed of subunits, or protomers, of one or more kinds, and the hypothesis states that the allosteric effect results from a symmetrical arrangement of the subunits in the molecule. In many instances, it has been possible to differentiate regulator binding sites, the sites at which nonsubstrate activators are bound, from substrate-binding sites. In one protein, the aspartic transcarbamylase of *E. coli*, these functionally different sites are present on different subunits of the molecule and can be physically separated from one another (40).

B. THE FOLDING OF SINGLE-CHAIN PROTEINS

With this background, we may now approach the major problem relating to protein synthesis and protein conformation: How does a newly synthesized polypeptide chain attain its final conformation? The general thesis has been that the folding of the protein is governed principally, if not entirely, by thermodynamic considerations and that special folding mechanisms or templates need not be invoked (*3, 15, 29*). Free energy is the thermodynamic quantity that defines the state of chemically reacting systems. When systems are in their most stable states, or at equilibrium, the total free energy will be at a minimum. Therefore, for a polypeptide chain that is folding into a native protein, it is presumed that the chain will fold in such a manner as to attain the state of lowest free energy. In the strictest sense, this thermodynamic hypothesis is only applicable to the conditions that exist at the time and place of protein synthesis, conditions which, for the most part, are impossible to define precisely. However, in its experimental applications the hypothesis has been extended to cover most *in vitro* situations that roughly approximate physiological conditions, and this extrapolation is valid as long as the proteins are able to change their conformations freely with changes in environment. It is conceivable that some proteins, particularly those with intramolecular cross-links, could, after their synthesis, enter environments in which they were no longer in their most stable states. They would then be in a metastable state, prevented from conformational alterations only by the time or activation energy necessary to effect such changes. While this is a possibility, the experimental evidence presently available does not, except possibly in a very few cases, indicate that such metastable molecules exist as native proteins.

The experimental verification of the thermodynamic hypothesis has relied upon the demonstration that native proteins, after being treated to disrupt all specific three-dimensional structures, are able to return spontaneously to their original conformations (Fig. 1). Under conditions of the experiments, the conformations attained after the proteins are allowed to refold are necessarily those of the lowest free energy. Therefore, if these conformations are indeed identical to those of the native materials, it may be inferred that the polypeptide chains are capable of folding spontaneously *in vivo*.

The assumption that proteins fold in a manner that is consonant with the lowest conformational free energy has recently been challenged (*73*). On the basis of kinetic arguments, it has been proposed that proteins follow a more or less predetermined pathway of folding that will, reproducibly, lead to the native but not necessarily most stable structures. Because of the absence of pathways, the latter are presumed to be unattainable. No evidence has as yet been presented in support of these contentions, and, for the present, attainment of minimal free energy is still

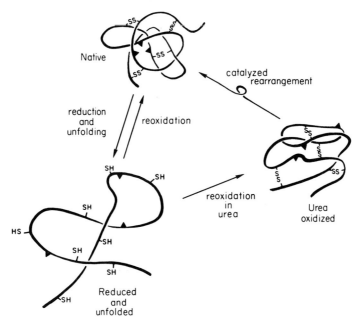

Fig. 1. General scheme for the reversible denaturation of proteins containing disulfide bonds. Treatment of native proteins with mercaptoethanol and urea or guanidine results in cleavage of the disulfide bonds and unfolding of the molecule. Following removal of these reagents and oxidation in air, the native molecule is regenerated. If reoxidation is permitted to take place in 8 M urea, the disulfide bonds do not form correctly. However, rearrangement to the correct conformation occurs if disulfide bond interchange is catalyzed by small amounts of thiols or the disulfide bond rearranging enzyme.

considered to be the driving force for the formation of native protein conformations.

1. *Proteins Containing Disulfide Bonds*

To facilitate the interpretation of experimental results, the original experiments were carried out with proteins that contained several intramolecular disulfide bonds. In native proteins these bonds are formed by specific pairs of sulfhydryl groups and thus provide useful indices of structure. Furthermore, once these bonds are formed, the protein is locked into a specific disulfide bonded configuration, rearrangement of which can be prevented during the assay procedure. The use of proteins with enzymic activity also facilitates the experiments because of the ease such activity affords in the detection of proper refolding.

The pioneering work with this approach was carried out with the enzyme bovine pancreatic ribonuclease (*98*). This protein, with 4 disulfide bonds and a molecular weight of 13,683, was reduced and unfolded by treatment with a reducing agent (β-mercaptoethanol) and concentrated urea (8–10 M) or guanidine hydrochloride (5–6 M). Although the β-mercaptoethanol actually breaks the disulfide bonds, the urea or guanidine is necessary to make these bonds easily accessible to the reducing agent. The precise modes of action of urea and guanidine are still being debated, but they are known to disrupt noncovalent interactions. If present in high enough concentrations, they cause the polypeptide chain to assume a random coil configuration, the existence of which is inferred from the fact that certain physical–chemical measurements reach the theoretical values for such a structure and do not change when the proteins are exposed to even more drastic treatments (*103*).

After treatment of ribonuclease with urea (or guanidine) and β-mercaptoethanol, these reagents were removed and the reduced unfolded protein allowed to refold. The polypeptide chain assumed a conformation that was no longer random and disulfide bonds reformed (reoxidized) through reaction with molecular oxygen. Under proper conditions of temperature, pH, solvent composition, and protein concentration, all the ribonuclease rapidly returned to an active form (*29*). By a variety of physical, chemical, and immunological techniques, the reoxidized protein could be shown to be identical with the starting material (*110*). Similar experiments have now been carried out with a large number of other disulfide bond containing proteins with measurable enzymic or binding activities. These proteins include egg white lysozyme, Taka-amylase A, pepsinogen, soybean trypsin inhibitor, kallikrein inactivator (basic trypsin inhibitor), ribonuclease T1, and cobrotoxin. They range in molecular weight from 10,000 to 50,000 and contain 2 to 4 disulfide bonds. While some of the reoxidized proteins have been recovered in high yields and have been well characterized, others have been examined solely in terms of activity or have been reactivated to only a limited extent. Nevertheless, the conclusion that they all appear to have been substantially restored to the native state does not appear to be seriously in question.

A few larger proteins, including thyroglobulin and serum albumin, have also been reduced and unfolded and then reoxidized. Restoration of various properties of the proteins has been observed, but the lack of specifically assayable activities and the large number of disulfide bonds have made difficult the direct demonstration that native conformation has been recovered. Furthermore, reoxidized thyroglobulin was found to be slightly but distinctly different from the native protein (*19*). The enzyme trypsin has also been reduced and reoxidized. In order to prevent autodigestion of the protein and proteolysis of unfolded protein by newly folded and active enzyme, the trypsin was immobilized on a carboxymethylcellulose column. By this procedure, it was possible to regenerate a small (4%) but significant amount of the original enzymic activity. Ribonuclease, similarly immobilized on carboxymethyl, cellulose, could, after reduction, be regenerated to the content of 40% as compared with the full regeneration obtained with soluble ribonuclease (27).

To verify further the fact that native proteins are indeed in their most stable forms, several reduced enzymes have been allowed to reoxidize in $8 M$ urea. This results in an improper pairing of disulfide bonds, and the reconstituted proteins thus formed are no more active than would be expected from the random formation of disulfide bonds. However, the addition of small amounts of β-mercaptoethanol (or of the disulfide-bond rearranging enzyme, see below) catalyzes the spontaneous rearrangement of these metastable urea-oxidized enzymes to the more stable active forms, in yields similar to those obtained under optimal conditions by direct reoxidation of the reduced proteins (46) (Fig. 1).

Because the *in vitro* conditions required for the most rapid reoxidation of unfolded reduced proteins were distinctly nonphysiological with regard to pH, and often to temperature (29), an attempt was made to determine whether a system might exist which would facilitate the reactivation process. Such a system has been found in extracts of rat and beef liver and of pigeon pancreas, and, with further purification, the responsible enzyme has been purified from beef liver microsomes. Although it was initially suspected that this enzyme might be functioning to catalyze sulfhydryl group oxidation, it now appears that it acts to accelerate the rearrangement of disulfide bonds from improper to proper configurations (41). The disulfide bond rearranging enzyme is nonspecific in its substrate requirements, and ribonuclease, egg white lysozyme, and soybean trypsin inhibitor all respond to its action. While the microsomal localization of this enzyme and its presence in many tissues indicate that it could be of general importance, it is still impossible to state whether this system has any related physiological function *in vivo*. Nevertheless, conditions can be established *in vitro* which allow reformation of presumably native protein conformations under relatively physiological conditions, at rates consistent with the estimated rates of protein synthesis *in vivo*.

2. Proteins without Disulfide Bonds

Experiments similar to those carried out with the disulfide bond containing proteins have also been performed with proteins devoid of internal linkage. Again, unfolding by concentrated solutions of urea and guanidine has been used to produce the random conformation of the polypeptide chain, and refolding has been allowed to take place after the removal of the denaturing agents. The addition of a reducing agent is often required to prevent the oxidation of sulfhydryl groups which may be protected in the native molecule. Among the proteins in this nondisulfide bond containing group that have been refolded or renatured are phosphoglyceric acid mutase, aldolase, enolase, α-amylase, phage lysozyme, carbonic anhydrase, malate dehydrogenase, lactate dehydrogenase, α-glycerophosphate dehydrogenase, fumarase, phosphoglucomutase, and tobacco mosaic virus protein. Several of these proteins are of particular interest in that they are multimeric, or composed of more than one subunit. The aggregation of these subunits, which may all be identical or of two, or rarely more, different kinds, gives rise to the so-called quaternary structure of the molecules. The successful *in vitro* denaturation and renaturation of multimeric proteins indicates that quaternary structure, like tertiary structure (the total conformation of the monomeric subunit), is also a function of the primary structure and that the potential for multimer formation is encoded in the amino acid sequence (*28*).

3. *Multiple Conformational States*

The methods used in these experiments provide a useful tool for the examination of a question raised earlier: Can a polypeptide chain assume more than one stable conformation in solution? Reports have recently appeared in which the existence of stable conformational variants has been claimed (*67*). Although the tacit assumption in the experiments discussed above is that a protein has only one conformation, it is theoretically possible that it could have more than one (*30*). This would occur if two or more different conformations were relatively similar in their conformational free energies but were separated by free energy barriers that prevented easy equilibration or interconversion (Fig. 2). In a thermodynamic sense, most such species would not be in the state of lowest conformational free energy and would be metastable rather than stable. Kinetically, however, they would appear to be stable.

Before it can be asserted that a set of proteins differ only in conformation, it must be proved that they are chemically identical and do not differ as a result of the differential binding of small ligands. Furthermore, if true conformational variants are completely unfolded under conditions that will not produce covalent changes, they will become indistinguishable and, on refolding, should give rise to the same set of proteins (Fig. 3). If they do not, it may be presumed that they are chemically different. Such tests have been applied to the several forms

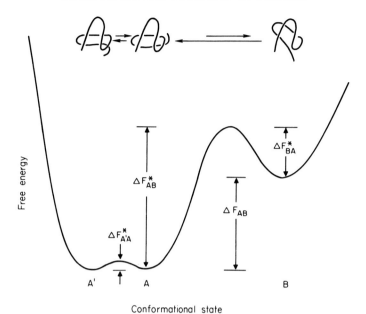

FIG. 2. Free energy relationships of hypothetical conformational variants of proteins. Molecules in states A and A′ are separated by only a small free energy of activation $(\Delta F^*_{A'A})$ and would readily interconvert. Molecules in state B would be considered metastable relative to those in states A and A′ since their free energy is higher by amount ΔF_{AB} than that of the latter. Because of the high free energy barriers ΔF^*_{AB} and ΔF^*_{BA} interconversion between B and A–A′ would be slow, and molecules in these 2 states could be considered stable conformational variants.

of crystalline sperm whale myoglobin and chicken heart mitochondrial malate dehydrogenase. These do not interconvert even after complete unfolding, and it is doubtful that they are conformational variants (30). It seems likely that many apparent conformational variants will turn out to be the same molecule with small chemical differences introduced either during or after synthesis or with small ligands bound in different amounts or in different chemical states.

C. The Formation of Multichain Proteins

Unlike the noncovalently linked multimeric proteins, proteins made of two or more polypeptide chains held together by disulfide bonds have presented a much greater problem, both experimentally and conceptually. Only in the cases of chymotrypsin and insulin is the manner of chain assembly understood. The former, an enzyme consisting of 3 disulfide-bond-linked polypeptide chains, is derived from the internally bonded

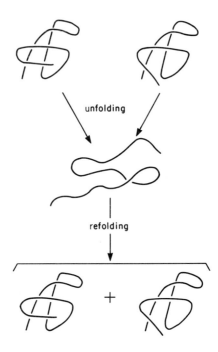

Fig. 3. A test for stable conformational variants. If 2 species of the same protein which differ solely in conformation are unfolded in urea or guanidine, they will assume equivalent unfolded structures. Upon refolding, the same species or set of species would be formed, regardless of the nature of the starting material. If, however, species differing in covalent structure are so treated, interconversion among them would not occur. Reprinted by permission (30).

single-chain zymogen chymotrypsinogen through the proteolytic activity of trypsin (85). In the conversion of chymotrypsinogen to chymotrypsin, 2 peptides are split out of the chain, and the resulting three-chain molecule, while enzymically active, is no longer in its most stable conformation (41). This result of activation should be contrasted with that of the conversion of trypsinogen to trypsin, an active single-chain disulfide-bonded protein. To a limited, but certainly greater than random, extent, chemical derivatives of the latter can be successfully reformed after reduction and unfolding (29).

1. Insulin

Several years ago it was proposed that the zymogen concept might be applicable to the biosynthesis of insulin (29). This view was strengthened by the observation that, although it was possible to obtain 60% or more of the theoretically expected amount of hormone by the *in vitro* re-combination of separated chains, the conditions were very far from physiological. An excess of reduced A-chain had to be mixed with the *S*-sulfonate form of B-chain at pH 10.6 for 18–22 hours at 2°C *(64)*. The yields with synthetic A- and B-chains were considerably lower. Further-more, it had been found that disulfide bond interchange and inactivation could be induced by treatment of insulin with mercaptoethanol or with the disulfide bond interchange enzyme discussed above *(41)*.

Steiner has recently isolated and characterized a protein, proinsulin, which appears to be the zymogen of insulin *(101)*. This protein consists of a single polypeptide chain, the N-terminal sequence of which is that of the B-chain. This B-chain sequence is linked to the C-terminal, A-chain sequence by a short polypeptide which presumably is split out during insulin biosynthesis (Fig. 4). Unlike native insulin, proinsulin can be successfully refolded in greater than 50% yield after reduction in 8 *M* urea under conditions similar to those employed in the reoxidation of other disulfide bond containing proteins.

2. *Immunoglobulins*

The immunoglobulins, as a group, are considerably more complex multichain proteins than is insulin. Immunoglobulin G (IgG, γ-globulin), the prototype of the group, is composed of 4 polypeptide chains—2 heavy (H) chains and 2 light (L) chains (Fig. 5). These chains combine to form 2 HL units or halves which are held together probably by a single disulfide bond and by noncovalent interactions between the H-chains. The H- and L-chains of each half of the molecule are similarly held together by a single disulfide bond and, in addition, each contains several intrachain disulfide bonds *(78)*. The manner by which IgG is assembled has been analyzed by the serial dissection of the molecule. By mild treatment with a reducing agent and acid, it is possible to split the molecule into halves which can be recombined by reoxidation of the inter-H-chain disulfide bond. These halves still possess the ability to combine with antigens, although they cannot produce precipitin reactions because of their univalent character. When the H- and L-chains of these univalent halves are separated by reduction of the interchain disulfide bond between them, they can in turn be recombined by reoxidation.

Even with the successful recombination of the various chains of γ-globulin, it is clear that none of these experiments is of direct relevance in determining whether the conformation of each IgG molecule is directly and solely determined by its amino acid sequence. This is so because all the fragments used in the recombination experiments still contain a high order of structure as a result of the presence of intact intrachain

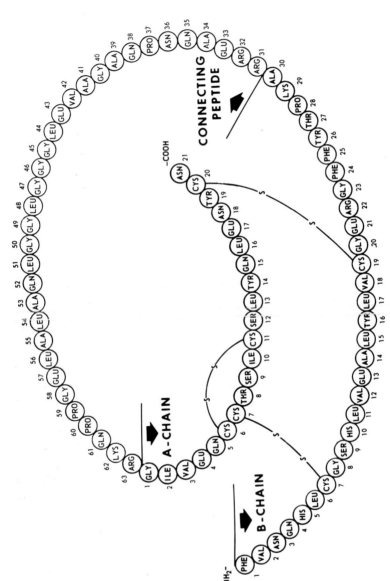

Fig. 4. The structure of porcine proinsulin, including the amino acid sequence of the connecting peptide that joins the B-chain to the A-chain (courtesy of Drs. Chance, Ellis, and Bromer, Eli Lilly Co., Indianapolis, Indiana, *11a*).

Fig. 5. The structure and reversible denaturation of immunoglobulin G (IgG). By successive reductions, the native molecule, H₂L₂, can be broken down to univalent halves, HL, and then to separate H- and L-chains. The isolated L-chains can be reconstituted after reduction and unfolding. Treatment of native IgG with papain produces a univalent fragment which lacks part of the H-chain. This fragment can also be successfully reformed after reduction and unfolding. In order to allow for reconstitution of unfolded IgG, it was found necessary to increase the solubility of reduced H-chains by the prior addition of chains of poly-DL-alanine to the exposed amino groups of the native molecule.

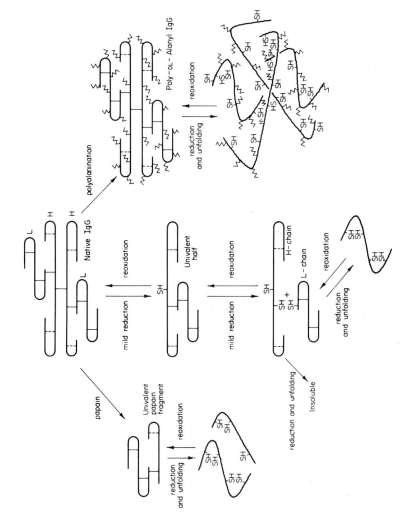

disulfide and noncovalent bonds. Because of its importance in deciding about the possible role of antigen in the folding of the molecule, attempts have been made to carry out the refolding and reoxidation of the molecule after complete reduction and unfolding (*37*).

That this approach might be feasible was indicated by the recovery of all antigenic sites after reoxidation of completely reduced and unfolded L-chains. However, completely reduced H-chains were found to be insoluble, and it was necessary to employ 2 special approaches (Fig. 5). The first was to use only a portion of the IgG molecule, a univalent fragment produced by treatment of intact IgG with papain, which contains an intact L-chain and a part of an H-chain. Following reduction and unfolding, this fragment could be reoxidized to yield material which resembled the starting material in 2 ways—antigenic properties and antibody-binding capacity. With regard to the latter, recoveries of binding capacity directed against specific antigens have been of the order of 17–56% of the soluble protein obtained after reoxidation (the soluble protein comprising 30–50% of the total protein reoxidized) (*45, 111*). The fact that the recoveries obtained are as high as they are is in a sense somewhat surprising since only a part of the IgG molecule was used. The inference from this is that the entire amino acid sequence of the H-chains is not necessary for the restoration of at least the antibody-combining site, and the explanation for this is probably twofold. First, the unfolded, intact L-chain appears capable of folding by itself. Second, it is possible that the combining part of the H-chain found in the papain fragment is more or less structurally independent of the remainder of the chain and, by virtue of this and the proximity of a folded L-chain, is able to refold properly.

The second way of circumventing the problem of the insolubility of reduced IgG chains has been to attach chains of poly-DL-alanine to the exposed amino groups of the molecule, a technique initially employed to solubilize reduced trypsin. The resulting poly-DL-alanyl γ-globulin, with over 800 added alanine residues per molecule, was soluble after unfolding and reduction of its 23 disulfide bonds, and the polyalaninated H- and L-chains could be separated. Under optimal conditions, in which the different chains were allowed to reoxidize separately and then combined, it was possible to recover soluble protein with all the antigenic sites of the starting material. Furthermore, when IgG with activity was used against specific antigens, it was found that the reconstituted material had from 25–50% of the antibody-combining capacity of the native protein (*37*). When the complexity of the protein being studied is considered, recoveries of this order are quite impressive, and any discrepancies from the theoretically expected yields probably result from the technical problems associated with the *in vitro* system. These results clearly indicate that the sequences of antibody molecules, like those of simpler proteins, can determine their three-dimensional structure and that the presence of antigen is not required for proper folding of the molecule. However, chemical heterogeneity has been observed even in antibodies prepared against single simple antigens. It appears, therefore, that more than one amino acid sequence can give rise to a relatively specific type of binding site, although it is probable that the binding sites of the various proteins differ somewhat in their actual conformations. This is not surprising in view of the results obtained with hemoglobin and myoglobin (see below) in which many amino acid sequences are compatible with the same three-dimensional structure and with the binding of heme.

D. THE ROLE OF PROSTHETIC GROUPS IN FOLDING

Many proteins contain noncovalently linked prosthetic groups, ranging in kind from simple cations to molecules as complex as heme, and it may be asked whether these groups are essential for the proper folding of the molecule. Few investigations have been carried out *in vitro* to examine this question, but one in particular deserves mention. Taka-amylase A is an enzyme with 4 disulfide bonds plus a free sulfhydryl group that requires calcium for activity. When care was taken to exclude calcium from the medium in which reduced and unfolded Taka-amylase A was allowed to reoxidize, no enzyme activity could be recovered (102). In the presence of calcium, however, significant recoveries of activity were obtained. Furthermore, inactive material obtained by reoxidation in the absence of calcium could be activated by treatment with calcium and catalytic amounts of reducing agents, indicating that the disulfide bonds may have been formed incorrectly. If these results are valid, the implication is that the prosthetic group must be available at the time of protein synthesis to insure proper folding and disulfide bond formation, or, if it is added later, that conditions must prevail for a disulfide bond interchange to allow for rearrangement to the correct conformation. However, recent reinvestigation of this question has revealed that the role of calcium may be more circumscribed. While it may be necessary to facilitate the oxidation of one disulfide bond, the proper reformation of the other three does not require its presence (39). It thus appears that the conformational effects of the calcium do not extend to the folding of the entire molecule.

With regard to proteins that do not contain disulfide bonds, the question of the role of prosthetic groups in folding is in some ways a trivial one. Even if an apoprotein (a protein without its prosthetic group) is not able to fold properly without the prosthetic group, the absence of covalent intrachain bonds should allow for the rapid rearrangement of the protein to the correct conformation once the prosthetic group is introduced. Nevertheless, it is possible that the folding of such a protein is facilitated by the prosthetic group, and the question concerning its role in folding is not without merit. Thus, it has recently been suggested that NAD is implicated in the folding of glyceraldehyde-3-phosphate dehydrogenase (18).

One protein without disulfide bonds that has been extensively studied is sperm-whale myoglobin. Apomyoglobin is produced by the treatment of myoglobin with cold acid acetone. Although this protein differs from the starting myoglobin in several respects, including its content of α-helix, it is probably not too different from myoglobin in the overall aspects of its structure. Apomyoglobin can be completely unfolded by treatment with urea or guanidine and, on removal of these reagents, can refold to produce apomyoglobin with all the properties of the starting material (49, 97). Furthermore, by addition of hematin, the refolded apoprotein can be converted to metmyoglobin—again with the properties of native metmyoglobin.

While it thus appears that heme is not required for the proper folding of apomyoglobin under normal conditions, it has, nonetheless, a profound effect on the stability of the protein and on its ability to fold in an adverse environment. When sperm-whale apomyoglobin is treated with increasing concentrations of urea, it loses all specific structure as determined by various physical–chemical parameters, at urea concentrations between 4 and 6 M. However, if hematin is added to the unfolded apomyoglobin while the latter is still in urea, it is specifically bound and there is a dramatic shift in the properties of the protein toward those of native metmyoglobin (97). Thus, in the presence of the prosthetic group heme, unfolded apomyoglobin can be refolded (to form metmyoglobin) under conditions which would otherwise result in virtually complete disruption of structure.

Substances other than prosthetic groups, such as substrates, inhibitors, other proteins, and ions, when studied in vitro can also affect the stability of proteins, and it may be asked whether they are able to influence the folding of these proteins. This has been investigated in a few instances. For example, the rate of reactivation of reduced ribonuclease is not affected by the presence of phosphate ions and substrate (29). Likewise, the presence of antigen during the refolding of the reduced papain fragment of specific antibody had little, if any, effect on the yield of active material (*111*). However, NADH increases the rate and extent of re-activation of unfolded fumarase, and malate and citrate improve the yield of refolded malate dehydrogenase (*12*). Furthermore, the association of the intact A- and B-chains of *E. coli* tryptophan synthetase appears to be enhanced by the presence of pyridoxal phosphate and serine (*51*). It is quite likely, therefore, that the ability of a protein to fold *in vitro* and *in vivo* may be influenced by many factors more or less directly related to the protein itself. Nevertheless, the general conclusion that it is principally the genetically controlled amino acid sequence of the protein that, in the final analysis, determines the conformation of the protein still appears valid.

II. THE PHYSICAL–CHEMICAL BASIS OF PROTEIN STRUCTURE

With the large mass of evidence connecting protein conformation to amino acid sequence, the question of how the two are related becomes one of great interest. Specifically, what is there about a sequence of amino acids that gives rise to a single (or few) unique three-dimensional conformation, rather than to a large number of different conformations? As has already been discussed, the answer is a thermodynamic one, and it can be considered as involving 2 important components—steric factors and interactions among amino acid side chains, peptide backbones, prosthetic groups (when present), and the solvent.

A. STERIC FACTORS

The steric factors involved in protein conformation are of several kinds. Perhaps the most obvious is that more than 1 atom cannot occupy a single position in space at the same time, or, stated another way, that the van der Waals' radii of 2 atoms that are not connected cannot overlap. For polypeptides of small size, such a restriction may severely limit the number of conformations available to the chain, but with proteins containing 100 or more amino acid residues, this number still remains enormous. However, within a tightly folded protein, the nature of an amino acid might become critical because of such steric considerations. For example, if 2 parts of the amino acid chain are to cross each other in close contact, only very small side chains may be acceptable at the point of crossing. This was thought to be the case with residues B6 and

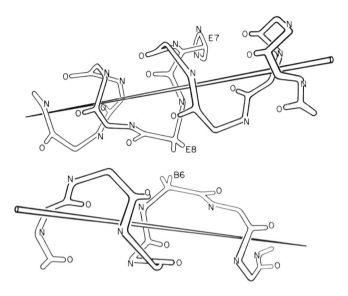

Fig. 6. The crossing of the B and E helices of sperm-whale myoglobin. Residues B6 and E8 at the point of crossing are both glycines; in rare instances, one of these is replaced by alanine. The insertion of larger residues at this point would result in marked distortion of the molecule—distortion which would be particularly serious because of the proximity of the histidine residue E7 to the prosthetic group heme. This has proved to be the case in hemoglobin Riverdale-Bronx (*93b*).

E8 of the heme proteins, the former being invariantly glycine and the latter glycine or alanine in all normal hemoglobins and myoglobins (Fig. 6) (*89*). However, the mutant hemoglobin Riverdale-Bronx (B6 Gly → Arg) has recently been described and, as would be expected, is unstable (*93b*).

The next group of steric restrictions is imposed by the existence of energetically preferred orientations for bond angles, particularly in the polypeptide chain backbone but also in the amino acid side chains and even in disulfide bonds. While the members of the peptide bond and the adjacent C^α carbon atoms must remain planar or nearly so (i.e.,

$$C^\alpha—\overset{\overset{\text{H}}{|}}{\underset{\overset{||}{\text{O}}}{C'}}—N—C^\alpha—\text{is planar}),$$ rotation is possible about the $N—C^\alpha$ bond

(angle of rotation denoted as φ) and the $C^\alpha—C'$ bond (angle of rotation ψ) of the backbone, as well as about the $C^\beta—C^\alpha$, $C^\gamma—C^\beta$, . . . bonds (χ_1, χ_2) in the side chains (Fig. 7). On the basis of theoretical calculations of the energies involved in the various orientations of the $N—C^\alpha$ and $C^\alpha—C'$ bonds, certain sets of values of ψ and φ have been found that have minimum steric energies and would be expected, therefore, to be preferred over any other (72). Similar calculations have been carried out for the side-chain bonds χ. Experimental verification of these concepts has been obtained from the determination of bond angle relationships found in protein structures as deduced by X-ray crystallography.

One of the peptide bond energy minima corresponds to values of ψ and φ that determine the formation of the α-helix. This is the polypeptide chain backbone configuration that has been designated as secondary structure. Other minima correspond to different helices, including the 3_{10}, α_{II}, and π helices. In the past it was thought that the first step in the folding of proteins was the formation of regions of α-helix, and that the chain then folded further to attain the final conformation. More recently it has been shown that the α-helix is just a component of a three-dimensional structure, albeit a somewhat special one, and the helical content of many proteins has been found in most instances, computed on the basis of optical rotatory dispersion measurements, to be quite low. Nevertheless, proposals advancing the primacy of the α-helix are again being advanced, and attempts are being made to predict the locations of α-helices in proteins of known sequence (87). The difficulty with this approach is that the factors that determine the existence or absence of such helices are only poorly understood. It is well known that

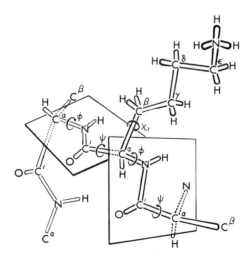

Fig. 7. The orientation of bond angles in the polypeptide chain backbone and side chains of proteins. Two planar peptide bonds which share a common carbon (C^{α}) are indicated. The orientation between the 2 planes is determined by the angles ψ(C'—C^{α}) and φ(N—C^{α}), while the orientation of the beginning of the side chain (of lysine) is determined by angle χ.

proline can exist only at the NH_2-terminal end of a helix; its rigid bond angles would otherwise make it a helix breaker. Other amino acid residues (glutamic acid, aspartic acid, and histidine) also appear to be correlated with the disruption of α-helices in proteins of known structure, although they do not always disrupt them. However, residues which are usually incompatible with α-helix formation in polyamino acids (valine, isoleucine, serine, and threonine) do not appear to be helix-breakers in proteins. Rules for the prediction of helical regions have been formulated from observations on metmyoglobin and hemoglobin, and their application has been fairly successful in the prediction of the positions of such regions. However, one difficulty with these rules and their tests is the possibility that there may be a significant difference in helical content between metmyoglobin and apomyoglobin (17). The former has been used in all calculations, but it seems that the heme is playing a major role in determining how "much" helix the protein is to have. Therefore, if the rules are to be generally applicable, a protein such as metmyoglobin is perhaps not the best one to use as their basis.

B. SIDE-CHAIN POLARITY

The recent crystallographic results of the groups of Kendrew and of Perutz have strikingly substantiated earlier proposals by Kauzmann and others that the distributions of polar and nonpolar residues in protein molecules are of paramount importance in determining their structures (65). While hydrogen bonding and the electrostatic and dipole interactions among various parts of the molecule contribute significantly to the free energy of a protein, a very large contribution to this term is made by the entropy involved in the interactions of solvent with the amino acid side chains. In an aqueous environment, nonpolar residues are thought to decrease the entropy of the surrounding water molecules by causing their repulsion and leading, therefore, to a more highly ordered local water structure. From this point of view, a substantial increase in entropy, leading to a decrease in free energy (since the contribution of a change in entropy, ΔS, to free energy is equal to $-T\Delta S$), will result from the removal of hydrophobic groups from the aqueous environment. Conversely, the tendency of the electrical potential energies of charged or otherwise polar groups to increase free energy are minimized by surrounding such groups with a medium of high dielectric constant—such as water. Both these requirements can best be satisfied by having all hydrophobic groups in the interior of the molecule (i.e., not exposed to solvent) and all polar groups on the surface. The distinctions between interior and surface must be interpreted broadly. For example, a molecule such as hemoglobin has a central cavity which is lined by polar residues and appears to be filled with water molecules. The lining of this cavity is considered to be part of the surface of the molecule.

With relatively few exceptions, the orientations of side-chain residues in the interiors of hemoglobin and myoglobin have conformed to the proposed distribution (88, 89). The same is also true of egg white lysozyme and ribonuclease, but with a somewhat greater number of exceptions (Fig. 8) (6, 63). In the heme proteins there are 33 interior residues, of which 30 are invariably hydrophobic and the other 3 are either serine or threonine and are probably internally hydrogen bonded. Of the 120 residues in the surface of the molecule or in surface crevices, 25 are always polar, 10 are always nonpolar, and the greater proportion, about two-thirds, may be either polar or nonpolar. Very often, however, the nonpolar surface residues are either glycine or alanine—in sperm-whale myoglobin these 2 residues account for nearly half of this group—and it is likely that these residues have much less effect on water structure than do the nonpolar residues with larger side chains. It thus appears that the surface of the molecule may vary considerably, in terms of side-chain polarity, as compared with the interior of the molecule, but there does appear to be a limit. Proteins with a higher proportion of hydrophobic residues than can be tolerated on the surface of the molecule seem to aggregate into multimers, thereby reducing the number of exposed hydrophobic groups (34).

On the basis of theoretical considerations and experimental data, it has been postulated that the folding of a protein is governed, to a large extent, by the distribution of polar and nonpolar residues in the amino acid sequence (29). (This hypothesis could be expanded to take into account such considerations as steric factors and α-helix makers and breakers, but this will be deferred for the present.) It has been predicted, therefore, that the alteration of the polar residues on the surface of a protein molecule should not interfere with the proper folding of the protein. This prediction was tested with several modifications of ribonuclease and was found, in general, to be valid (29). Poly-DL-alanyl ribonuclease, ribonuclease with long chains of poly-DL-alanine attached to the exposed α- and ε-amino groups, was able to refold properly after reduction and unfolding, as were several other derivatives of ribonuclease prepared by esterification of the side-chain carboxyl groups and/or reaction of exposed amino groups with various acid anhydrides. Likewise, poly-DL-alanyl ribonuclease T1 could be reduced and reoxidized with a 70% yield (70), and poly-DL-alanyl trypsin gave about 4% recovery of activity; the results with poly-DL-alanyl gamma globulin have already

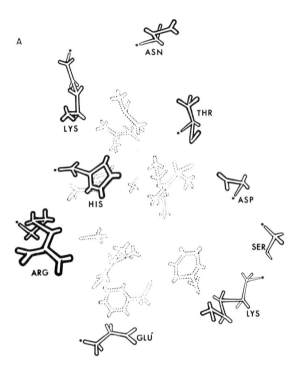

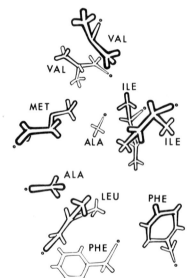

Fig. 8. A view of the location of amino acid side chains in one part of the molecule of egg white lysozyme. (A) The surface is composed entirely of polar residues, while the interior of the molecule consists of residues with hydrophobic side chains. (B) An enlargement of the interior of the molecule. The polypeptide chain backbone has been omitted for clarity.

been discussed. However, it should be noted that the diiodination of the 3 exposed tyrosine residues of bovine pancreatic ribonuclease did prevent the proper refolding of the molecule (*38*). It has not as yet been possible to carry out the converse experiments—those of modifying hydrophobic groups and, of greatest interest, of converting interior hydrophobic residues to polar residues and then testing the ability of the unfolded proteins to fold properly. It would of course be necessary to insure that the modifications did not introduce steric interference with folding, but if they did not, experiments with such altered proteins would provided the most direct evidence for the role of side-chain polarity in protein folding.

C. The Effects of Mutations on Structure

In addition to the experimental manipulation of proteins, another source of information about the effects of amino acid alterations on the folding and conformation of proteins has been provided by the study of those mutant proteins, both natural and induced, that can be isolated and studied. Although the full understanding of these effects will require more knowledge about the forces that influence conformation, much has been learned about their nature and diversity. A change of 1 amino acid in a protein may cause conformational alterations that range from being virtually indiscernible to those causing gross alterations in stability and function. The numerous electrophoretic variants of a wide variety of proteins provide examples of the former. To illustrate the latter, different amino acid substitutions, all of which decrease activity, in the A protein of *E. coli* tryptophan synthetase have been found to have quite different effects on the sensitivity of the protein to heat denaturation (*51*). When an arginine or valine residue is substituted for glycine, sensitivity to denaturation is increased; a glutamic acid residue, however, decreases sensitivity. Furthermore, the effects of the latter substitution on both activity and stability are reversed by a substitution of a cysteine for a tyrosine residue, 36 amino acid residues distant (Fig. 9). Interestingly enough, the effects of a different mutation, glycine to valine, 2 amino acids away from the mutations just discussed, can be reversed by a second mutation, leucine to arginine, again 36 residues away. These results indicate the complex effects on conformation of even single amino acid changes and are not readily explained by any simple theory. Further explanations will have to await the elucidation of the three-dimensional structure of the protein.

Of the many known mutations of human hemoglobin, most appear to

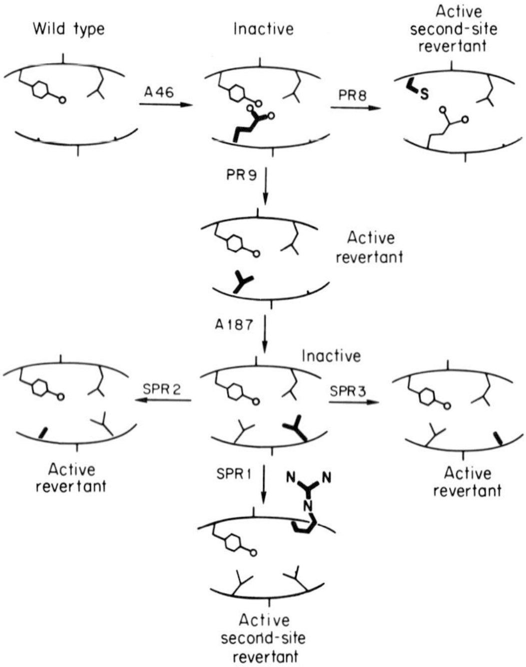

FIG. 9. A series of mutations affecting tryptophan synthetase (51). Substitution of glutamic acid for glycine (A46) results in an inactive protein, but the effects of this mutation can be ameliorated either by substituting cysteine for tyrosine in another part of the molecule (PR8) or by substituting valine for the mutant glutamic acid (PR9). The latter is inactivated by a mutation which substitutes a valine for a second nearby glycine (A187), and these effects can be reversed either by replacement of one of the valines by alanine (SPR2, SPR3), or by replacement of a leucine by arginine (SPR1). One possible explanation for these results is depicted in this diagram. Two regions of the chain are presumed to be in close proximity. Mutations causing overcrowding of the region of contact result in loss of activity, while those allowing more space result in regain of activity. To explain the regain of activity following the substitution of arginine for leucine, it is postulated that the polar guanido group pulls the side chain of arginine out of the interior of the molecule.

affect only electrophoretic mobility. However, a few are known to affect protein conformation. Some of these cause the sickling of erythrocytes upon deoxygenation. This results from the polymerization of the hemoglobin molecules into long stiff aggregates which deform the red blood cells. The substitution of valine for glutamic acid in the sixth position of the β-chain (hemoglobin S) is the most common mutation causing sickle hemoglobin, but sickling has also been observed with hemoglobins I (α^{16} Lys $\rightarrow$ Glu), D Punjab (β^{121} Glu $\rightarrow$ Gln), and C Georgetown and C

Harlem (84). The latter two are most interesting since they have been found to represent double mutations of the β-chain—the regular S mutation (β^6 Glu $\rightarrow$ Val) plus, for C Harlem, a second mutation at β^{73} (Asp $\rightarrow$ Asn) (7). It has also been found that the intensity of the sickling phenomenon produced by the S mutation is decreased by the simultaneous presence of an α-chain mutation, α^{23} Glu $\rightarrow$ Gln (69). It has been proposed that the S mutation allows for a localized conformational alteration which, in turn, permits stacking of the hemoglobin molecules (84). No explanation is yet available for the effects of the other mutations. One other interesting mutation affecting hemoglobin conformation is that in hemoglobin Köln (β^{98} Val $\rightarrow$ Met). The methionine, probably because of its greater size and critical location close to the heme, results in a markedly decreased protein stability, both *in vivo* and *in vitro* (11).

Some effects of more complex alterations in amino acid sequence have been studied (see also Chapters I and II). Three reading-frame shift mutations (mutations involving the addition or deletion of 1 or 2 nucleotide bases) were induced with acridine in the lysozyme gene of phage T4 (86). None of the mutant phages had any lysozyme activity, either due to the long carboxyl-terminal region of altered amino acids or the premature termination of the polypeptide chain, or both. Two of the acridine-induced mutant phages were each crossed to a third mutant, and the resulting pseudowild double-mutant proteins were isolated. One protein was shown to contain an altered sequence of 5 amino acids due to the deletion and insertion of single nucleotide bases about 15 nucleotides apart (Fig. 10). Although it has not been characterized physically, the double-mutant protein was found to have a specific activity about 50% of normal. The second pseudowild protein also had an altered sequence of amino acids with the insertion of an extra residue. It ap-

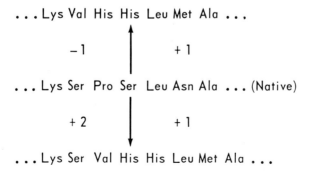

FIG. 10. Pseudowild phage lysozymes resulting from the recombination of mutant genes carrying frame-shift mutations (from Okada *et al.*, 86).

pears to be the result of a double and a single-base insertion, again about 15 residues apart. These results indicate that changes in the sequence of 4 or 5 amino acids and even the addition of an amino acid may, in some circumstances, be compatible with the proper or nearly proper folding of the molecule.

III. THE FOLDING OF PROTEINS *IN VIVO*

While the foregoing discussion analyzes the experimental and theoretical bases for our present ideas concerning the folding of proteins *in vitro* and, by inference, *in vivo*, it does not provide information about the actual process of protein synthesis. It is not known, for example, whether the polypeptide chains fold during the sequential addition of amino acids, or whether folding takes place only after completion of the polypeptide chain. The former has been postulated to occur, for example, in the biosynthesis of egg white lysozyme (*90*), but evidence for this mechanism is lacking. And, if folding does take place before completion of the chain, is it the proper folding, and does it significantly affect the rate or efficiency of the folding process? In higher organisms, at least, the presence of the disulfide-bond interchange system could make possible the rapid reshuffling of incorrectly formed disulfide bonds, and their formation would not necessarily interfere with the rapid attainment of the correct conformation. Evidence obtained from the study of the rate of synthesis of hemoglobin chains in a reticulocyte ribosome system has been interpreted to suggest that folding is taking place during chain synthesis (*113*). A significant decrease in the rate of synthesis was found to occur at about the hundredth residue of the α- and β-chains, and it was postulated that this could result from conformational changes introduced by the binding of heme to the chain at this time. Such binding would be feasible since the 2 histidine residues thought to be involved in heme binding occur in the first 100 residues. It is possible, of course, that the change in rate is related to the presence of some rate-determining codon rather than to conformational alterations. Furthermore, data on the incorporation of ^{59}Fe into reticulocyte ribosomes and the finding of free globin in reticulocytes would be more compatible with the introduction of the heme into hemoglobin after the chains have been released (*32*).

Active and latent, or nascent, completed proteins have frequently been found in combination with ribosomes, and it has been inferred that the folding of these proteins has taken place while the polypeptide chain was still connected to or in contact with these particles. While this explanation is probably correct in some instances, there is danger in assuming that all ribosome-bound proteins have been newly synthesized, since nonspecific binding of proteins to ribosomes can occur.

Studies with proteins composed of more than 1 polypeptide chain have indicated that there may be preferential release of specific chains from the ribosomes, and that the released chain(s) might be of importance in determining the folding and/or release of the other chain or chains from their ribosomes. Two systems have been of particular interest in this regard. One is the synthesis of the chains of IgG by mouse plasma cell (myeloma) tumors (99). The L- and H-chains of the myeloma protein are synthesized on polysomes of different size, each class of polysomes being commensurate in size with the length of the chain. The L-chains were found to be synthesized in a number in excess of the number of H-chains and appeared to attach to the H-chains while the latter were still attached to their polysomes. Although the methodology does not allow determination of the conformations of the individual chains, it seems likely that a high degree of folding has probably occurred even before chain combination. The finding that L-chains can refold by themselves *in vitro* would support this idea.

A similar situation may exist in the synthesis of the α- and β-chains of rabbit hemoglobin (14). Incubations of reticulocyte systems with radioactive amino acids have demonstrated that β-chains are relatively more highly labeled than are α-chains during the early phases of incubation. These results have been interpreted as indicating the preferential release of newly synthesized β-chains and their combination, on or near the polysomes, with already synthesized but ribosomally bound α-chains. However, the data could also be interpreted as showing the combination of released α-chains with newly synthesized β-chains, and other data favor this conclusion (98a). In either case, it has been postulated that the globin chains form $\alpha\beta$ dimers which, after combining with heme, give rise to the $\alpha_2\beta_2$ tetramers of hemoglobin (32).

Although heme does not appear to be required for the release of globin chains from polysomes, heme has been shown to stimulate the synthesis of globin chains *in vitro* (74, 109). The basis for this stimulation is not understood, but one interpretation is that the binding of heme to the polypeptide chain causes release of the chain from the polysomes and allows further synthesis to take place (74). However, the observation of a shift in the distribution of ribosomes to larger polysomes, the stabilizing effects of heme on polysomes *in vitro,* and the failure of protoporphyrin (which will bind to globin) to stimulate incorporation suggest that the effect may not involve protein conformation but, rather, some action at the ribosomal level (*109*). In support of the latter hypothesis, hemin has recently been shown to inhibit erythroid cell nuclease (*9b*).

IV. STRUCTURAL ALTERATIONS FOLLOWING
POLYPEPTIDE CHAIN SYNTHESIS

In many instances, the synthesis of a protein does not stop with the completion of the amino acid sequence or even with the folding of the polypeptide chain. Several classes of proteins are known in which further changes in the molecule are accomplished either by alterations of the amino acids themselves, by the covalent addition of various types of substituents, or by the noncovalent binding of other nonprotein molecules. These changes may occur either during or immediately after the synthesis of the polypeptide chain or later.

A. Modifications of the Amino Acid Side Chains

Changes in the amino acid residues themselves are particularly well illustrated by the biosynthetic steps involved in the synthesis of collagen. Collagen contains 2 amino acids that are not commonly found in other proteins—hydroxyproline and hydroxylysine. Neither amino acid is incorporated as such into the protein, but each is produced by hydroxylation of already incorporated proline or lysine. The formation of hydroxyproline has been extensively studied, and the results indicate that the incorporation of proline into peptide linkages and the hydroxylation of proline are separable processes (*106*). In a soluble system derived from chick embryos, the incorporation of ^{14}C-proline into collagen is completed before ^{14}C-hydroxyproline is found. If hydroxylation is blocked by anaerobiasis, a deficiency of ascorbic acid, or the destruction of the hydroxylating enzyme (an oxygenase which uses O_2 as the oxygen source), a collagen deficient in hydroxyproline will be produced. However, hydroxyproline formation is not prevented by treatment with puromycin, and even a synthetic polypeptide (Pro-Gly-Pro)$_n$ will serve as a substrate for the soluble hydroxylating system. The sequence X-Gly-Pro appears to be required for hydroxyproline formation. With *in vitro* microsomal systems, hydroxylation takes place while the collagen polypeptide is still attached to the ribosomes. It is thought that the same is also true *in vivo,* and that, under optimal conditions, hydroxylation keeps pace with the growth of the polypeptide chain.

The hydroxylation of lysine in collagen is thought to occur by a similar mechanism, but the details of the process have not been worked out. Furthermore, lysine has been implicated in the formation of intramolecular interchain cross-links, with the δ-semialdehyde of α-amino adipic acid, derived from lysine, postulated as an intermediate (*91*). Many other types of atypical linkages including γ-glutamyl and ester bonds have been claimed to exist in collagen and related structural proteins.

Another type of covalent alteration in an amino acid, one which results in the masking of the terminal NH_2 group, has been observed in some of the immunoglobulin chains, including the λ type of human L-chain, and H-chains of rabbit IgG and of a human myeloma protein. In each of these, the NH_2-terminal residue glutamine appears to cyclize spontaneously to form the cyclic pyrrolidone carboxylic acid (*112*).

B. Covalently Bound Substituents

Many types of substituents covalently bound to protein have been identified and are discussed in detail below, but several others deserve mention. These include carboxyl terminal amides, serine-0-phosphates, tyrosine-0-sulfates, and Schiff's bases of pyridoxal and ε-amino groups. Some of these may not require specific enzyme systems for their attachment, or they may become attached as a consequence of the mechanism of action of the enzyme in question. For example, a phosphate group can be reversibly transferred from glucose 1,6-diphosphate to the active serine of phosphoglucomutase. Because its precise identity is unknown, 1 substituent of particular interest is the thiol (RSH) in disulfide linkage with one of the cysteines of streptococcal proteinase (*33*).

1. *N-Acetyl Groups*

N-Acetyl groups have been recognized in many proteins. Several plant viruses contain such groups attached to a variety of different amino acids (not including proline). Acetylation of the NH_2-terminal residue occurs in melanophore-stimulating hormones, vertebrate cytochromes c, egg albumin, and histones. The β-chains of chicken hemoglobin and the α-chains of carp hemoglobin are *N*-acetylated, as are the γ-chains of human fetal hemoglobin F_1 (*57*). *N*-Formyl proline has been found in lamprey hemoglobin (*8*). As has been discussed in other chapters,

N-formyl methionine has been implicated as the amino acid which corresponds to the initiating codon, and it may be enzymically removed following synthesis of some proteins. However, in the vertebrates and probably also in the viruses, the N-acetyl group (and probably the N-formyl group as well) appears to be the product of acetylation of the completed protein. An acetyl transferase or acetokinase has been isolated from chicken reticulocytes, and is capable of transferring acetyl groups from acetyl-CoA to several proteins in a relatively nonspecific manner (76). The most marked transfer is to chicken hemoglobin and human fetal hemoglobin F_2. It is not presently known when acetylation occurs *in vivo*, whether it is only this enzyme system that carries out N-acetylation *in vivo*, or what factors determine the acetylation of some proteins and not others.

2. Carbohydrate

A large number of proteins contain covalently bound carbohydrate (31). It is found attached to many circulating serum proteins (or, more properly, glycoproteins), including haptoglobin, immunoglobulins, and fibrinogen, as well as to bovine ribonuclease B, ovalbumin, fetuin, and human chorionic gonadotropin. There also exists the complex group of mucopolysaccharides which, although they have a higher carbohydrate content, seem to contain specific proteins. The sugars usually found with proteins are galactose, mannose, fucose, glucosamine, N-acetylglucosamine, N-acetylgalactosamine, and N-acetylneuraminic acid. Several types of sugar-protein linkages have been recognized. These include O-glycosidic linkages to serine and threonine, glycosylamine linkages to asparagine, and possibly ester linkages to the β and γ carboxyl of aspartic and glutamic acids (43). A unique glycoprotein is the β-chain of hemoglobin A_{Ic} in which a hexose is combined with the N-terminal valine as a Schiff's base (6a, 54).

Considerable effort has recently been directed to the study of the biosynthesis of glycoproteins and mucopolysaccharides, and the evidence indicates that the sugars are added, one by one, to the protein by a series of specific enzymic reactions. In chondroitin sulfate, the sequence of sugars at the site of attachment to the protein moiety has been found to be serine-O-xylose-galactose-galactose-glucuronic acid (104). Although a microsomal system has been implicated in the attachment of the carbohydrate, protein synthesis and carbohydrate attachment are, in analogy to proline hydroxylation, dissociable. The attachment of xylose, the proximal sugar in the carbohydrate chain of heparin, has been carried out enzymically in a completely soluble system in which the protein is not associated with ribosomes (44). Again, as is the case with the other types of constituents, it is not known whether a particular amino acid sequence or protein conformation, or both, is required for the attachment of carbohydrate, or whether such attachment normally occurs during or only after polypeptide chain synthesis.

3. Heme

A substituent of particular interest is the covalently bound heme ring of the cytochromes c. This group is bound to the protein by thioether linkages to 2 cysteines which are separated from one another by 2 amino acid residues. Nothing is known about the mechanism by which this linkage is established *in vivo*, especially whether it is spontaneous, or enzymically stimulated. However, the process has been studied *in vitro*. The heme can be removed from cytochrome c by treatment of the protein with silver sulfate. When the apo-cytochrome c thus prepared is reacted with protoporphyrinogen (a heme precursor) at pH 3.5 and then with iron, a small amount of active cytochrome c can be reconstituted (*95*). Because of the conditions employed *in vitro* for the coupling process, it is not clear whether the addition of the heme is dependent on a specific amino acid sequence plus a particular protein conformation, or whether the specific sequence of amino acids is all that is required.

C. Noncovalently Bound Substituents

The combination of proteins with noncovalently bound prosthetic groups has already been alluded to, and, as has been mentioned, groups of all types may be found. However, a somewhat different type of interaction appears to be present in the lipoproteins (*96*). These proteins, which are present in several distinct forms, are capable of binding large quantities of neutral and charged lipids including cholesterol, phospholipids, and triglycerides. It has been possible to remove most, if not all, of the lipids from the protein moieties by treatment of lipoproteins with lipid solvents, but attempts to reconstitute truly native lipoproteins *in vitro* have had only limited success. Although the nature of the interactions between lipid and protein has not been well elucidated, it does appear that a combination of electrostatic and nonpolar bonds are involved—the former between polar amino acid residues and the choline and phosphate moieties of phospholipids, and the latter between hydrophobic amino acid residues and the hydrocarbon portions of the lipid. Investigations of the amino acid composition of soluble serum α- and β-lipoproteins have not revealed abnormally high contents of nonpolar amino acids, but elevations have been found in structural lipoproteins.

Likewise, physical studies on native and almost completely delipidated α-lipoproteins have not as yet indicated any special physical properties.

D. MACROMOLECULAR COMPLEXES

Before concluding this section, brief mention must be made of the fact that many proteins are now known to exist in the form of macromolecular complexes with other proteins and nonprotein constituents. The electron transport system in mitochondria is the prototype of this type of aggregation of functionally related but distinct proteins, and similar systems have been found in other sites (94). For example, the pyruvate dehydrogenase complex from E. coli, with a molecular weight of about 4.8×10^6, appears to contain several molecules each of pyruvate decarboxylase, dehydrolipoyl dehydrogenase, and lipoyl reductase-transacetylase. These 3 enzymes have been separated from one another by treatment with urea and chromatography, and can be recombined under appropriate conditions to form an active complex with the properties of the native complex. Similar results have been obtained with the triple enzyme complex of E. coli α-ketoglutarate dehydrogenase and with the mitochondrial electron-transport complexes. Thus, despite the fact that the complexes have not been reconstituted from completely unfolded proteins, it appears reasonable to assume that the assembly of even the largest protein molecules and aggregates, with their very complex quaternary structures, can occur spontaneously and in accord with the thermodynamic hypothesis.

V. THE CHEMICAL EVOLUTION OF PROTEINS

Proteins, like all other constituents and characteristics of organisms, have arrived at their present state by a long process of evolution. It is, perhaps, improper to speak of proteins as having themselves evolved, since, in the strictest sense, only organisms evolve. The evolution of organisms is brought about by changes in their content of genetic information, and these changes are presumed to have been preserved by the processes of natural selection because of advantages they conferred on the host. Since the expression of genetic information is mediated principally through proteins, the evolution of an organism will often be reflected by alterations in its proteins—alterations which may be qualitative, quantitative, or temporal. The latter two are related principally, although not exclusively, to systems of control and will not be discussed in any detail here. Qualitative alterations may, however, be analyzed in terms of the structures of the proteins, and evolutionary changes in these structures can be readily perceived. For the purposes of discussion, it is convenient to speak of the evolution of proteins as distinct from the evolution of the organisms that carry them.

Many methods have been employed in the study of protein evolution. Many years ago, respiratory pigments (hemoglobins and related proteins) from many species of animals were compared on the basis of their gross crystal structures, but this approach did not have general applicability. These and other types of proteins have also been compared by analysis of their functional properties—substrate requirements, kinetic constants, and so on. Such comparisons are quite useful for understanding the effects of structural changes on the activities of the proteins and on the physiology of their hosts. Another method of comparison, perhaps the earliest that was used in any extensive and systematic way, has been the immunochemical comparison of proteins from different species. Although the initial work was relatively crude, more recent approaches, including such techniques as immunodiffusion, immunoelectrophoresis, and quantitative complement fixation reactions, has allowed for quite sensitive comparisons. In general, these methods respond to both the three-dimensional conformations and the sequences of proteins and provide information about alterations in these properties. Changes of even a single amino acid residue in a hemoglobin chain can be detected by suitably sensitive techniques. Unfortunately, the information derived from immunochemical comparisons cannot be simply interpreted in terms of structural changes, and attention has turned principally to the analysis of amino acid sequences for insight into the details of protein evolution.

A. Types of Mutations

1. Point Mutations

Evolutionary changes in proteins are brought about by a series of mutational events which may be of 2 types: point mutations which affect single nucleotide bases in the structural genes, and larger chromosomal aberrations. A single-base change may result in either no apparent alteration (if the new codon codes for the same amino acid as the old), a substitution of one amino acid for another, or the termination of the polypeptide chain (if the new codon is of the *amber* or *ochre* variety). The first of these, the so-called isosemantic substitution, will not affect protein structure, but the possibility has been raised that the base change could result in a change in the rate of polypeptide chain synthesis (*59*).

Single-base substitutions resulting in the replacement of one amino acid by another have been recognized in the evolution of many groups of proteins. In proteins from closely related species, there may be one or only a few such substitutions, and these will usually involve single-base changes. With more widely separated species, proteins may differ by many amino acids and the substitutions will involve double and even triple-base changes. If it is assumed that mutations occur at random throughout the structural gene, the occurrence of multiple mutations within single triplets will become more likely as the number of base substitutions increases. Thus, in comparing the sequences of the cytochromes c of different species, it has been found that the number of base changes required to explain the amino acid substitutions increased from 15 (9 one base and 3 two base) when the human and horse cytochromes were compared to 70 (25 one base, 21 two base, and 1 three base) for tuna versus *Neurospora* cytochrome (*62*). These numbers must, of course, represent minima since a large number of isosemantic substitutions and some multiple mutations that produced substitutions compatible with single-base changes have probably also occurred. The former could amount to a quarter or more of all base changes (see below), and may in part explain why the ratio of one- to two-base substitutions is less than would be expected on the basis of random base substitutions governed by the Poisson distribution (*62*).

2. *Chromosomal Accidents*

The simplest type of chromosomal accident is the addition or deletion of a single nucleotide base resulting in a translational frame shift. Although such mutations have been induced with mutagens of the acridine type (see Chapter I), there is no evidence that this type of mutation occurs commonly, if at all, spontaneously *in vivo*. One reason for its uncommonness, or at least for the failure to observe it, could be the drastic effects which a frame shift would have on the amino acid sequence of the protein unless the mutation occurred very near to the C-terminal end. Otherwise, the large change in sequence would probably result in a protein which was nonfunctional, nonsynthesizable (by chain termination), or unrecognizable.

In the natural evolution of proteins, base substitutions which result in chain termination and thereby in a carboxyl-terminal deletion, cannot be distinguished from chromosomal accidents (Fig. 11). However, the deletion of one or more amino acids from within a protein or at the amino-terminal end can be attributed to a chromosomal accident since it is highly improbable that 3 adjacent bases would be simultaneously or even sequentially removed as the result of single-base deletions.

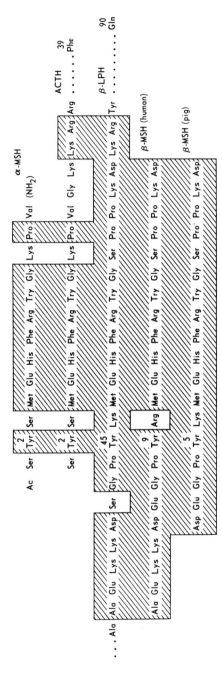

Fɪɢ. 11. Homologous proteins from the anterior pituitary. All hormones possess the hexapeptide Met-Glu-His-Phe-Arg-Try-Gly, and several have larger regions of homology. However, they all differ in the lengths of their NH₂- and COOH-terminal portions.

Furthermore, many internal deletions involve considerably more than one amino acid, and the likelihood is even less that a series of single-base deletions are involved. Likewise, the addition of amino acids within the protein or to its ends, must be attributed to a chromosomal accident, although it is not always possible to determine whether the differences in the lengths of 2 proteins results from an addition of amino acids to one or their deletion from the other.

An example of a deletion of a single amino acid may be the mutant human hemoglobin β-chain found in hemoglobin Freiburg (60). From a determination of the amino acid composition and a limited analyses of the sequence of the abnormal peptide from the mutant β-chain, it has been concluded that a valine normally present at position 23 has been deleted:

$$
\begin{array}{cc}
& 20 \qquad\qquad 25 \\
\text{Normal} & \text{Val-Asp-Glu-Val-Gly-Gly-Glu} \\
& 20 \qquad\; \downarrow \quad\; 25 \\
\text{Freiburg} & \text{Val-Asp-Glu——Gly-Gly-Glu}
\end{array}
$$

A similar situation appears to have occurred in the evolution of the hemoglobin α-chain. Sometime after the divergence of the progenitors of man and the carp, a single amino acid was deleted from the α-chain of the former at the position between residues 46 and 47 of the present human α-chain (8):

$$
\begin{array}{ll}
\qquad\quad 40 \qquad\qquad\qquad\qquad\qquad\qquad 50 \\
\text{Human Lys-Thr-Tyr-Phe-Pro-His-Phe——Asp-Leu-Ser-His} \\
\text{Carp Lys-Thr-Tyr-Phe-Ala-His-Trp-Ala-Asp-Leu-Ser-His} \\
\qquad\quad 40 \qquad\qquad\qquad\qquad\qquad\qquad 50
\end{array}
$$

A quite dramatic mutation has recently been found in hemoglobin Gun Hill, the β-chain of which has a deletion of 5 amino acids (probably F9 to FG4) (7a). The resulting β-subunits, while capable of folding and combining with α-chains, are unable to bind heme, and the resulting hemoglobin tetramer contains only 2 heme groups.

The chromosomal events responsible for additions or deletions may be thought of as involving unequal crossing over during mitosis or meiosis. Other mechanisms can also be postulated but there is little evidence to favor one over another. If 2 chromosomes, or DNA strands carrying the same structural gene, pair in such a way that the bases are not in one-to-one correspondence with each other, then a crossover between them can produce structural genes bearing either deletions or additions at any place (Fig. 12). (In this discussion, we are only speaking figuratively since the precise mechanisms of crossing-over are not known.) Depending on the amount of mismatching, the resulting alterations in the structural gene can range in size from the addition or subtraction of a single base

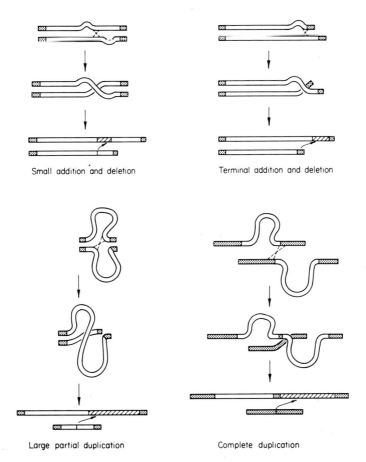

Small addition and deletion

Terminal addition and deletion

Large partial duplication

Complete duplication

Fig. 12. The generation of additions, deletions, and duplications of structural genes by nonhomologous crossing over. Misalignment of homologous genes during DNA replication is visualized as giving rise to pairs of genes, one of which contains an additional segment of the gene while the other reciprocal product lacks this segment.

to the addition or subtraction of a series of many bases and, as a result, of many amino acids in the protein. It is also possible that crossing-over could occur between 2 unrelated genes, in which case the material added to 1 gene would bear no relationship to material already present in that gene. However, if the crossing-over takes place between 2 chromosomes bearing the same gene, the added material will result in a duplication of material already present and will, as long as a frame-shift has not occurred, be read as such during protein synthesis. Depending on how the genes are matched, the duplication could range from a small partial duplication to a virtually complete tandem duplication. Our ability to recognize such events depends, of course, on the product of the partially duplicated gene being a polypeptide which is capable of folding into a protein which is itself recognizable as related to the original protein.

B. GENE DUPLICATIONS

The classic example of an almost complete gene duplication is that of the haptoglobin 2α gene (20). It has been shown by amino acid sequence analysis that this gene results from the partial duplication of the haptoglobin 1α gene. The latter has 2 alleles, $1\alpha F$ and $1\alpha S$, and these are both represented in the 2α gene—the $1\alpha F$ gene at the end corresponding to the amino terminal portion of the protein and the $1\alpha S$ gene to the carboxyl terminal end. Even more remarkable is the finding of a heptoglobin α-chain, the Johnson α-chain, which from its molecular weight appears to represent a partial triplication of the 1α-chain. Although it is not surprising that a triplication should occur, it is remarkable that the product should, despite its length, be both recognizable and functional as a haptoglobin. The same is, of course, also true for the 2α protein.

Gene duplications may be complete as well as partial. The former will occur when the 2 structural genes are completely mismatched. In either situation, the presence of a gene duplication will predispose to the recurrence of the same type of genetic accident that caused it and, therefore, to the formation of hybrid genes and of triplications and even higher states of gene multiplication (Fig. 13). A striking example of the formation of hybrid genes is known. A pairing between the 2 chromosomes carrying the linked genes for the β- and δ-chains of human hemoglobin presumably occurred in such a way that the δ gene of one strand was paired with the β gene of the other. The result of a crossover in this region was the formation of a hybrid (δ–β) gene which, in turn, codes for a hybrid (δ–β) polypeptide chain known as hemoglobin Lepore (71). In addition, the normal β and δ genes were deleted, and, as a result, individuals with this chromosome show defective hemoglobin synthesis. The reciprocal product of the crossover, one with the genetic constitution δ–(β–δ)–β, has not been recognized, probably because the presence of the normal δ and β genes makes the individual hematologically normal and, therefore, unlikely to be investigated. Because of the great similarity in sequence between the δ and β chains, the hybrid chain is structurally quite similar to the normal hemoglobin chains.

The amino acid sequences of several proteins have recently been found to be compatible with the conclusion, tentative in some cases, that their

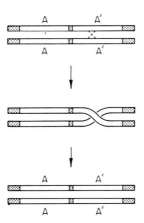

Normal crossing over with correct
matching of duplicate genes

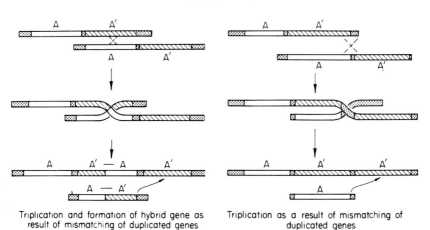

Triplication and formation of hybrid gene as   Triplication as a result of mismatching of
result of mismatching of duplicated genes   duplicated genes

Fig. 13. Propagation of gene duplication and formation of hybrid genes by recombination between homologous but nonidentical genes or gene segments (A and A').

genes were built up by one or a series of repeated partial duplications. Despite the ease with which partial duplication was recognized in the haptoglobins, the derivation of such evidence for most proteins has been based on the analysis of sequences by computers, and 2 principal types of programs have been written. One makes use of the actual amino acids in the sequence and systematically compares one part of the molecule with all others, attempting to maximize the number of amino acids which occupy homologous positions in two or more portions of the chain. If it is possible to find orientations in which this number is sufficiently greater

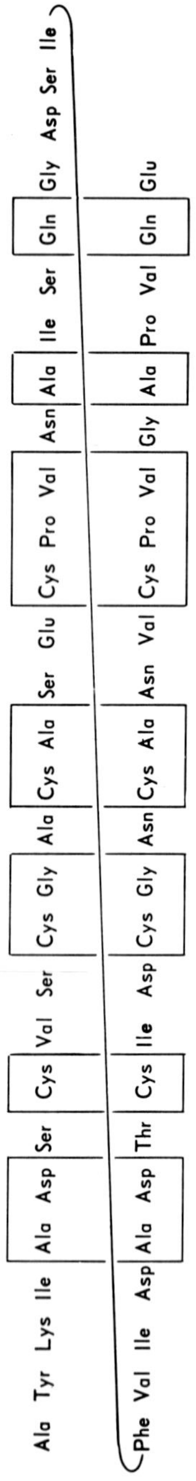

FIG. 14. Internal duplication in the sequence of Clostridial ferredoxin (23). Residues 1-26 are homologous to residues 30-55.

than would be expected to occur at random, the finding is taken as evidence for the occurrence of a partial duplication. The difficulties in this type of analysis are twofold: defining what constitutes random interchanges, and accounting for the further evolutionary alterations (base changes, additions, and deletions) which may have taken place after the duplication event and obscured the sequence homologies. Despite these problems, when the sequence of *Clostridium pasteureanium* ferredoxin is divided into 2 halves, residues 1–29 and 30–55 and these 2 parts are paired from their amino-terminal residues, 12 of the 26 pairs of amino acids have identical residues in the 2 peptides (Fig. 14) (*23*). It has also been proposed, on the basis of further analyses, that the duplicate halves of the ferredoxin molecule were originally derived from a tetrapeptide which was repeatedly duplicated until a gene with 7 sequential segments was built up. The latter proposal is quite tentative, but it seems reasonable to conclude that the ferredoxin gene is the product of one, if not a large series, of partial gene duplications.

The second method for the internal comparison of amino acid sequences by computer analyses involves the calculation of the minimum number of nucleotide base substitutions required to convert one amino acid sequence into another, again taking into account additions and deletions (Fig. 15) (*10, 35*). If 2 unrelated sequences are compared, the average number of required base substitutions found is of the order of 1.5 to 1.6 per residue. When the sequences of the human hemoglobin chains are compared internally by this method, again taking one segment of nucleotide bases and matching it with all others, evidence for

	Residue differences	Base changes per residue
...|Lys. Asp Val Leu His Gly|...		
...Phe Arg|Tyr Glu **Val** Asn Ala Pro|...	5	1.50
...Phe|Arg Tyr Glu Val Asn Ala|Pro ...	6	1.00
...|Phe Arg Tyr Glu Val Asn|Ala Pro ...	6	2.00
Random	~6	1.61

FIG. 15. Detection of relatedness of amino acid sequences by comparison of base sequences. The possible base sequences corresponding to a pair of amino acid sequences are compared, and the minimal number of mutations that would be required to convert one sequence into the other calculated. If the number of required base changes is significantly less than the 1.61 per codon that would be expected when unrelated sequences are compared, it is assumed that the 2 base sequences and, therefore, the 2 amino acid sequences are evolutionarily related. In theory, this method can detect relationships that may not be apparent from comparisons of amino acid sequences. In practice, considerably longer sequences than are shown here are used.

either a single duplication or for as many as 5 partial duplications may be obtained. Similarly, the internal comparisons of the sequences of cyto-chromes c from *Neurospora crassa* and *Pseudomonas* C-551 have been interpreted as showing several partial duplications. While these con-clusions must still be considered largely tentative, they do raise the interesting possibility that many present day genes have been derived from much smaller primitive genes by a process of multiple partial duplications.

Unlike the relatively limited evidence for repeated partial gene dupli-cations, the evidence for multiple complete gene duplications is quite strong, and new data are constantly being accumulated. Most of the evidence is based on visual comparisons of the amino acid sequences of different proteins, but here again, computer approaches involving both amino acid and nucleotide base sequences are now being applied. The best-known and still most impressive example of multiple complete gene duplications is found in the heme proteins, hemoglobin and myo-globin (*58*). In man, at least 6 and possibly even more homologous poly-peptide chains with genetically independent and different sequences have been recognized: myoglobin and the α-, β-, γ-, δ-, and ϵ-chains of hemo-globin. Other independent duplications may have occurred in other mam-mals: for example, 3 β-chains are found in sheep and 2 in mice.

One factor that has aided in the comparison of diverse protein sequences has been the finding of similar or identical sequences of amino acids in the vicinity of functionally important residues in functionally related proteins. Thus, the finding of a common sequence about the single DFP-reactive serine of trypsin, chymotrypsins A and B, elastase, and other enzymes with esterase activity has led to the observation that the sequences and disulfide bonds of these proteins are generally homol-ogous (*9a, 50, 85a*). It now appears, therefore, that, rather than having obtained similar or identical active center peptides by the convergent evolution of originally different proteins, these and probably several more enzymes are derived from the divergent evolution of duplicates of a common ancestor. In a similar manner, several dehydrogenases with an active cysteine residue appear to be related (Fig. 16). These include lactate dehydrogenase, alcohol dehydrogenase, and, possibly, triose phos-phate dehydrogenase (*36*). The peptide sequence around the active system of the latter also bears some resemblance to similar peptides in luciferase and ATP–creatine phosphotransferase (*105*). All these enzymes are similar in having the cysteine residue separated by 2 amino acid residues from aspartic acid (in the first 3) or asparagine (in the last 3). The vertebrate lactate dehydrogenase cited above is itself a tetramer composed of 2 or more different polypeptide chains which are, them-

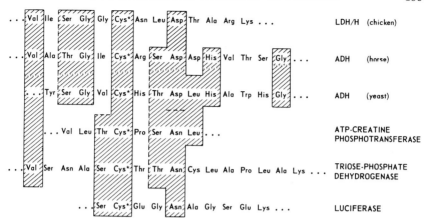

Fig. 16. Sequence homologies among peptides containing the active-center cysteines (Cys*) of several dehydrogenases and related enzymes.

selves, presumably products of duplicate (and, in the trout, genetically linked) genes. Three dehydrogenases from *Bacillus subtilis* have also been examined—alanine, malic, and lactate dehydrogenase (*117*). The first and second and the second and third are immunologically cross reactive; the first and third are weakly reactive. Peptide mapping indicates that the first and second have 4 identical peptide sequences, of which 2 are also present in the third. Thus, even without more direct evidence from sequences, it appears that these 3 enzymes are related gene duplication products.

Other proteins which appear to be the result of complete gene duplications are: the α_1- and α_2-chains of collagen; the protein hormones, secretin and glucacon; the enzymes, phosphorylase and UDP glucose: α-1,4-glucan α-4-glucosyl transferase, and fructose-diphosphate aldolase 1A and 1B; growth hormone and placental lactogen; and several anterior pituitary hormones (*20, 91*). The last include ACTH, α-MSH, β-MSH, and β-LPH (β-lipotropic hormone). LPH is a protein 90 residues in length which contains, within its sequence, all the sequence of human β-MSH, unchanged except for 2 amino acid substitutions (Fig. 11) (*75*). The posterior pituitary gland is also the site of origin of several hormones which appear to be the result of gene duplications. In mammals, the 2 principal hormones are the octapeptides (considering cystine as a single residue), oxytocin, and vasopressin. These hormones correspond, probably by simple evolution rather than by further duplications, to the mesotocin, vasotocin, isotocin, and glumitocin of lower vertebrates (Fig. 17) (*1*). In addition, 2 other posterior pituitary peptides have recently been identified. One is the β-corticotropin releasing factor (β-CRF), a cyclic

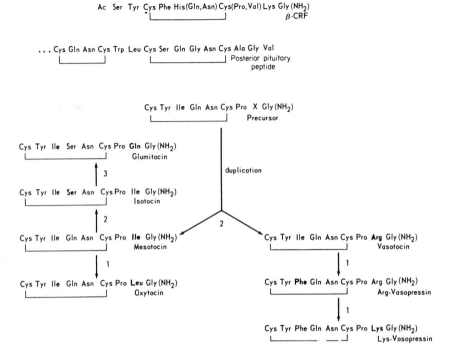

FIG. 17. Sequence homologies among polypeptide hormones from the posterior pituitary gland. The precursor peptide and the evolutionary scheme shown in the lower part of the figure are hypothetical: the numbers adjacent to the arrows indicate the minimum number of base changes required to convert the gene for one peptide hormone into the gene for another. The 2 peptides at the top of the figure are of interest because they also contain the internal disulfide ring structure (with an asparagine residue adjacent to the second cysteine) found in the peptide hormones.

peptide with the same type of ring structure as the others plus 3 additional residues. The other is an unnamed peptide, 48 residues in length, which also contains a similar ring structure and, like oxytoxin, vasopressin, and possibly β-CRF, contains an asparagine residue just preceding the second half cystine in the ring (Fig. 17) (93).

Interesting structural homologies have been found between hen egg white lysozyme and mammalian α-lactalbumin, a component of lactose synthetase (9). The former is an enzyme involved in carbohydrate chain degradation while the latter is involved in chain synthesis. If the 2 proteins are, in fact, evolutionarily related, it remains to be determined whether they are gene duplication products or result from the divergent evolution of the same gene in the 2 classes of vertebrates.

It has been proposed that the A- and B-chains of insulin also represent the products of gene duplication. If several deletions are postulated, it is possible to align the A- and B-chains so that 9 out of 17 residues are identical (*62*). However, this evidence must, at present, be considered as only suggestive, and it is much less firmly supported than the other examples that have been presented.

While they are not proteins, the demonstration that several soluble RNA's are related in structure provides further evidence for the generality of gene duplications (*61*). Common sequences have been found in base residues 64–72 (by the numbering of Jukes) in serine, tyrosine, and alanine sRNA's, and in residues 22–33 of the first 2 and, in part, the third. The 2 serine sRNA's of yeast differ by only 3 bases. Furthermore, if the sequences are compared with regard to their ability to form the complex secondary structure of helices and loops postulated for sRNA, all are compatible with a common structure despite gross differences in sequence. This type of result is similar to that obtained with the different globin chains which, although they differ greatly in sequence, still share a common tertiary structure (see below). It is likely that many, if not all, of the sRNA's may be evolutionarily related by gene duplications, and the same may also prove to be the case for the amino acid activating enzymes.

C. The Evolution and Synthesis of Immunoglobulins

To complete this survey of gene duplications, it is important to discuss the fascinating but complicated situation of the immunoglobulins. Two types of human L-chains are known and have been designated κ and λ. Analyses of 2 large peptides from the invariant portions of these chains have revealed extensive sequence homologies (Fig. 18), and these 2 types of chains appear to be the products of a gene duplication (*79*). Furthermore, a comparison of the carboxyl-terminal peptides of the κ and γ chains of human IgG indicates definite sequence homologies. Still another area of homology is present in the midportion of these chains (*100*). Even more impressive homologies are found when the invariant portions of the sequences of the Fc (carboxyl terminal) portion of the H-chain of rabbit IgG and of an L-chain Bence-Jones protein are compared. Because the H-chain is twice as long as the L-chain, the possibility has been raised that the former was derived from the latter by an almost complete gene duplication. If some deletions and substitutions are postulated, the same amino acid sequences are present in 2 and possibly even in 3 portions of the H-chain of rabbit IgG (Fig. 18) (*52*). The ag-

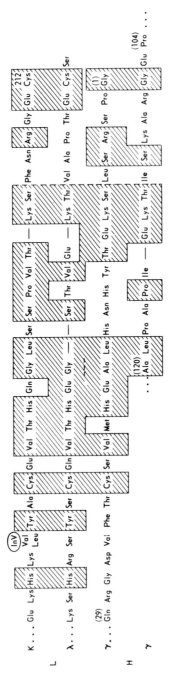

Fig. 18. Sequence homologies among the COOH-terminal portions of human κ and λ L-chains and the rabbit γ H-chain. The lowest line shows a possibly homologous peptide found within the sequence of the rabbit chain. These findings indicate a common origin of the genes for κ, λ, and γ chains and the possible occurrence of partial duplication in γ. Modified from Hill et al. (52).

gregate of the evidence is, therefore, compatible with the hypothesis that the H-chain was derived from the L-chain, with the major homologies in sequence being preserved in the invariant carboxyl terminal portions of the chains. Furthermore, it has also been suggested that the L-chain, with about 215 residues, is itself the product of the duplication of a gene originally coding for 110–120 residues (*93a*). In analogy with the β- and δ-chains of human hemoglobin and the diffuse (double) β-chains in the mouse, the finding of close genetic linkage among the H-chains of the mouse immunoglobulins γA, γG (γG-Be1), and γH (γG-Be2), suggests that the different H-chains are also gene duplication products (*92*). This conclusion is substantiated by the finding of several identical peptides in γG and γH.

The question of how many duplications must be postulated to explain the existence of the very large number of antibodies that may be evoked is one of the major problems of modern immunology. Present evidence indicates that antibodies with different specificities have different amino acid sequences (*68*), and it is possible that there will be as many sequences as there are different antigens. This number could be even higher since there appears to be a heterogeneity of antibodies with even a single known specificity. If it is assumed that H- and L-chains both contribute residues to the antibody combining site then, to have n different antibodies, the minimum number of different H- and L-chains required would be $\sqrt{n}$ for each chain. For $n = 10^4$, 100 H- and 100 L-chain sequences would be required for just 1 class of antibody. If, however, combination with antigen is principally mediated by only a single chain, the total number of different sequences required would be even greater. Considering the various different classes of antibodies (IgG, IgA, IgM, etc.), the total numbers, in either case, become much higher.

If each different amino acid sequence is coded for by a different gene, then a number of H- and L-chain genes commensurate with the number of sequences would be necessitated. However, there are certain facts which militate against the assumption of such a large number of independent structural genes. One is the close linkage of the mouse H-chain genes. While nothing quantitative is known about the relationship of linkage map distances or recombination frequencies to gene distances in terms of nucleotide bases, it does not seem likely that 100 or more of each type of H-chain genes could all be so closely linked. A second complication is that for all of the L-chain sequences studied, from immunoglobulins, myeloma proteins, and human Bence-Jones proteins, the carboxyl terminal half of the molecule is virtually invariant in sequence while the amino-terminal half exhibits considerable variability (Fig. 19) (*53, 100*). The same is probably true of H-chains. With the evidence

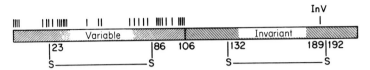

Fig. 19. Variable and invariant regions in the sequences of human κ L-chain, Bence-Jones proteins. The vertical lines indicate positions at which differences in sequence have been observed. InV indicates the position in the invariant region at which 2 different genetically controlled residues are found.

presently available, the variability in the amino-terminal portion appears to be the result of multiple, single-site mutations and not of larger chromosomal alterations. However, even the variable portions do not seem to have unlimited sequence possibilities. In the NH₂-terminal peptides of the L-chains of pooled or antigen-specific rabbit, mouse, and human IgG's and Bence-Jones proteins, only a few amino acids are found at each position in the sequence, and a few positions contain only a single residue (21). The nature of the latter appears to be different in different species. Yet, the invariant portions of the L-chains are not totally invariant. An amino acid substitution at position 189 (in Bence-Jones protein) has been identified and is thought to be responsible for the antigenic difference between L (κ)-chains which are Inv (a+) and Inv (a−). The former has a leucine at this position, while the latter has a valine, and this substitution is controlled by an autosomal locus with 2 or more alleles (53). A similar situation has been found at position 190 of human L (λ)-chains, producing the Oz locus (the nature of which is still uncertain), and has been claimed for the Gm antigen locus in human H (γ)-chains.

In the case of the Bence-Jones proteins, it is possible to ascertain how many kinds of L-chains a single type of neoplastic cell is making, and the answer has been 1 (100). Therefore, it must be assumed that only a single messenger for L-chains is being made, or, at least, is operative in these cells. For normal antibody-producing cells, the evidence is not clear whether a cell can make 1 or several different antibodies simultaneously.

Many hypotheses, in addition to that involving multiple duplicated H- and L-chain genes, have been advanced to explain the diverse findings outlined above (56, 92, 100, 100a), and they touch on issues of importance for both protein synthesis and protein evolution. One is that a single hypermutable L-chain gene exists and gives rise, by somatic point mutations, to many different genes in different precursors of antibody-producing cells. This notion has been objected to because it would appear to allow for a wider variety of substitutions than has actually been found and would not exclude mutations in the invariant portion of the chain. A second hypothesis is that there is again a single gene, but that certain base triplets in the portion of the gene corresponding to the NH$_2$-terminal part of the chain could be translated differently, but consistently, under different conditions. Why such translational variability, presumably related to changes in tRNA species or to similar mechanisms, should not affect the sequences of other proteins is not clear. The third possibility is that the variable and invariant portions of the L-chain are controlled by 2 separate genes or sets of genes (55). These genes could produce a single messenger RNA for both by a copy-choice form of transcription prior to polypeptide chain synthesis, or they could produce 2 messenger RNA's and, therefore, 2 polypeptide chains. The latter would then have to be coupled to form the whole L-chain. The gene corresponding to the variable NH$_2$-terminal portion of the molecule is visualized as having been duplicated many times, with the duplicates then having undergone divergent evolution. This hypothesis requires, therefore, that a large number of genes exist for the variable portion of the L-chain, but does not explain why certain residues do not show variability in sequence. It would be possible, of course, to visualize the existence of just 2 separate genes, with the one for the variable portion being, for some reason, hypermutable. Still another idea is that there are a limited number of genes which give rise to a wide variety of sequences by somatic recombination. All these hypotheses are also applicable to the H-chain system, and the tight linkage of the different H-chain genes could be explained by postulating separate genes for the 2 ends of the chain, with the genes for the invariant portions of the different H-chains being clustered together. At present, none of the hypotheses completely explains all the facts known about the various immunoglobulin chains, and further work will be required to prove or disprove them.

D. Convergent Evolution of Proteins

It has frequently been questioned whether proteins with different origins may evolve to states in which they are functionally or structurally similar. For the evolution of such proteins to be considered as having been truly convergent, it would have to be proven that their phylogenetic precursors were originally functionally unrelated and that they evolved independently to yield proteins with the same or similar functions. While it is likely that convergent evolution for function has occurred, it has not yet been determined whether convergence in activity has been accompanied by convergence of amino acid sequences. Many instances which were initially thought to represent examples of the latter have, on further investigation, turned out to represent the divergent evolution of proteins derived from duplicated genes; the several enzymes in the trypsin–chymotrypsin–elastase group illustrate this point well (50). However, instances are also known in which a given enzymic activity may be represented, although not necessarily in the same organism, by 2 or more groups of enzymes, each with distinct functional and physical properties, and the problem of deciding whether they are or are not evolutionarily related is a difficult one. An example of this difficulty is provided by the enzyme phosphoglucomutase which exists in 3 phylogenetic types—that found in vertebrates, that of *E. coli*, and that of *M. lysodeikticus* and *B. cereus* (47). The first and third types differ particularly with regard to their kinetic properties and to their ability to be phosphorylated, while the second has an intermediate type of behavior. Were it not for the fact that the same pentapeptide has been found in the active site of the *E. coli* and vertebrate enzymes, it might be postulated that the bacterial and vertebrate enzymes are not at all related. The sequence evidence suggests that they are. Whether the third group of enzymes is also related is not known.

Convergent evolution has been suggested to explain why bacterial and mold proteinases differ from the vertebrate proteinases in disulfide-bond content and in the amino acid sequences about the DFP-reactive serines that both groups possess, and to explain why fructose diphosphate aldolases from animals, protozoa, plants, and green algae differ so markedly in physical and catalytic properties from the aldolases of bacteria, yeast, fungi, and blue-green algae (20). While convergent evolution may in fact be the case, especially for the aldolases, full sequence data on the various groups of proteins will be necessary before a final decision can be made, and even then the decision may not be unequivocal (26).

VI. SELECTIVE FORCES IN PROTEIN EVOLUTION

All the examples used to illustrate the various aspects of protein evolution have been of proteins that have survived the evolutionary process. They represent, therefore, products of evolution more or less closely related to their ancient precursors which, for reasons that are mostly unclear to us, have been changed to and preserved in their present forms.

Some of the kinds of mutational events that may have occurred have already been discussed in detail. Others, such as gene inversions, translocations, and chromosomal accidents involving unrelated genes have probably also taken place, although no specific evidence from protein sequences has yet been obtained. The important fact is that proteins have evolved and continue to evolve, not by one mechanism but by a combination of these events. Thus, a complete duplication does not in itself result in the formation of a new protein, although it could, by allowing for more mRNA synthesis, alter and possibly even double the rate of synthesis of the original protein. New proteins do not arise until further changes, both large and small, have occurred in 1 and probably both the duplicate genes. Since the organism still possesses the original gene from which the duplications were derived, it is possible to tolerate major alterations in 1 of the duplicate genes while retaining the original function of the other (118).

A. The Randomness of Mutations and Nonrandomness of Amino Acid Replacements

It is generally believed that mutations are random in occurrence, both in time and in location. However, this statement must be qualified in 2 ways. First, it applies principally to point mutations and possibly to small additions and deletions, since, as has already been pointed out, the existence of 1 tandem gene duplication, whether partial or complete, will predispose to further duplications and deletions. Second, the randomness of even point mutations has not been proven. It has long been known that induced mutations in lower organisms such as bacteriophage may not be random, and that there may be mutational hot spots. With spontaneous mutations, it has been difficult to amass enough information to permit any definitive conclusions. However, some idea of the randomness of the mutational process may be derived from an examination of the human hemoglobin variants. Most of these are detected because of differences in electrophoretic mobility and mutations involving neutral hydrophobic residues in the interior of the molecule would thus be likely to be missed. It is assumed that selection has not been operative for the rarer mutations, although the mutations must, of course, have been compatible with synthesis of identifiable proteins. With these considerations in mind, a cursory examination of the sites at which amino acid replacements have been found indicates that they are distributed throughout the length of the chain (Fig. 20) (4). This observation is extended by the finding of differences in amino acid sequence at nearly every

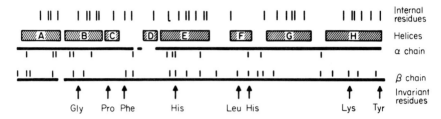

Fig. 20. Variants of human hemoglobins. The vertical lines in the α- and β-chains indicate positions at which single amino acid substitutions have been found. Those residues which are found in all the globins are shown at the bottom of the figure, while those present in the interior of the molecule are at the top. The helical portions of the chain are identified by the rectangles containing the letters used to designate the helices.

position in the naturally evolved hemoglobin chains (Fig. 20) (*89*), but the possible existence of mutational hot or cold spots still cannot be assessed. It is of interest that 2 mutations have been found at certain sites: the mutations causing hemoglobins S and C at β6, San Jose and Siriraj at β7, Mexico and Shimonoseki at α54, and Zurich and M Saskatoon at β63 (*4*). Furthermore, more than 1 mutation may be present in the same chain. Both situations may merely be the results of 2 random mutations occurring either at the same site or in the same gene. However, the former might also indicate the existence of a hot spot, while the latter could have resulted from a crossover between 2 allelic genes carrying different mutations. Thus the double-mutant hemoglobin C Harlem may represent either the product of a crossover between the genes for hemoglobin S (β6) and for a mutation at β73, the result of a second mutation in a hemoglobin S gene, or a mutation at β6 in a gene carrying the β73 substitution (*7*). The first possibility is strengthened by the recent finding of the mutant hemoglobin Korle-Bu, a hemoglobin with the same β73 (Asp → Asn) substitution as C Harlem (*67a*).

Assuming that single-site mutations are essentially random, how can one explain the observation that amino acid sequence changes occurring during protein evolution are often distinctly nonrandom in character? This nonrandomness may range from the occurrence of lengths of amino acid sequence that are completely or nearly completely unaltered to apparent restrictions in the kinds of amino acid residues that are found at specific positions. A well-known example of the former is cytochrome c which in addition to having 28 invariant residues (at the time of writing) has an invariant region encompassing residues 70–80—a total of 39 invariant residues in all. Of these, 9 are glycines and 7 are lysines (*77*). By contrast, no more than 8 residues are invariant in the sequences of normal globins and 2 of these are the heme-linked histidines which have been found to be altered in the mutant hemoglobins M. Nevertheless, even with such variability in sequence, the various globin chains and proteins appear to have similar, if not identical, conformations. As was noted earlier, this is probably related to the fact that certain regions of the molecule and hence of the amino acid sequence contain only hydrophobic groups (89).

B. The Conservative Nature of the Amino Acid Code

The factors that operate to limit the range of amino acids at specific sites may be divided into 2 groups—those that are inherent in the mutation process itself and those that are controlled by the processes of selection. Considering only random single-site nucleotide base substitutions, an examination of the amino acid "code" (Fig. 21) indicates that single-base mutations will not result in the random replacement of 1 amino acid by another (24). Rather, there will be a preferential substitution of an amino acid by itself about 25% of the time and by another amino acid with the same polarity about 40% of the time. If the former are eliminated from consideration because they would not be recognizable, then almost 60% of detectable single-site mutations would not change the polarity of the affected residue. For 2-step mutations, this value is closer to 50%. Furthermore, the arrangement of codons is such that single-site mutations would account for interchanges of such similar amino acids as serine and threonine; alanine and glycine; leucine, isoleucine, and valine; glutamic and aspartic acids; and tyrosine and phenylalanine. Expressed in terms of the polarity of the side chains, the amino acid code as presently constituted favors isopolar substitutions and the interconversion of structurally similar amino acids and thereby exerts a conservative force on the nature of the substitutions that are allowed.

C. The Evolution of the Code

Several writers have commented on the particular arrangement of the codons and have concluded that the code may have evolved to its present arrangement to minimize the deleterious effects of mutations and/ or of mistakes in codon transcription and translation (24, 42, 114). That mistakes, presumably in translation, do occur is well illustrated by the α-chain of rabbit hemoglobin in which more than 1 amino acid residue has been found at several sites in the molecule (108). Another proposal

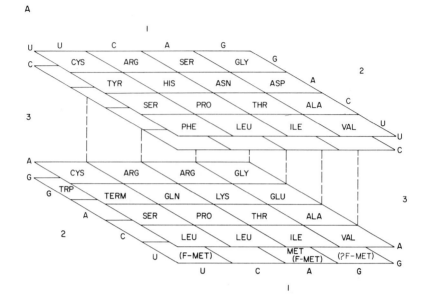

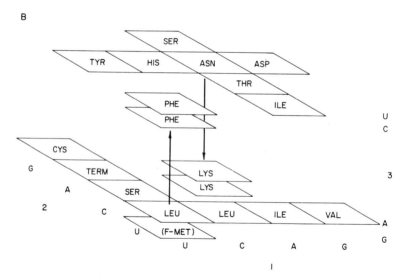

Fig. 21. (A) The general organization of the genetic code. The bases of each codon are represented on the 3 axes of the grid. Because of the high degree of degeneracy in the third position, the sets of planes representing the 2 pyrimidine bases and the 2 purine bases have been collapsed together (term = chain termination). (B) Possible outcomes of single-base changes in the codons for asparagine (upper right) and leucine (lower left). The high degree of preservation of polarity is apparent in each case.

that has been made about the origin of the code is that its present organization is the consequence of amino acid-codon assignments that were established when the number of different amino acids was fewer and the codons were perhaps of a 1- or 2-base type (16). According to this hypothesis, the expansion of the code from its primitive state was to some extent influenced by the need to minimize the deleterious effects of the introduction of new amino acids and codons. As a result, related amino acids acquired related codons. However, at some critical point in its evolution, the code is thought to have become locked into an organization which precluded further major alterations, and optimization for mutational and translational stability could not and did not occur. If enough were known about the actual effects of amino acid substitutions on proteins, it might be possible to determine whether the present code is or is not really optimal in these regards, but unfortunately, this decision cannot presently be made.

One other hypothesis for the basis of the codon assignments is that the codons are chemically related to the amino acid residues because of an actual amino acid-codon interaction (115). While such interactions are not believed to occur, it is conceivable that, sometime in the distant past, they did exist and, following the intervention of sRNA, the amino acid–codon relationships persisted.

D. Selection for Conformation and Other Properties of Proteins

Particular attention has been paid to the effects of mutations on the polarity of the amino acid residues because of the assumption, discussed earlier, that side-chain polarity is a critical factor in determining the conformations of proteins. If it is also assumed that preservation of the function of an evolving protein will require preservation of conformation, it can be predicted that maintenance of a specific polarity distribution of amino acid residues would be observed in homologous proteins. This certainly appears to be the situation in the case of the globins (25, 89) and probably also for several other proteins including subtilisin, tobacco mosaic virus protein, and pancreatic ribonuclease. It is of course conceivable, although probably unlikely, that major conformational changes could occur and be compatible with retention of function as long as the proper geometry were preserved in the regions required for activity.

Although it can be demonstrated that amino acid substitutions in homologous proteins are nonrandom with respect to polarity, it remains to be proven that this nonrandomness is the result of natural selection rather than merely being a consequence of the conservative organization of the code itself. As an approach to this problem, a numerical scale has been devised to express the average changes in polarity that result from sets of amino acid substitutions (25). These values could then be compared with the average changes to be expected from random unselected mutations, taking into account the organization of the amino acid code. Such analyses have been carried out for several of the globins and for pairs of cytochromes c, subtilisins, tobacco mosaic virus proteins, pancreatic ribonucleases, and trypsinogen and chymotrypsinogen, and the results indicate that the changes in polarity are even more conservative than the amino acid code itself would insure (25). When a distinction is made between interior and exterior residues, as can be done for the globlins, it is quite apparent that polarity is much more rigorously preserved for the former. Except for one instance (myoglobin in comparison with the hemoglobin β-chain), however, selection for polarity of the external residues is also demonstrable.

Aside from the folding aspects of conformation, it is possible to conceive of selection as having been directed to other properties of proteins, both conformational and nonconformational. It has been suggested, for example, that the covalent binding of carbohydrate to proteins, to form glycoproteins, was important for the development of secretory mechanisms for proteins which were to function extracellularly (31). Likewise, the evolution of multienzyme systems and of subcellular particles and organelles was dependent on changes in proteins which allowed them to interact with one another and with membranes (83). The development of multimeric proteins has resulted in several properties which can be considered desirable. Of these, probably the two most important are the allosteric control systems, already discussed, and complementation. The latter is a process in which conformational alterations in 1 subunit may be partially or completely corrected by interaction with a normal region in another subunit (13). By such a mechanism, an active enzyme could be synthesized by organisms that are heterozygous for 2 different mutant alleles that would produce an inactive enzyme if present in the homozygous state. In some instances it is even possible that the effects of a mutation present in the homozygous state would be ameliorated by complementation. Complementation may, therefore, represent a kind of mechanism for self-correction for proteins containing more than 1 subunit.

Evolutionary changes may also affect the synthesis and degradation of proteins. Aside from specific control systems for induction, repression, and related phenomena, the amount of a protein within a cell or in an extracellular compartment is dependent, to a first approximation, on the balance of these 2 rates. For proteins being made at a constant rate and degraded by a process that follows first-order kinetics, the steady state concentration of the protein will be equal to K_1/K_2, where K_1 and K_2 are the respective rate constants for synthesis and degradation (5). A change in either rate will affect the final concentration of the protein. For example, a decrease in the resistance of the protein to degradation, either because the insertion of an amino acid makes the protein susceptible to attack by proteases or because the formation of a less stable conformation, will result in an increase in K_2 and, therefore, in a decrease in the concentration of the protein. Lowered intracellular protein concentrations resulting from such mutationally induced decreases in stability have been observed for 2 human red blood cell enzymes, catalase and glucose-6-phosphate dehydrogenase (2, 66). These effects on degradation may, to a large extent, be considered as being conformational in nature. Changes of the rate of synthesis by mutations are, however, probably related more to translational mechanisms than to conformational effects. Several mutant human hemoglobins, including hemoglobins S (Glu → Val) and C (Glu → Lys), are found in lower than normal amounts in red blood cells. It has been proposed, although certainly not proven, that the decrease in the rate of synthesis occurs because the mutant codons are such that their combination with the appropriate sRNA's take place at a slower rate than usual—possibly because they require sRNA's that are present only in small amounts (59).

E. The Role of the Host in Protein Evolution

Although the preceding discussion has been framed in terms of proteins as the subjects of evolutionary changes, it must again be emphasized that the ultimate criterion for selection does not involve the proteins themselves. Rather, it is the effect that a change in a protein has on the ability of the host organism to reproduce that determines whether selection will act for or against the mutation that produced the change. While it is possible to detail the ways in which mutations affect the properties of protein molecules, an understanding of how changes in these properties affect the fertility of the organism is, except in the most obvious cases, almost completely lacking. This is particularly true for higher organisms and is well illustrated by 3 human mutations that have been extensively studied with regard to selection: sickle-cell hemoglobin (hemoglobin S), β-thalassemia (a deficiency in the production of hemoglobin β-chains), and deficiency of glucose-6-phosphate dehydrogenase (81). Despite the deleterious effects of these mutations on protein structure or synthesis, all have survived in man and are found with relatively high frequencies in some populations. The explanation that has been advanced is that these mutations, when present in the heterozygous state, protect their carriers against malaria. Therefore, the loss of the mutant genes by the death of homozygous or severely affected hemizygous individuals in the population, is counterbalanced by the increased fertility of those less severely affected. However, the evidence for the importance of malaria in preserving some or all of these mutations is not accepted by all workers. Furthermore, even if malaria is involved, nothing is actually known about how the changes in the proteins affect the susceptibility of their hosts to this disease, although glucose-6-phosphate dehydrogenase deficient erythrocytes in female heterozygotes have been claimed to have fewer malarial parasites than do normal erythrocytes (75a).

The existence in a population of 2 or more forms of a protein, the structures of which are determined by allelic genes, is referred to as polymorphism. Polymorphisms are distinguished from rare single mutations in that the frequencies of the variant alleles are higher than can be explained by simple mutation alone. Several systems of polymorphic human proteins are known in which the only discernible differences among the proteins are in electrophoretic mobility (48). These proteins include placental alkaline phosphatase, some glucose-6-phosphate dehydrogenase variants, and red blood cell phosphoglucomutase and adenylate kinase. Additional polymorphisms, involving differences in protein concentration or enzyme activity, have been found for red blood cell acid phosphatase, liver acetyl transferase, and serum pseudocholinesterase (48). Again, nothing is now known about the selective forces that maintain these polymorphisms, although it is possible that some of these variants represent selectively neutral mutations which became established by genetic drift (26, 65a).

Polymorphic systems such as those listed above are of considerable interest to students of evolution. They represent, as it were, evolution occurring in the present and thus provide a point of departure for the analysis of the mechanisms by which mutation and selection interact. Any insights that can be gained into the origin and maintenance of these polymorphisms will significantly advance our understanding of the processes of evolution.

REFERENCES

1. R. Acher, Evolutionary aspects of the structure of proteins. *Angew. Chem. Intern. Ed. Engl.* **5**, 798 (1966).

2. H. Aebi and M. Cantz, Über die cellüläre Verteilung der Katalase im Blut homozygoter und heterozygoter Defektträger (Akatalasie). *Humangenetik* **3**, 50 (1966).

3. C. B. Anfinsen, The tertiary structure of ribonuclease. *Brookhaven Symp. Biol.* **15**, 184 (1962).

4. D. Beale and H. Lehmann, Abnormal hemoglobins and the genetic code. *Nature* **207**, 259 (1965).

5. C. M. Berlin and R. T. Schimke, Influence of turnover rates on the responses of enzymes to cortisone. *Mol. Pharmacol.* **1**, 149 (1965).

6. C. C. F. Blake, G. A. Mair, A. C. T. North, D. C. Phillips, and V. R. Sarma, On the conformation of the hen egg-white lysozyme molecule. *Proc. Roy. Soc.* **B167**, 365 (1967).

6a. R. M. Bookchin and P. M. Gallop, Structure of hemoglobin A_{1c}: Nature of the N-terminal β chain blocking group. *Biochem. Biophys. Res. Commun.* **32**, 86 (1968).

7. R. M. Bookchin, R. L. Nagel, and H. M. Ranney, Structure and properties of hemoglobin C Harlem, a human hemoglobin variant with amino acid substitutions in two residues of the β-polypeptide chain. *J. Biol. Chem.* **242**, 248 (1967).

7a. T. B. Bradley, Jr., R. C. Wohl, and R. F. Rieder, Hemoglobin Gun Hill. Deletion of five amino acid residues and impaired heme binding. *Science* **157**, 1581 (1967).

8. G. Braunitzer, Phylogenetic variation in the primary structure of hemoglobins. *J. Cellular Physiol.* **67**, Suppl. 1, 1 (1966).

9. K. Brew, T. C. Vanaman, and R. L. Hill, Structural studies on α-lactalbumin, the B protein of lactose synthetase. *Federation Proc.* **26**, 724 (1967).

9a. J. R. Brown, D. L. Kauffman, and B. S. Hartley, The primary structure of porcine pancreatic elastase. The N-terminus and disulfide bridges. *Biochem. J.* **103**, 497 (1967).

9b. E. R. Burka, Hemin: An inhibitor of erythroid cell ribonuclease. *Science* **162**, 1287 (1968).

10. C. R. Cantor and T. H. Jukes, The repetition of homologous sequences in the polypeptide chains of certain cytochromes and globins. *Proc. Natl. Acad. Sci. U. S.* **56**, 177 (1966).

11. R. W. Carrell, H. Lehmann, and H. E. Hutchison, Haemoglobin Köln (β 98 valine $\rightarrow$ methionine): An unstable protein causing inclusion-body anemia. *Nature* **210**, 915 (1966).

11a. R. E. Chance, R. M. Ellis, and W. W. Bromer, Porcine proinsulin: Characterization and amino acid sequence. *Science* **161**, 165 (1968).

12. O. P. Chilson, G. B. Kitto, J. Pudles, and N. O. Kaplan, Reversible inactivation of dehydrogenases. *J. Biol. Chem.* **241**, 2431 (1966).

13. A. Coddington, J. R. S. Fincham, and T. K. Sundaram, Multiple active varieties of *Neurospora* glutamate dehydrogenase formed by hybridization between two inactive mutant proteins *in vivo* and *in vitro*. *J. Mol. Biol.* **17**, 503 (1966).

14. B. Colombo and C. Baglioni, Regulation of haemoglobin synthesis at the polysome level. *J. Mol. Biol.* **16**, 51 (1966).

15. F. H. C. Crick, On protein synthesis. *Symp. Soc. Exptl. Biol.* **13**, 138 (1958).

16. F. H. C. Crick, Origin of the genetic code. *Nature* **213**, 119 (1967).

17. M. J. Crumpton and A. Polson, A comparison of the conformation of sperm whale metmyoglobin with that of apomyoglobin. *J. Mol. Biol.* **11**, 722 (1965).

18. W. C. Deal and S. M. Constantinides, Possible control of glyceraldehyde-3-phosphate dehydrogenase through NAD⁺ requirement for folding. *Federation Proc.* **26**, 348 (1967).

19. B. de Crombrugghi and H. Edelhoch, The properties of thyroglobulin. XIV. The structure of reduced thyroglobulin. *Biochemistry* **5**, 2238 (1966).

20. G. H. Dixon, Mechanisms of protein evolution. *In* "Essays in Biochemistry" (P. N. Campbell and G. D. Greville, eds.), Vol. 2, pp. 148–204. Academic Press, New York, 1966.

21. R. F. Doolittle, The amino-terminal amino acid sequences of rabbit immunoglobulin light chains. *Proc. Natl. Acad. Sci. U. S.* **55**, 1195 (1966).

22. M. S. Doscher and F. M. Richards, The activity of an enzyme in the crystalline state. *J. Biol. Chem.* **238**, 2399 (1963).

23. R. V. Eck and M. O. Dayhoff, Evolution of the structure of ferredoxin based on surviving relics of primitive amino acid sequences. *Science* **152**, 363 (1966).

24. C. J. Epstein, Role of the amino-acid "code" and of selection for conformation in the evolution of proteins. *Nature* **210**, 25 (1966).

25. C. J. Epstein, Non-randomness of amino-acid exchanges in the evolution of homologous proteins. *Nature* **215**, 356 (1967).

26. C. J. Epstein, Structural and genetic aspects of protein evolution. *In* "Homologous Enzymes and Biochemical Evolution" (N. V. Thoai and J. Roche, eds.), pp. xxi–xlix. Gordon and Breach, New York, 1968.

27. C. J. Epstein and C. B. Anfinsen, The reversible reduction of disulfide bonds in trypsin and ribonuclease coupled to carboxymethyl cellulose. *J. Biol. Chem.* **237**, 2175 (1962).

28. C. J. Epstein, M. M. Carter, and R. F. Goldberger, Reversible denaturation of rabbit-muscle lactate dehydrogenase. *Biochim. Biophys. Acta* **92**, 391 (1964).

29. C. J. Epstein, R. F. Goldberger, and C. B. Anfinsen, The genetic control of tertiary protein structure: Studies with model systems. *Cold Spring Harbor Symp. Quant. Biol.* **28**, 439 (1963).

30. C. J. Epstein and A. N. Schechter, An approach to the problem of conformational isozymes. *Ann. N. Y. Acad. Sci.* **151**, 85 (1968).

31. E. H. Eylar, On the biological role of glycoproteins. *J. Theoret. Biol.* **10**, 89 (1965).

32. L. Felicetti, B. Colombo, and C. Baglioni, Assembly of hemoglobin. *Biochim. Biophys. Acta* **129**, 380 (1966).

33. W. Ferdinand, W. H. Stein, and S. Moore, An unusual disulfide bond in streptococcal proteinase. *J. Biol. Chem.* **240**, 1150 (1965).

34. H. F. Fisher, A limiting law relating the size and shape of protein molecules to their composition. *Proc. Natl. Acad. Sci. U. S.* **51**, 1285 (1964).

35. W. M. Fitch, Evidence suggesting a partial internal duplication in the ancestral gene for heme-containing globins. *J. Mol. Biol.* **16**, 17 (1966).

36. T. P. Fondy, J. Everse, G. A. Driscoll, F. Castillo, F. E. Stolzenbach, and N. O. Kaplan, The comparative enzymology of lactic dehydrogenase. IV. Function of sulfhydryl groups in lactic dehydrogenases and the sequence around the essential group. *J. Biol. Chem.* **240**, 4219 (1965).

37. M. H. Freedman and M. Sela, Recovery of specific activity upon reoxidation of completely reduced polyalanyl rabbit antibody. *J. Biol. Chem.* **241**, 5225 (1966).

38. M. E. Friedman, H. A. Scheraga, and R. F. Goldberger, Structural studies on

ribonuclease. XXVI. The role of tyrosine 115 in the refolding of ribonuclease. *Biochemistry* **5,** 3770 (1966).

39. T. Friedmann and C. J. Epstein, The role of calcium in the reactivation of reduced Taka-amylase. *J. Biol. Chem.* **242,** 5131 (1967).

40. J. C. Gerhart and H. K. Schachman, Distinct subunits for the regulation and catalytic activity of aspartate transcarbamylase. *Biochemistry* **4,** 1054 (1965).

41. D. Givol, F. de Lorenzo, R. F. Goldberger, and C. B. Anfinsen, Disulfide interchange and the three-dimensional structure of proteins. *Proc. Natl. Acad. Sci. U. S.* **53,** 676 (1965).

42. A. L. Goldberg and R. E. Wittes, Genetic code: Aspects of organization. *Science* **153,** 420 (1966).

43. A. Gottschalk and E. R. B. Graham, The basic structure of glycoproteins. *In* "The Proteins" (H. Neurath, ed.), 2nd ed., Vol. 4, pp. 95–151. Academic Press, New York, 1966.

44. E. E. Grebner, C. W. Hall, and E. F. Neufeld, Glycosylation of serine residues by a UDP-xylose:protein xylosyltransferase from mouse mastocytoma. *Arch. Biochem. Biophys.* **116,** 391 (1966).

45. E. Haber, Recovery of antigenic activity after denaturation and complete reduction of disulfides in a papain fragment of antibody. *Proc. Natl. Acad. Sci. U. S.* **52,** 1099 (1964).

46. E. Haber and C. B. Anfinsen, Side chain interactions governing the pairing of half-cystine residues in ribonuclease. *J. Biol. Chem.* **237,** 1839 (1962).

47. P. Handler, T. Hashimoto, J. G. Joshi, H. Dougherty, K. Hanabusa, and C. del Río, Phosphoglucomutase—evolution of an enzyme. *Israel J. Med. Sci.* **1,** 1173 (1965).

48. H. Harris, D. A. Hopkinson, and J. Luffman, Enzyme diversity in human populations. *Ann. N. Y. Acad. Sci.* **151,** 232 (1968).

49. S. C. Harrison and E. R. Blout, (1965). Reversible conformational changes of myoglobin and apomyoglobin. *J. Biol. Chem.* **240,** 299 (1965).

50. B. S. Hartley, J. R. Brown, D. L. Kauffman, and L. B. Smillie, Evolutionary similarities between pancreatic proteolytic enzymes. *Nature* **207,** 1157 (1965).

51. D. R. Helinski and C. Yanofsky, Genetic control of protein structure. *In* "The Proteins" (H. Neurath, ed.), 2nd ed., Vol. 4, pp. 1–93. Academic Press, New York, 1966.

52. R. L. Hill, K. Delaney, R. E. Fellows, Jr., and H. E. Lebovitz, The evolutionary origins of the immunoglobulins. *Proc. Natl. Acad. Sci. U. S.* **56,** 1762 (1966).

53. N. Hilschmann and L. C. Craig, Amino acid sequence studies with Bence-Jones proteins. *Proc. Natl. Acad. Sci. U. S.* **53,** 1403 (1965).

54. W. R. Holmquist and W. A. Schroeder, A new N-terminal blocking group involving a Schiff base in hemoglobin A1c. *Biochemistry* **5,** 2489 (1966).

55. L. Hood and D. Ein, Immunoglobulin lambda chain structure: Two genes, one polypeptide chain. *Nature* **220,** 764 (1968).

56. L. E. Hood, W. R. Gray, and W. J. Dreyer, On the mechanism of antibody synthesis: A species comparison of L-chains. *Proc. Natl. Acad. Sci. U. S.* **55,** 826 (1966).

57. E. R. Huehns and E. M. Shooter, The properties and reactions of haemoglobin F and their bearing on the dissociation equilibrium of haemoglobin. *Biochem. J.* **101,** 852 (1966).

58. V. M. Ingram, Gene evolution and the haemoglobins. *Nature* **189,** 704 (1961).

59. H. A. Itano, Genetic regulation of peptide synthesis in hemoglobins. *J. Cellular Physiol.* **67**, Suppl. 1, 65 (1966).

60. R. T. Jones, B. Brimhall, T. H. J. Huisman, E. Kleihauer, and K. Betke, Hemoglobin Freiburg: Abnormal hemoglobin due to a deletion of a single amino acid residue. *Science* **154**, 1024 (1966).

61. T. H. Jukes, Indications for a common evolutionary origin shown in the primary structure of three transfer RNAs. *Biochem. Biophys. Res. Commun.* **24**, 744 (1966).

62. T. H. Jukes, "Molecules and Evolution." Columbia Univ. Press, New York, 1966.

63. G. Kartha, J. Bello, and D. Harker, Tertiary structure of ribonuclease. *Nature* **213**, 862 (1967).

64. P. G. Katsoyannis, Synthesis of insulin. *Science* **154**, 1509 (1966).

65. W. Kauzmann, Some factors in the interpretation of protein denaturation. *Advan. Protein Chem.* **14**, 1 (1959).

65a. J. L. King and T. H. Jukes, Non-Darwinian evolution. *Science* **164**, 788 (1969).

66. H. N. Kirkman, P. R. McCurdy, and J. L. Naiman, Functionally abnormal glucose-6-phosphate dehydrogenases. *Cold Spring Harbor Symp. Quant. Biol.* **29**, 391 (1964).

67. G. B. Kitto, P. M. Wassarman, and N. O. Kaplan, Enzymatically active conformers of mitochondrial malate dehydrogenase. *Proc. Natl. Acad. Sci. U. S.* **56**, 578 (1966).

67a. F. I. D. Konotey-Ahulu, E. Gallo, H. Lehmann, and B. Ringelhann, Haemoglobin Korle-Bu (β73 aspartic acid $\rightarrow$ asparagine). *J. Med. Genetics* **5**, 107 (1968).

68. M. E. Koshland, Primary structure of immunoglobulins and its relationship to antibody specificity. *J. Cellular Physiol.* **67**, Suppl. 1, 33 (1966).

69. L. M. Kraus, T. Miyaji, I. Iuchi, and A. P. Kraus, Characterization of $\alpha^{23\ Glu\ NH_2}$ in hemoglobin Memphis. Hemoglobin Memphis/S, a new variant of molecular disease. *Biochemistry* **5**, 3701 (1966).

70. Y. Kuriyama, Enzymatic activity of polyalanyl ribonuclease T_1 and polyglutamyl ribonuclease T_1. *J. Biochem. (Tokyo)* **59**, 596 (1966).

71. D. Labie, W. A. Schroeder, and T. H. J. Huisman, The amino acid sequence of the δ-β chains of hemoglobin Lepore$_{Augusta}$ = Lepore$_{Washington}$. *Biochim. Biophys. Acta* **127**, 428 (1966).

72. S. J. Leach, G. Némethy, and H. A. Scheraga, Computation of the sterically allowed conformations of peptides. *Biopolymers* **4**, 369 (1966).

73. C. Levinthal, Are there pathways for protein folding? *J. Chim. Physi.* **65**, 44 (1968).

74. R. D. Levere and S. Granick, Control of hemoglobin synthesis in the cultured chick blastoderm. *J. Biol. Chem.* **242**, 1903 (1967).

75. C. H. Li, L. Barnafi, M. Chrétien, and D. Chung, Isolation and amino acid sequence of β-LPH from sheep pituitary glands. *Nature* **208**, 1093 (1965).

75a. L. Luzzatto, E. A. Usanga, and S. Reddy, Glucose-6-phosphate dehydrogenase deficient red cells: Resistance to infection by malarial parasites. *Science* **164**, 839 (1969).

76. G. Marchis-Mouren and F. Lipmann, On the mechanism of acetylation of fetal and chick hemoglobins. *Proc. Natl. Acad. Sci. U. S.* **53**, 1147 (1965).

77. E. Margoliash and A. Schejter, Cytochrome *c*. *Advan. Protein Chem.* **21**, 113 (1966).

78. C. Milstein, Immunoglobulin κ-chains. Comparative sequences in selected stretches of Bence-Jones proteins. *Biochem. J.* **101**, 352 (1966).

79. C. Milstein, Comparative peptide sequences of kappa and lambda chains of human immunoglobulins. *J. Mol. Biol.* **21**, 203 (1966).

80. J. Monod, J. Wyman, and J.-P. Changeux, On the nature of allosteric transitions. *J. Mol. Biol.* **12**, 88 (1965).

81. A. G. Motulsky, Hereditary red cell traits and malaria. *Am. J. Trop. Med. Hyg.* **13**, Part 2, 147 (1964).

82. H. Muirhead and M. F. Perutz, Structure of reduced human hemoglobin. *Cold Spring Harbor Symp. Quant. Biol.* **28**, 451 (1963).

83. K. D. Munkres and D. O. Woodward, On the genetics of enzyme locational specificity. *Proc. Natl. Acad. Sci. U. S.* **55**, 1217 (1966).

84. M. Murayama, Tertiary structure of sickle cell hemoglobin and its functional significance. *J. Cellular Physiol.* **67**, Suppl. 1, 21 (1966).

85. H. Neurath, Mechanism of zymogen activation. *Federation Proc.* **23**, 1 (1964).

85a. H. Neurath, K. A. Walsh, and W. P. Winter, Evolution of structure and function of proteases. *Science* **158**, 1638 (1967).

86. Y. Okada, E. Terzaghi, G. Streisinger, J. Emrich, M. Inouye, and A. Tsugita, A frame-shift mutation involving the addition of two base change of pairs in the lysozyme gene of phage T4. *Proc. Natl. Acad. Sci. U. S.* **56**, 1692 (1966).

87. P. F. Periti, G. Quagliarotti, and A. M. Liquori, Recognition of α-helical segments in proteins of known primary structure. *J. Mol. Biol.* **24**, 313 (1967).

88. M. F. Perutz, Structure and function of haemoglobin. I. A tentative atomic model of horse oxyhaemoglobin. *J. Mol. Biol.* **13**, 646 (1965).

89. M. F. Perutz, J. C. Kendrew, and H. C. Watson, Structure and function of haemoglobin. II. Some relations between polypeptide chain configuration and amino acid sequence. *J. Mol. Biol.* **13**, 669 (1965).

90. D. C. Phillips, The hen egg-white lysozyme molecule. *Proc. Natl. Acad. Sci. U. S.* **57**, 484 (1967).

91. K. A. Piez, G. R. Martin, A. H. Kang, and P. Bornstein, Heterogeneity of the α chains of rat skin collagen and its relation to the biosynthesis of cross-links. *Biochemistry* **5**, 3813 (1966).

92. M. Potter, R. Lieberman, and S. Dray, Isoantibodies specific for myeloma γG and γH immunoglobulins of BALB/c mice. *J. Mol. Biol.* **16**, 334 (1966).

93. E. C. Preddie, Structure of a large polypeptide of bovine posterior pituitary tissue. *J. Biol. Chem.* **240**, 4194 (1965).

93a. F. W. Putnam, Immunoglobulin structure: Variability and homology. *Science* **163**, 633 (1969).

93b. H. M. Ranney, A. S. Jacobs, L. Uden, and R. Zalusky, Hemoglobin Riverdale-Bronx an unstable hemoglobin resulting from the substitution of arginine for glycine at helical residue B6 of the β polypeptide chain. *Biochem. Biophys. Res. Commun.* **33**, 1004 (1968).

94. L. J. Reed and D. J. Cox, Macromolecular organization of enzyme systems. *Ann. Rev. Biochem.* **35**, 57 (1966).

95. S. Sano and K. Tanaka, Recombination of protoporphyrinogen with cytochrome c apoproteins. *J. Biol. Chem.* **239**, PC3109 (1964).

96. A. M. Scanu, Factors affecting lipoprotein metabolism. *Advan. Lipid Res.* **3**, 63 (1965).

97. A. N. Schechter and C. J. Epstein, Spectral studies on the denaturation of myoglobin. *J. Mol. Biol.* **35**, 567 (1968).

98. M. Sela, F. H. White, Jr., and C. B. Anfinsen, Reductive cleavage of disulfide bridges in ribonuclease. *Science* **125**, 691 (1957).

98a. J. R. Shaefer, P. K. Trostle, and R. F. Evans, Rabbit hemoglobin biosynthesis: Use of human hemoglobin chains to study molecule completion. *Science* **158**, 488 (1967).

99. A. L. Shapiro, M. D. Scharff, J. V. Maizel, Jr., and J. W. Uhr, Polyribosomal synthesis and assembly of the H and L chains of gamma globulin. *Proc. Natl. Acad. Sci. U. S.* **56**, 216 (1966).

100. S. J. Singer and R. F. Doolittle, Antibody active sites and immunoglobulin molecules. *Science* **153**, 13 (1966).

100a. O. Smithies, Antibody variability. *Science* **157**, 267 (1967).

101. D. F. Steiner and J. L. Clark, The spontaneous reoxidation of reduced beef and rat proinsulins. *Proc. Natl. Acad. Sci. U. S.* **60**, 622 (1968).

102. T. Takagi and T. Isemura, Necessity of calcium for the renaturation of reduced Taka-amylase A. *J. Biochem. (Tokyo)* **57**, 89 (1965).

103. C. Tanford, K. Kawahara, and S. Lapanji, Proteins in 6 M guanidine hydrochloride. Demonstration of random coil behavior. *J. Biol. Chem.* **241**, 1921 (1966).

104. A. Telser, H. C. Robinson, and A. Dorfman, The biosynthesis of chondroitin-sulfate protein complex. *Proc. Natl. Acad. Sci.* **54**, 912 (1965).

105. J. Travis and W. D. McElroy, Isolation and sequence of an essential sulfhydryl peptide at the active site of firefly luciferase. *Biochemistry* **5**, 2170 (1966).

106. S. Udenfriend, Formation of hydroxyproline in collagen. *Science* **152**, 1335 (1966).

107. D. W. Urry and H. Eyring, The role of the physical sciences in biomedical research. *Perspectives Biol. Med.* **9**, 450 (1966).

108. G. von Ehrenstein, Translational variations in the amino acid sequence of the α-chain of rabbit hemoglobin. *Cold Spring Harbor Symp. Quant. Biol.* **31**, 705 (1966).

109. H. S. Waxman and M. Rabinowitz, Control of reticulocyte polyribosome content and hemoglobin synthesis by heme. *Biochim. Biophys. Acta* **129**, 369 (1966).

110. F. H. White, Jr., Regeneration of native secondary and tertiary structures by air oxidation of reduced ribonuclease. *J. Biol. Chem.* **236**, 1353 (1961).

111. P. L. Whitney and C. Tanford, Recovery of specific activity after complete unfolding and reduction of an antibody fragment. *Proc. Natl. Acad. Sci. U. S.* **53**, 524 (1965).

112. J. M. Wilkinson, E. M. Press, and R. R. Porter, The N-terminal sequence of the heavy chain of rabbit immunoglobulin IgG. *Biochem. J.* **100**, 303 (1966).

113. R. M. Winslow and V. M. Ingram, Peptide chain synthesis of human hemoglobins A and A₂. *J. Biol. Chem.* **241**, 1144 (1966).

114. C. R. Woese, On the evolution of the genetic code. *Proc. Natl. Acad. Sci. U. S.* **54**, 1546 (1965).

115. C. R. Woese, D. H. Dugre, S. A. Dugre, M. Kondo, and W. C. Saxinger, The fundamental nature and evolution of the genetic code. *Cold Spring Harbor Symp. Quant. Biol.* **31**, 723 (1966).

116. J. A. Yankeelov, Jr. and D. E. Koshland, Jr., Evidence for conformational changes induced by substrates of phosphoglucomutase. *J. Biol. Chem.* **240**, 1593 (1965).

117. A. Yoshida, Structural and serological similarity of three dehydrogenases of Bacillus subtilis. *Biochim. Biophys. Acta* **105**, 70 (1965).

118. E. Zuckerkandl and L. Pauling, Molecular disease, evolution, and genic heterogeneity. *In* "Horizons in Biochemistry" (M. Kasha and B. Pullman, eds.), pp. 189–225. Academic Press, New York, 1962.

GENERAL REFERENCES

A discussion of the chemistry and biology of immunoglobulins. *Proc. Roy. Soc.* **B166**, 113–243 (1966).

C. B. Anfinsen, "The Molecular Basis of Evolution." Wiley, New York, 1959.

V. Bryson and H. J. Vogel, eds., "Evolving Genes and Proteins." Academic Press, New York, 1965.

M. O. Dayhoff and R. V. Eck, "Atlas of Protein Sequence and Structure 1969." Natl. Biomed. Res. Founda., Silver Springs, Maryland, 1969.

C. J. Epstein, and A. G. Motulsky, Evolutionary origins of human proteins. *Progr. Med. Genet.* **4**, 85–127 (1965).

J. R. S. Fincham, "Genetic Complementation." Benjamin, New York, 1966.

W. M. Fitch and E. Margoliash, Construction of phylogenetic trees. *Science* **155**, 279 (1967).

C. Nolan and E. Margoliash, Comparative aspects of primary structures of proteins. *Ann. Rev. Biochem.* **37**, 727 (1968).

V. M. Sarich and A. C. Wilson, Immunological time scale for hominid evolution. *Science* **158**, 1200 (1967).

Symposium on differentiation and growth of hemoglobin- and immunoglobin-synthesizing cells. *J. Cellular Physiol.* **67**, Suppl. 1 (1966).

Author Index

Numbers in parentheses are reference numbers and indicate that an author's work is referred to although his name is not cited in the text. Numbers in italics show the page on which the complete reference is listed.

A

Abe, M., 56(1), *118*

Abelson, I., 203(1a), *204*

Abelson, J. N., 23(40), *38*, 149(67), 150 (67), 158(67), 161(67), 180(98, 150), 202(150), 203(67, 98, 150), *207, 209, 212*

Acher, R., 409(1), *424*

Acs, G., 190(187), *214*, 307(99), *365*

Adams, A., 144(1, 57), 172(1, 57), *204*

Adams, J. M., 192(2), *204*

Adloff, E., 277(26), *359*

Aebi, H., 423(2), *424*

Afanaseva, T. P., 96(198), 103(198), 116 (198), *127*

Akinrimisi, E. O., 109(2), 111(2), *118*

Alcock-Howden, J., 257(82), 309(82), *363*

Allende, C. C., 188(3), 195(3), *204*

Allende, J. E., 188(3), 195(3), *204*

Allfrey, V. G., 111(3), *118*, 253(52), 261 (1), 262(1), 264, 265(1), 266(8, 75), 267(75), 268(1, 69), 272(53), 273(29), 275(70), 276(61, 70, 74), 277(61), 278 (61), 279, 280(61a), 281(61), 282(61), 283(7), 291(4, 21), 292(11), 295(11), 296(11), 297(5), 299(9, 11, 70), 300 (6), 301(6), 302(6), 303(6), 304(6, 69), 307(2), 308(2, 11, 28), 309(9, 28, 75), 310(2), 311(2), 312(2), 313(9, 11, 85), 314(9, 11, 85), 318, 322, 323(8), 324(10, 21c), 332(76), 334(76), 335 (8), 336(74), 337(52), 338(52), 340 (53), 341(3, 9, 10), 343(3), 344(32a), 345(3), 347(3), 348(3, 32a), 350(29, 47a, 47b), 351(47c), *358, 359, 360, 361, 362, 363, 364*

Alpers, D. H., 26(1), 30(1), *36*, 87(5, 6), *118*

Ames, B. N., 25, 29(2), 31, 33(3, 78, 83), *36, 39, 40*, 90(7), *118*, 181(148), 186 (141), 187(141), *211, 212*, 234(1), *241*

Anders, M., 100(104), 106(166), *123, 126*, 182(87, 88), *208, 209*

Anderson, W. F., 234(2, 64a), 240(2), *241, 244*

Andoh, T., 202(4), *204*

Anfinsen, C. B., 370(3, 29), 372(29, 98), 373(27, 29, 41, 46), 376(29, 41), 377 (41), 382(29), 387(29), 395(3), *425, 426, 427, 430,' 431*

Anthony, D. D., 102(8), 103(8), *118*

Apgar, J., 145(6, 44), 158(5, 84), 165 (84), 166(5, 6, 44), *204, 206*

Apirion, D., 24, 25, *36*, 80(236), *129*

Aposhian, H. V., 62(9), 64(8a, 9, 297), 71(297), *118, 132*

Appelman, A. W. M., 287(48), *361*

Arias, I. M., 267(99a), 298(99a), 308(99a), 309(99a), 312(99a), *365*

Asano, K., 81(257), 108(10), *118, 130*

Ascione, R., 144(57), 172(57), *207*

Astbury, W. T., 215, *241*

Astrachan, L., 217, *245*

Attardi, B., 81(13), *118*

Attardi, G., 30(6), 32(6), *36*, 77(12), 78 (12, 130), 79, 81(12, 13), 82(12), 83 (130), 90(11), *118, 124*, 269(41), 339 (41), *361*

Atwood, K. C., 180(140), *211*

August, J. T., 65(164), 106(164), *126*, 315(43), 323, 332(43), 333, *361*

433

444 AUTHOR INDEX

Kurland, C. G., 77(213), 78(131), 83(131), *124, 128*

L

Labie, D., 404(71), *428*
Lagerquist, U., 157(16, 96), 184(94), 196(95), *205, 209*
Lake, J. A., 170(97), *209*
Lambert, L., 58(89), *122*
Lamberti, A., 111(81), *121*
Landy, A., 23(40), *38,* 149(67), 150(67), 158(67), 161(67), 180(98), 203(1a, 67, 98), *204, 207, 209*
Lang, D., 67(204), *128*
Langan, T. A., 350(50b), *361*
Langendorf, H., 304(89), *364*
Langridge, R., 252, *130*
Lanka, E., 202(168, 169), *213,* 231(94), *246*
Lapanji, S., 372(103), *430*
Lark, C., 46(214), *128*
Lark, K. G., 46(215), 50(216), 51, 52, 79, *122, 128*
Last, J. A., 232(55), *244*
Laszlo, J., 257(60), 265(60), 267(60), 299(60), 305(60), 306(60), 309(60), 313(60), *362*
Latham, H., 328(83), 329(83), *364*
Lavallée, R., 86(218), *128*
Lazzarini, R. A., 80(374), *136*
Leach, S. J., 385(72), *428*
Lebovitz, H. E., 411(52), 412(52), *425*
Lebowitz, J., 50(360), 110(360), *136*
Lebowitz, P., 149(99), *209*
Leder, P., 83(327), *134,* 221(65), 222, *244*
Lederberg, J., 58(106), *123*
Lehman, I. R., 62(80), 68(294), 71(294), *121, 132*
Lehmann, H., 391(11), 417(4), 418(4, 67a), *425, 428*
Lehrer, G. M., 268(51), 282, *361*
Leive, L., 83(219), 85(219, 220), 86(219, 220), 87, *128*
Lembach, K. J., 117(221), *128*
Leng, M., 109(222), 111(222), *129*
Levere, R. D., 393(74), *428*
Levin, J. G., 31(15), *36,* 112(31, 50), *119, 120,* 228(11, 18), 230(18), 234(64a), *241, 244*
Levinthal, C., 77(313), 80(391), 83(225,

313), 84(313), 85(225, 313, 391), 86(96, 223, 225, 391), 88(223), 96(224), *122, 129, 133, 137,* 233(57), *244,* 370(73), *428*
Lewis, E. B., 14(64), *39*
Lezzi, M., 295, *361*
Li, C. H., 409(75), *428*
Li, L., 158(11), 166(12), *204*
Liau, M. C., 330(51b), *362*
Lieberman, R., 413(92), 414(92), *429*
Lielausis, A., 62(93), 96(93), 116(93), *122*
Lindahl, T., 76(226), *129,* 144(1, 57), 172(1, 57), *204*
Lindblow, C., 170(54), *207*
Lipmann, F., 140(36), 141(36), 190(187), 199(14, 166), *204, 206, 213, 214,* 221(57a), 233(91), *244, 245,* 307(99), *365,* 396(76), *428*
Lipschitz, R., 111(68), *121*
Lipsett, M. N., 151(100, 101), *209*
Liquori, A. M., 385(87), *429*
Littau, V. C., 253(52), 266(75), 267(75), 272(53), 297(5), 309(75), 332(76), 334(76), 337(52), 338(52), 340, *358, 362, 363*
Littau, U. Z., 97(73), *121,* 150(102), 183(102, 138), *209, 211*
Littlefield, J. W., 47(61), *120*
Live, T. R., 63(295), *132*
Loebel, J. E., 200(91), *209,* 225(33, 51), 229(51), *242, 243*
Loening, U., 310(96a), *365*
Loftfield, R. B., 195(103, 104), *209*
Lohrmann, R., 200(153), *212,* 225(78, 79), *245*
Loken, M. R., *128*
London, I. M., 82(316), *133*
Lovlie, A., 298(93), 331(93), *365*
Lowenstein, W. R., 258(54), 334(54), *362*
Lucas, Z. J., 64(126), 66(126), *124*
Lucy, J. A., 111(227), *129*
Luffman, J., 424(48), *427*
Lunt, M. R., 104(49), *120*
Luria, S. E., 90(150), 103(228), *125, 129*
Luvin, S. E., 32(47), *38*
Luzzati, D., 116(206), *128*
Luzzatto, L., 424(75a), *428*

M

Maaløe, O., 51, 75(231), 76(231), *129, 133*

Roth, J. R., 33(78), *39*, 186(141), 187 (141), *211*

Roth, T. F., 50(309), *133*

Rottman, F. M., 222(10), 230, 231(71), 234(64a), *241, 244*

Rouiller, C., 265(17), *359*

Rouvière, J., 30(6), 32(6), *36*, 88(265), 90(11), *118, 131*

Rozencwajg, R., *361*

Rushizky, G. W., 160(142), *211*

Russell, R., 203(1a), *204*

Ruttenberg, G. J. C. M., 12(63), *39*, 50 (359), *136*

Rymo, L., 196(95), *209*

Ryter, A., 58(310), *133*

S

Sabatini, D., 310(81), *364*

Sadowski, P. D., 257(82), 309(82), *364*

Sagik, B. P., 81(311), *133*

St. Lawrence, P., 238, *246*

Sakabe, K., 58(274), 59(273, 274), 62 (274), *131*

Salas, M., 232(55), *244*

Salser, W. A., 62(32), 77(313), 83(313), 84, 85(313), 97(32, 312), 99(32), 116 (312), *119, 133*

Sanchez, C., 26(54), *38*

Sandeen, G., 109(98), *122*

Sanger, F., 77(46a), *120*, 160(143), 161 (27, 143), 164(27), 166(10a), 170 (10a), 192(111), *204, 210, 212*, 229 (58), *244*

Sano, S., 397(95), *429*

Sarabhai, A. S., 16, 19(79), *39*, 231, *245*

Sarich, V. M., *431*

Sarid, S., 97(73), *121*

Sarin, P. S., 170(144), 172(145), 190(146), *212*

Sarma, V. R., 368(6), 387(6), 395(6), *425*

Sarnat, M. T., 82(357), 103(111, 112), 107(111, 112), 115(64, 111, 112, 357), 116(111), *121, 123, 135*

Sato, K., 26(51), *38*, 82(168, 169), 90 (168, 169), 94(169), 115(168, 169), *126*

Saxinger, W. C., 421(115), *430*

Scaife, J., 114(173), *126*

Scanu, A. M., 397(96), *429*

Schachman, H. K., 369(40), *427*

Schaechter, M., 51(314), 58(85), 83(315), 86(315), *122, 133*

Schandl, E., 333(92c), 355(92c), 357 (92c), *364*

Scharff, M. D., 393(99), *430*

Schechter, A. N., 374(30), 375(30), 376 (30), 382(97), *426, 429*

Schejter, A., 418(77), *428*

Scheraga, H. A., 385(72), 389(38), *426, 428*

Scherberg, N. H., 97(365a), *136*

Scherrer, K., 82(316), *133*, 328, 329(83), *364*

Schildkraut, C., 64(297), 67(172, 296, 317), 68(317), 71(297), *126, 133*, 267 (56), 269(56), 260(56), 339(56), *362*

Schimke, R. T., 423(5), *425*

Schlessinger, D., 25, *36*, 80(235, 236), 83 (23), *119, 129*, 217, *245*

Schlessinger, S., 33(80), *39*

Schneider, W. C., 264(37), *360*

Scholtissek, C., *136*

Schopf, J. W., 235(5), *241*

Schroeder, W. A., 396(54), 404(71), *427, 428*

Schultze, B., 293, *364*

Scolnick, E., 232(22, 62, 73), *242, 244, 245*

Scott, J. F., 170(144), 189(81), *208, 212*, 266(36), *360*

Seaman, E., 170(54), *207*

Sekiguchi, M., 96(318), *133*

Sela, M., 372(98), 380(37), *426, 430*

Seno, T., 144(124), 145(124), *210*

Sentenac, A., 107(74), *121*

Setlow, R. B., 61(319, 321), *133, 134*

Shaefer, J. R., 393(98a), *430*

Shapiro, A. L., 393(99), *430*

Shearer, R. W., 353(84a), *364*

Sheldrick, P., 104(351), 105(351), 116 (206), *128, 135*

Shemyakin, M. F., 89(197), 96(198, 199), 103(198), 116(198, 199), *127*

Shepherd, W. M., 181(156), *212*

Sheppard, D. E., 34(81), *40*

Shigeura, H. T., 103(322), *134*

Shimizu, Y., 318(29a), *360*

Shin, D. H., 112, *134*, 230(74), *245*

Shooter, E. M., 395(57), *427*

Shub, D., 96(224), *128*

Sibatani, A., 313(85), 314(85), *364*

Stretton, A. O. W., 16(79), 19(12, 79), 23(12), *36*, *39*, 202(24), *205*, 231(15), *242*, *245*
Stuart, A., 149(135), 155(135, 136), *211*
Studier, F. W., 45(345), *135*
Stulberg, M. P., 145(92), *209*
Suback-Sharpe, H., 181(73, 157), 202 (157), *208*, *212*
Subramanian, A. R., 232(83), *245*
Sueoka, N., 47(346), 51(347), 54(279), 55(271), 58, 77(272), 79, 97(348), *120*, *131*, *135*, *137*, 144(63), 145(158, 159), 172(63), 197(177), 198(176, 177), *212*, *213*, 234(84), *245*
Sugimoto, K., 58(274), 59(248a, 273, 274), 62(274), *131*, *135*
Sugimura, T., 318(29a), *360*
Sugino, A., 58(274), 59(274), 62(274), *131*
Summers, W., 115(349), *135*
Sundaram, T. K., 422(13), *425*
Suskind, S. R., 22, *40*
Susman, M., 62(93), 96(93), 116(93), *122*
Suzuki, H., 101(187), *127*
Szer, W., 239(85), *245*
Szybalski, W., 47(82), 104(211, 277, 351), 105(351), 115(211, 349, 350), 116 (350), *121*, *128*, *131*, *135*

T

Takagi, T., 381(102), *430*
Takagi, Y., 101(187), *127*
Takanami, M., 232(86), *245*
Takemura, S., 158(160a), 166(160b), *213*
Tanaka, K., 396(95), *429*
Tanaka, N., 241, *245*
Tanford, C., 372(103), 380(111), 383 (111), *430*
Tashiro, Y., 310(81), *364*
Tata, J. R., 332(91), 349, 352, *364*
Tatum, E. L., 10, *36*
Taylor, A. L., 12(91), 13(91), *40*, 180 (161), 181(161), 187(161), *213*
Taylor, J. H., 47(146, 352, 353, 354), *125*, *135*, 333(92c), 355, 357(92a,b,c), *364*
Taylor, K., *135*
Telser, A., 396(104), *430*
Tener, G. M., 144(65), 145(65), 152(13), 160(163), *204*, *213*
Terzaghi, E., 17(89), 18(71, 89), *39*, *40*, *212*, 391(86), *429*

Thach, R. E., 222, *245*
Thedford, R., 151(70), *207*
Thiebe, R., 145(162), *213*
Thomas, R., 58(89), *122*
Thompson, L. R., 356(92d), *364*
Thorpe, D., 22(104), *41*
Tissières, A., 88(356), *135*, 217, *245*
Tocchini-Valentini, G. P., 82(357), 103 (112), 107(112), 111(81, 365a), 115 (64, 112, 357), *121*, *123*, *135*
Tomizawa, J., 56(1), *118*
Tomkins, G. M., 26(1), 30(1), 31, *36*, *40*, 87(5, 6), *118*
Tomlinson, R. V., 152(13), 160(163), *204*, *213*
Tompkins, R., 232(22, 73), *242*, *245*
Toschi, G., 115(14, 15), *118*
Travers, A. A., 101(48, 357a), 117(48), *120*, *136*, 333(92e), *365*
Travis, J., 408(105), *430*
Tremblay, G. Y., 58(85), *122*
Trostle, P. K., 393(98a), *430*
Trotter, C. D., 12(91), 13(91), *40*
Ts'o, P. O. P., 109(2, 358), 111(2), *118*, *119*, *136*
Tsuboi, M., 109(358), *136*
Tsugita, A., 4(93), 17(89), 18(71, 89), *39*, *40*, *212*, 391(86), *429*
Tsunakawa, S., 151(21), *205*
Tsuzuki, J., 302(92f), *365*
Turner, M. K., 292(79), *363*

U

Uden, L., 384(93b), *429*
Udenfriend, S., 394(106), *430*
Ueda, K., 286, 318(77c), 319(69a, 92h), *363*, *365*
Uhr, J. W., 393(99), *430*
Ukita, T., 177(120), *210*
Ullman, A., 19(57), *38*
Umezawa, H., 111(255, 255a), *130*, 241 (87), *245*
Urry, D. W., 368(107), 369(107), *430*
Usanga, E. A., 424(75a), *428*

V

Vanaman, T. C., 410(9), *425*
Van Bruggen, E. F. J., 12(63), *39*, 50 (359), *136*
Vaughan, M. H., 322(90a), 353(90a), *364*
Vendrely, C., 334, *365*

Subject Index

A

Acetokinase, occurrence of, 396

N-Acetyl groups, of protein, 395–396

Acetyl transferase, occurrence of, 396

Acridine, mutation of lysozyme gene of phage T4 by, 391

Actinomycin D, inhibition of RNA synthesis by, 83, 88, 110

Adaptor hypothesis, 139–140

S-Adenosylmethionine, in methylation of unusual bases, 182

Alanine tRNA, nucleotide sequence of, 158, 165

Allelic genes, in polymorphism, 424

Allosteric proteins, 369

Amber mutations, 16
operon transcription in, 95

Amber suppressor strain su_I^+, *amber* triplet of, 202

Amber suppressor su_{III}^+, tyrosine tRNA and, 181

Amber suppressors, 23

Amino acid(s)
acceptor activity of E. coli tRNA, 143
activation
in nucleus, 304
tRNA and, 139–214
activation of, tRNA and, 139–214
analogs, use by aminoacyl-tRNA synthetases, 196

Amino acid acceptor site, of tRNA, 176

Amino acid code, conservative nature of, 419

Amino acid-dependent ATP-P³²P exchange, enzymes catalyzing, 187–188

Amino acid hydroxamates, formation of, 184, 188

Amino acid starvation, effect on mRNA, 88

Aminoacyladenylate complex, formation of, 183, 187–189

Aminoacyl-tRNA
attachment to ribosomes, 219–224
structure of ester, 189–190, 192
synthesis of specificity, 194–199

3′-Aminoacyl-tRNA, in peptide bond formation, 191

Aminoacyl-tRNA complex, chloramphenicol and puromycin lack of effect on, 306–307

Aminoacyl-tRNA synthetases
in amino acid activation, 140, 141, 183–187
amino acid analog use by, 196–197
mutants of, 187
structural genes for, 185–187

Aminoacyl-tRNA synthetase recognition sites, 194
of tRNA, 176, 179

2-Aminopurine, as chemical mutagen, 16

Anthranilate synthetase, in tryptophan pathway, 90, 91

Anticodon, of tRNA, 176

Apomyoglobin, conformation studies on, 382

Aspartic transcarbamylase, regulator binding sites on, 369

ATP
nuclear type, 274
synthesis in nucleus, 275–288

Autoradiography, in nuclear protein studies, 292–295, 303

B

Bacillus subtilis
chromosome map of, 54, 56
genes of, ordered replication of, 55, 56
histidase of, in mRNA studies, 87–88
parameters of messenger metabolism in, 85

Spores, of bacteria, germination of, 53–54
Streptomycin, in phenotypic curing, 25
Sucrose, nuclei isolation in, 264–266
Suppression, following mutation, 20–24, 28
Suppressor gene, 203
Suppressor mutations, 22–23, 28–29
Suppressor tRNA's, 202–204
Swivel, 74
 in double-strand DNA, 49

T

Temperature-sensitive mutants, in study of protein and nucleic acid synthesis, 185
Template, definition of, 66
template RNA, amino acid activation and, 139, 140
Thermodynamic hypothesis, 398
Thiogalactoside transacetylase in *lac* operon transcription, 94
Thymine dimers, from DNA irradiation, 60
Tobacco mosaic virus (TMV), RNA of, 156
 infectivity of, 4
T₁-ribonuclease, specificity of, 159
T₁RNase, in analysis of tRNA, 157
Trans dominance, 34
Transitions and transversions, description of, 16–17

Transforming substance as DNA, 1
Translational frame shift, in protein evolution, 400
Trypsin
 DFP-reactive serine of, 408
Tryptophan synthetase, conformation studies on, 389, 390
trp operon, 27
 transcription of, 89–95
Tryptophan pathway, *try* operon transcription and, 89–95
Tryptophan synthetase
 gene studies using, 16
 in tryptophan pathway, 91
Tyrosine tRNAs, nucleotide sequences of, 161

U

Ultraviolet, damaged DNA, repair of, 60
Universality, of genetic code, 197, 199, 232–236

V

Vaccinia virus, transcription in, 97
Vasopressin, in hormone evolution, 409
Vasotocin, in hormone evolution, 409
Viral replication, 62–63

W

Watson-Crick model of double-strand DNA, 7
Wobble hypothesis, 178, 200